Cow's Milk and Allergy

Cow's Milk and Allergy

Special Issue Editors

Joost van Neerven
Huub Savelkoul

MDPI • Basel • Beijing • Wuhan • Barcelona • Belgrade

MDPI

Special Issue Editors
Joost van Neerven   Huub Savelkoul
Wageningen University   Wageningen University
The Netherlands   The Netherlands

Editorial Office
MDPI
St. Alban-Anlage 66
4052 Basel, Switzerland

This is a reprint of articles from the Special Issue published online in the open access journal *Nutrients* (ISSN 2072-6643) from 2018 to 2019 (available at: https://www.mdpi.com/journal/nutrients/special_issues/milk_allergy).

For citation purposes, cite each article independently as indicated on the article page online and as indicated below:

LastName, A.A.; LastName, B.B.; LastName, C.C. Article Title. *Journal Name* **Year**, *Article Number*, Page Range.

ISBN 978-3-03928-028-5 (Pbk)
ISBN 978-3-03928-029-2 (PDF)

Contents

About the Special Issue Editors

Joost van Neerven Professor in Mucosal Immunity at Wageningen University and Sr. Research Specialist at FrieslandCampina. Trained as a biologist, Prof. Joost van Neerven received his Ph.D. in 1995 at the University of Amsterdam in the Netherlands, discussing a thesis on the role of T cells in allergies. He then joined the biopharmaceutical company ALK-Abello in Denmark, where he studied the application of allergens for immunotherapy and the underlying immunological mechanisms. In 1999, he returned to the Netherlands, where he subsequently worked in several biotechnology companies that worked on monoclonal antibodies for therapy of immune mediated diseases. In 2003, he cofounded Bioceros BV, a biotechnology company that develops and manufactures therapeutic monoclonal antibodies. In 2006, he joined FrieslandCampina and, in 2013, he was appointed special Professor of Mucosal Immunity at Wageningen University. His research interests are (mucosal) immunology, allergy, nutrition, and milk.

Huub Savelkoul Professor in Cell Biology and Immunology. Trained as a biologist with majors in Biochemistry, Cell Biology, and Genetics, Prof Huub Savelkoul received his PhD cum laude in immunology from the Medical Faculty of the Erasmus University in Rotterdam, discussing a thesis on IgE formation and regulation in mouse models of allergy. He was then a postdoctoral fellow at the DNA Research Institute of Molecular and Cellular Biology in Palo Alto, California, where he studied cytokine-based immunoregulation in allergies. Since 2000, he has been a full professor at Wageningen University, where he co-founded the Allergy Consortium Wageningen. Prof. Savelkoul's main research interests are the regulation of IgE antibody formation in allergies, the immunogenicity and allergenicity of dietary components, the basic immune-mediated mechanisms in food allergies, the immunomodulation by food and feed, and the development of allergy-linked immunodiagnostics.

Preface to "Cow's Milk and Allergy"

Dear reader, we are pleased to present to you the printed version of the Special Issue of *Nutrients* on "Cow's Milk and Allergies" that is in your hands now. This Special Issue describes not only the epidemiology, management and prevention of cow's milk allergy, but also focuses on the roles that (heat) processed milk as well as unprocessed milk may have in the initiation or prevention of allergies—issues that are currently important focus areas in allergy research. We are very pleased that leading scientists in this field have contributed to this book offering a comprehensive overview of the current status in the field of cow's milk allergies. This book will hopefully help to provide an up-to-date overview on the link between cow's milk and allergies.

We hope you will enjoy reading this book!

Joost van Neerven, Huub Savelkoul
Special Issue Editors

MDPI

Editorial

The Two Faces of Cow's Milk and Allergy: Induction of Cow's Milk Allergy vs. Prevention of Asthma

R. J. Joost van Neerven [1,2,*] and Huub F. J. Savelkoul [1,3]

1   Cell Biology and Immunology Group, Wageningen University & Research,
    6708 WD, Wageningen, The Netherlands
2   FrieslandCampina, 3818 LE, Amersfoort, The Netherlands
3   Allergy Consortium Wageningen, Wageningen University & Research,
    6708 WD, Wageningen, The Netherlands
*   Correspondence: joost.vanneerven@wur.nl

Received: 15 August 2019; Accepted: 16 August 2019; Published: 19 August 2019

Abstract: Cow's milk has been consumed by humans for over 5000 years and contributed to a drastic change in lifestyle form nomadic to settled communities. As the composition of cow's milk is relatively comparable to breast milk, it has for a very long time been used as an alternative to breastfeeding. Today, cow's milk is typically introduced into the diet of infants around 6 months, except when breastfeeding is not an option. In that case, most often cow's milk based infant formulas are given. Some children will develop cow's milk allergy (CMA) during the first year of life. However, epidemiological evidence also suggests that consumption of unprocessed, "raw" cow's milk is associated with a lowered prevalence of other allergies. This Special Issue of *Nutrients* on "Cow's Milk and Allergy" (https://www.mdpi.com/journal/nutrients/special_issues/milk_allergy) is dedicated to these two different sides of cow's milk and allergy, ranging from epidemiology of CMA, clinical presentation and sensitization patterns, treatment and prevention, effects of milk processing, and current management guidelines for CMA, but also the epidemiological evidence linking cow's milk to lower asthma prevalence as well as the tolerance-inducing effect of raw cow's milk in food allergy models. In this editorial, we discuss these issues by highlighting the contributions in this Special Issue.

Keywords: cow's milk allergy; milk; hydrolysate; asthma; processing; tolerance

1. Cow's Milk Allergy

The first recorded food allergy to dairy was described by Hippocrates (460–375 BC) that mentioned how some people could eat cheese without problems, but that eating cheese in some people causes pain [1]. As described in (Mucosal Immunology), Besredka described already in 1909 the process we now know as oral tolerance induction by performing experiments in laboratory animals. He demonstrated that oral (or rectal) administration of cow's milk could prevent the development of anaphylaxis [2].

Cow's milk allergy in children is characterized by the fact that most children will overgrow the allergy and develop immunological tolerance to cow's milk before three to four years of age. Proper diagnosis and allergy management in early childhood is provided in the Diagnosis and Rationale for Action against cow's milk allergy (CMA) (DRACMA) guidelines [3]. Extensively hydrolyzed milk formulas and amino acid formulas are often used to treat CMA in infants [4]. In the Middle East, the use of partial hydrolysates as an intermediate, first step-down from the use of intact cow's milk formulas before introducing extensively hydrolyzed or amino acid formulas was recently discussed and approved during a consensus meeting [5].

In addition to CMA treatment—which in essence is the avoidance of ingestion of intact cow's milk allergens—there is an urgent need to accelerate immunological tolerance to cow's milk in these children. Recent progress in the understanding of the immunological basis of tolerance induction can aid in this development of novel sustainable therapies [4,6].

Oral tolerance is an active process of local and systemic immune unresponsiveness by which the immune system does not respond to an orally administered antigen such as food. The mechanism of this tolerance induction remains poorly understood but comprises the induction of regulatory T cell (Tregs) and also the induction of anergy or deletion of responding T-cell subsets [7].

Since these early descriptions of cow's milk allergy (CMA) and oral tolerance induction, the prevalence of CMA in infants and toddlers has increased to approximately 0.5%–3% at 1 year of age, both in Western countries as in the rest of the world [8].

However, the prevalence of CMA, as the most common food allergy in early childhood, varies considerably based on the definition (i.e., parent-reported CMA > doctor-diagnosed CMA > double blind placebo-controlled food challenge (DBPCFC)-confirmed CMA), so the real prevalence may actually be lower than 1%–2%. Some children and even adults remain allergic to cow's milk and often show severe reactions even to traces of milk.

In bovine milk, caseins are the most abundant protein group, while whey proteins are more abundant in human breastmilk. In particular, β-lactoglobulin and αS2-casein are present in bovine milk but absent in human breastmilk and are important allergens in cow's milk or preparations thereof.

The majority of patients with IgE-mediated CMA are sensitized to more than one milk allergen, with a great variability in the specificity and intensity of the induced IgE responses. Molecular-based allergy diagnosis allows associating each patient with a specific immune reactivity profile and potentially to identify different CMA phenotypes. Thus, different and novel diagnostic tests are needed to characterize the sensitization patterns of patients with different clinical pictures of CMA. Such tests comprise multiplex assays using purified (recombinant) allergen molecules as allergen arrays [4] to detect all potentially clinically relevant allergens and functional basophil activation tests to correlate to the severity of the allergic symptoms [9].

It is widely accepted that the prevalence of allergic sensitization and food allergy have increased over the last few decades. Still, it is unclear whether this also holds true for the development of CMA [8]. Moreover, high-risk children with CMA are more likely to also have multiple food allergies and suffer also from asthma, atopic dermatitis, and allergic rhinitis. These factors, in addition to the resolution of the CMA in many children, complicate the determination of the real prevalence. The susceptibility to developing CMA (and its persistence) also seems to be linked to differences in microbiota composition [10,11] and may also involve epigenetic components, which can explain the observed increase of food allergy prevalence worldwide. The described shared genetic etiology between CMA and asthma can thereby explain the associated sensitivity for subsequent development of asthma [12].

In contrast to sheep and goat's milk that are very similar to cow's milk, milk derived from camels and cow's milk have a low cross-reactivity, which is indicative of a low protein similarity. As such, camel milk may be an alternative to cow's milk-based hypoallergenic infant formulas for children at high risk of developing CMA [13]. In addition to the most frequent type I or IgE-mediated allergy against cow's milk, also non-IgE-mediated milk allergy exists. Often these patients show a delayed onset of disease after allergen exposure. The mechanism of this delayed-type hypersensitivity reaction is poorly understood, and a mouse model can help to improve our understanding [14].

2. Milk Processing and CMA

Until the industrial revolution and the founding of milk processing factories, milk was mostly consumed in its raw or boiled form. Since the 1880s, milk pasteurization and higher heat-treated milks have become the industrial standard to prevent the spread of milk-borne pathogens that can cause *a.o.* diarrhea. Various heat treatments, including pasteurization and ultra-high temperature, are widely used today to increase the safety of raw milk and extend its shelf life.

Thermal treatments will also lead to structural changes of the milk proteins, including protein aggregation and glycation. These protein alterations can thus modulate the cow's milk protein immunogenicity (the induction of specific IgE production) and/or allergenicity (the ability to cross-link cell-bound IgE on mast cells with specific allergens, provoking the release of mediators, including histamine) [15]. Such aggregated and glycated proteins can interact with multiple receptors on immune cells linking the allergic sensitization profile to individual clinical responsiveness when exposed to cow's milk allergens [16].

Surprisingly, it was found that most children with CMA can tolerate baked milk, and such children appeared to have lower β-lactoglobulin, and casein-specific IgE concentrations and higher numbers of regulatory T (Treg) cells present in their circulation [6,17].

Baked milk (as within the muffin matrix) might also promote formation of complexes with food components, inducing a modulation of the immune reactivity and reduction of allergenicity of milk allergens. Addition of baked milk products into the daily diet can accelerate the induction of tolerance to unheated raw milk rather than complete avoidance of the allergenic food [18]. Regular ingestion of baked milk products could thus drive a change in immune responsiveness, thereby inducing milk tolerance.

Allergy prevention was for a very long time based on allergy avoidance measures.

Currently, however, actively interfering with influencing immune tolerance based on novel insights into the (heat treatment modified) hypoallergenic allergen molecules, the use of (partially) hydrolysed formula's, the epitope-specificity of the IgE antibodies, the shifting balance between T-cell subsets, and the induction of Tregs are considered to be more effective in prevention of the development of allergic symptoms [6,17].

3. Raw Milk and Allergy: Evidence from Epidemiology and Animal Models

Interestingly, in the early 2000s, milk was associated with allergy in quite a different and unexpected way. Searching for an environmental link with asthma, children growing up on small farms were found to have a much lower risk for developing asthma and allergic rhinitis [19]. In follow-up studies, it was shown that this effect was independently linked to exposure to the farming environment (stables and animals) on the one hand and to consumption of unprocessed (raw) cow's milk on the other hand [20–22]. To date, more than 15 epidemiological studies have shown that the consumption farm milk, most of which is consumed as raw unpasteurized cow's milk, can reduce the risk of allergic diseases. This has been reviewed in Sozanska in this issue [23] and in [24].

These results have now been confirmed by causal evidence in mouse model systems [25,26]. In this issue, these authors demonstrated that the suppression of food allergic symptoms by raw cow's milk is retained after skimming but abolished after pasteurization of the milk, and subsequent addition of alkaline phosphatase might restore the allergy-protective effects. In a follow-up paper, the same group addressed the role of epigenetic modification as part of the mechanism of action [27]. This is in line with the finding that exposure to raw farm milk in pregnancy and the first year of life induces epigenetic changes in innate immunity receptor genes [28].

As a further mechanism, the consumption of immunomodulatory cytokines in unprocessed bovine milk may create the environment to promote the development of regulatory T cells, enabling establishment and maintenance of oral tolerance in the gut, a process in which bovine IgG may also be involved through the formation of allergen–IgG complexes [29,30]. Such IgG–allergen-immune complexes in murine milk have been shown to protect against allergies in experimental models [31,32].

4. Conclusions

We hope this Special Issue provides a state-of-the-art overview of the two-faced story of cow's milk and allergy, on the one hand, by describing how early introduction, milk processing, milk protein hydrolysates, and new immune-therapeutic approaches can help to prevent and manage CMA in the clinic, and on the other hand, highlighting how consumption of raw or minimally processed milk might become relevant in future preventive strategies for respiratory and possibly food allergies (see Figure 1 for overview).

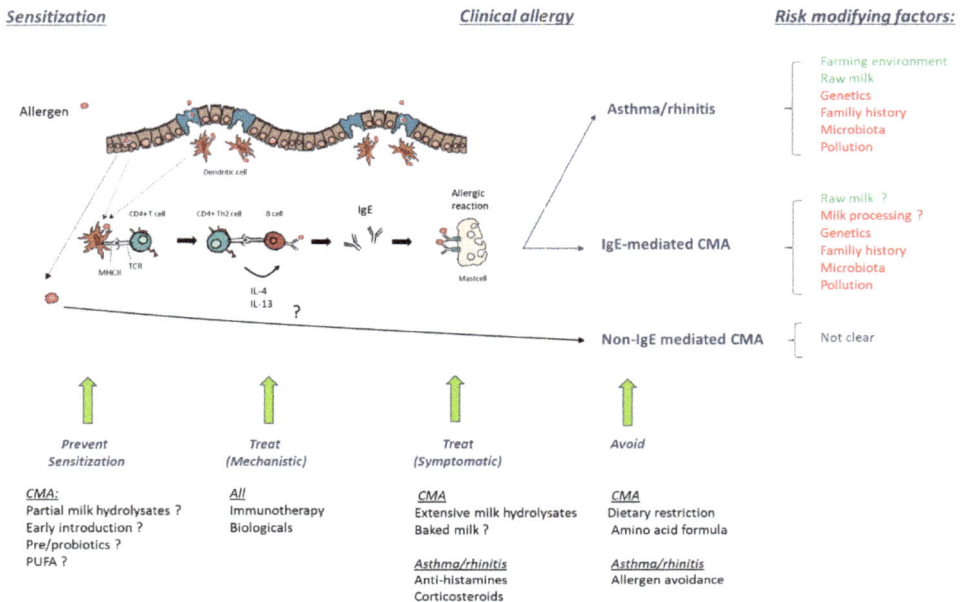

Figure 1. Sensitization, risk factors, and current strategies to prevent and treat allergies (cow's milk allergy (CMA) as well as inhalation allergies).

Author Contributions: R.J.J.v.N. and H.F.J.S. both contributed to the writing of this editorial.

Funding: This research received no external funding.

Conflicts of Interest: R.J.J.v.N. is employed by of FrieslandCampina.

References

1. Schadewaldt, H. Magen-Darmallergie. In *Geschichte de Allergie*; Schadewaldt, H., Ed.; Deisenhofen Dustri: Oberhaching, Germany, 1983; Volume 4, pp. 37–51.
2. Mestecky, J.; McGhee, J.R.; Bienenstock, J.; Lamm, M.E.; Strober, W.; Cebra, J.J.; Mayer, L.; Ogra, P.L.; Russell, M.W. Historical Aspects of Mucosal Immunology. In *Mucosal Immunology*, 4th ed.; Mestecky, J., Strober, W., Russell, M.W., Kelsall, B.L., Cheroutre, H., Lambrecht, B.N., Eds.; Elsevier: Amsterdam, The Netherlands, 2015; pp. 23–43.
3. Fiocchi, A.; Brozek, J.; Schünemann, H.; Bahna, S.L.; von Berg, A.; Beyer, K.; Bozzola, M.; Bradsher, J.; Compalati, E.; Ebisawa, M.; et al. World Allergy Organization (WAO) Special Committee on Food Allergy. World Allergy Organization (WAO) Diagnosis and Rationale for Action against Cow's Milk Allergy (DRACMA) Guidelines. *Pediatr. Allergy Immunol.* **2010**, *21* (Suppl. 21), 1–125. [CrossRef]
4. Linhart, B.; Freidl, R.; Elisyutina, O.; Khaitov, M.; Karaulov, A.; Valenta, R. Molecular approaches for diagnosis, therapy and prevention of cow's milk allergy. *Nutrients* **2019**, *11*, 1492. [CrossRef]
5. Vandenplas, Y.; Al-Hussaini, B.; Al-Mannaei, K.; Al-Sunaid, A.; Helmi Ayesh, W.; El-Degeir, M.; El-Kabbany, N.; Haddad, J.; Hashmi, A.; Kreishan, F.; et al. Prevention of Allergic Sensitization and Treatment of Cow's Milk Protein Allergy in Early Life: The Middle-East Step-Down Consensus. *Nutrients* **2019**, *11*, 1444. [CrossRef]
6. Knol, E.F.; de Jong, N.W.; Ulfman, L.H.; Tiemessen, M.M. Prevention and treatment of cow's milk allergy, what are the options? *Nutrients* **2019**, in press.

7.   Commins, S.P. Mechanisms of Oral Tolerance. *Pediatr. Clin. N. Am.* **2015**, *62*, 1523–1529. [CrossRef]

8.   Flom, J.D.; Sicherer, S.H. Epidemiology of Cow's Milk Allergy. *Nutrients* **2019**, *11*, 1051. [CrossRef]

9.   Schocker, F.; Kull, S.; Schwager, C.; Behrends, J.; Jappe, U. Individual Sensitization Pattern Recognition to Cow's Milk and Human Milk Differs for Various Clinical Manifestations of Milk Allergy. *Nutrients* **2019**, *11*, 1331. [CrossRef]

10. Bunyavanich, S.; Shen, N.; Grishin, A.; Wood, R.; Burks, W.; Dawson, P.; Jones, S.M.; Leung, D.Y.M.; Sampson, H.; Sicherer, S.; et al. Early-life gut microbiome composition and milk allergy resolution. *J. Allergy Clin. Immunol.* **2016**, *138*, 1122–1130. [CrossRef]

11. Berni Canani, R.; De Filippis, F.; Nocerino, R.; Paparo, L.; Di Scala, C.; Cosenza, L.; Della Gatta, G.; Calignano, A.; De Caro, C.; Laiola, M.; et al. Gut microbiota composition and butyrate production in children affected by non-IgE-mediated cow's milk allergy. *Sci. Rep.* **2018**, *8*, 12500. [CrossRef]

12. Jansen, P.R.; Petrus, N.C.M.; Venema, A.; Posthuma, D.; Mannens, M.M.A.M.; Sprikkelman, A.B.; Henneman, P. Higher Polygenetic Predisposition for Asthma in Cow's Milk Allergic Children. *Nutrients* **2018**, *10*, 1582. [CrossRef]

13. Maryniak, N.Z.; Bech-Hansen, E.; Ravn-Ballegaard, A.-S.; Sancho, A.I.; Lindholm-Bøgh, K. Comparison of the Allergenicity and Immunogenicity of Camel and Cow's Milk—A Study in Brown Norway Rats. *Nutrients* **2018**, *10*, 1903. [CrossRef]

14. Wąsik, M.; Nazimek, K.; Nowak, B.; Askenase, P.W.; Bryniarski, K. Delayed-Type Hypersensitivity Underlying Casein Allergy Is Suppressed by Extracellular Vesicles Carrying miRNA-150. *Nutrients* **2019**, *11*, 907. [CrossRef]

15. Teodorowicz, M.; van Neerven, J.; Savelkoul, H. Food Processing: The Influence of the Maillard Reaction on Immunogenicity and Allergenicity of Food Proteins. *Nutrients* **2017**, *9*, 835. [CrossRef]

16. Zenker, H.E.; Ewaz, A.; Deng, Y.; Savelkoul, H.F.J.; van Neerven, R.J.J.; De Jong, N.W.; Wichers, H.J.; Hettinga, K.A.; Teodorowicz, M. Differential effects of dry vs. wet heating of β-lactoglobulin on formation of sRAGE binding ligands and sIgE epitope recognition. *Nutrients* **2019**, *11*, 1432. [CrossRef]

17. D'Auria, E.; Salvatore, S.; Pozzi, E.; Mantegazza, C.; Sartorio, M.; Pensabene, L.; Baldassarre, M.E.; Agosti, M.; Vandenplas, Y.; Zuccotti, G.V. Cow's milk allergy: Immunomodulation by dietary intervention. *Nutrients* **2019**, *11*, 1399. [CrossRef]

18. Bavaro, S.; De Angelis, E.; Barni, S.; Pilolli, R.; Mori, F.; Novembre, E.; Monaci, L. Modulation of milk allergenicity by baking milk in foods: A proteomic investigation. *Nutrients* **2019**, *11*, 1536. [CrossRef]

19. Riedler, J.; Braun-Fahrländer, C.; Eder, W.; Schreuer, M.; Waser, M.; Maisch, S.; Carr, D.; Schierl, R.; Nowak, D.; Von Mutius, E.; et al. Exposure to farming in early life and development of asthma and allergy: A cross-sectional survey. *Lancet* **2001**, *358*, 1129–1133. [CrossRef]

20. Perkin, M.R.; Strachan, D.P. Which aspects of the farming lifestyle explain the inverse association with childhood allergy? *J. Allergy Clin. Immunol.* **2006**, *117*, 1374–1381. [CrossRef]

21. Waser, M.; Michels, K.B.; Bieli, C.; Flöistrup, H.; Pershagen, G.; Von Mutius, E.; Ege, M.; Riedler, J.; Schram-Bijkerk, D.; Brunekreef, B.; et al. Inverse association of farm milk consumption with asthma and allergy in rural and suburban populations across Europe. *Clin. Exp. Allergy* **2007**, *37*, 661–670. [CrossRef]

22. Loss, G.; Apprich, S.; Waser, M.; Kneifel, W.; Genuneit, J.; Büchele, G.; Weber, J.; Sozanska, B.; Danielewicz, H.; Horak, E.; et al. The protective effect of farm milk consumption on childhood asthma and atopy: The GABRIELA study. *J. Allergy Clin. Immunol.* **2011**, *128*, 766–773. [CrossRef]

23. Sozańska, B. Raw Cow's Milk and Its Protective Effect on Allergies and Asthma. *Nutrients* **2019**, *11*, 469. [CrossRef]

24. Von Mutius, E.; Vercelli, D. Farm living: Effects on childhood asthma and allergy. *Nat. Rev. Immunol.* **2010**, *10*, 861–868. [CrossRef]

25. Abbring, S.; Verheijden, K.A.T.; Diks, M.A.P.; Leusink-Muis, A.; Hols, G.; Baars, T.; Garssen, J.; van Esch, B.C.A.M. Raw Cow's Milk Prevents the Development of Airway Inflammation in a Murine House Dust Mite-Induced Asthma Model. *Front. Immunol.* **2017**, *8*, 1045. [CrossRef]

26. Abbring, S.; Ryan, J.T.; Diks, M.; Hols, G.; Garssen, J.; van Esch, B. Suppression of food allergic symptoms by raw cow's milk in mice is retained after skimming but abolished after heating the milk – a promising contribution of alkaline phosphatase. *Nutrients* **2019**, *11*, 1499. [CrossRef]

27. Abbring, S.; Wolf, J.; Ayechu Muruzabal, V.; Diks, M.; Alashkar Alhamwe, B.; Alhamdan, F.; Harb, H.; Renz, H.; Garn, H.; Garssen, J.; et al. Raw cow's milk suppresses allergic symptoms in a murine model for food allergy—A potential role for epigenetic modifications. *Nutrients* **2019**, *11*, 1721. [CrossRef]

28. Loss, G.; Bitter, S.; Wohlgensinger, J.; Frei, R.; Roduit, C.; Genuneit, J.; Pekkanen, J.; Roponen, M.; Hirvonen, M.R.; Dalphin, J.C.; et al. Prenatal and early-life exposures alter expression of innate immunity genes: The PASTURE cohort study. *J. Allergy Clin. Immunol.* **2012**, *130*, 523–530. [CrossRef]

29. Van Neerven, R.J.J.; Knol, E.F.; Heck, J.M.L.; Savelkoul, H.F.J. Which factors in raw cow's milk contribute to protection against allergies? *J. Allergy Clin. Immunol.* **2012**, *130*, 853–858. [CrossRef]

30. Ulfman, L.H.; Leusen, J.H.W.; Savelkoul, H.F.J.; Warner, J.O.; van Neerven, R.J.J. Effects of Bovine Immunoglobulins on Immune Function, Allergy, and Infection. *Front. Nutr.* **2018**, *5*, 52. [CrossRef]

31. Mosconi, E.; Rekima, A.; Seitz-Polski, B.; Kanda, A.; Fleury, S.; Tissandie, E.; Monteiro, R.; Dombrowicz, D.D.; Julia, V.; Glaichenhaus, N.; et al. Breast milk immune complexes are potent inducers of oral tolerance in neonates and prevent asthma development. *Mucosal Immunol.* **2010**, *3*, 461–474. [CrossRef]

32. Ohsaki, A.; Venturelli, N.; Buccigrosso, T.M.; Osganian, S.K.; Lee, J.; Blumberg, R.S.; Oyoshi, M.K. Maternal IgG immune complexes induce food allergen-specific tolerance in offspring. *J. Exp. Med.* **2018**, *215*, 91–113. [CrossRef]

nutrients

MDPI

Review

Epidemiology of Cow's Milk Allergy

Julie D. Flom * and Scott H. Sicherer

Elliot and Roslyn Jaffe Food Allergy Institute, Division of Allergy, Department of Pediatrics, Kravis Children's Hospital, Icahn School of Medicine at Mount Sinai, New York, NY 10029, USA; scott.sicherer@exchange.mssm.edu
* Correspondence: julie.flom@mountsinai.org; Tel.: +212-241-5548; Fax: +212-426-1902

Received: 16 April 2019; Accepted: 6 May 2019; Published: 10 May 2019

Abstract: Immunoglobulin E (IgE)-mediated cow's milk allergy (CMA) is one of the most common food allergies in infants and young children. CMA can result in anaphylactic reactions, and has long term implications on growth and nutrition. There are several studies in diverse populations assessing the epidemiology of CMA. However, assessment is complicated by the presence of other immune-mediated reactions to cow's milk. These include non-IgE and mixed (IgE and non-IgE) reactions and common non-immune mediated reactions, such as lactose intolerance. Estimates of prevalence and population-level patterns are further complicated by the natural history of CMA (given its relatively high rate of resolution) and variation in phenotype (with a large proportion of patients able to tolerate baked cow's milk). Prevalence, natural history, demographic patterns, and long-term outcomes of CMA have been explored in several disparate populations over the past 30 to 40 years, with differences seen based on the method of outcome assessment, study population, time period, and geographic region. The primary aim of this review is to describe the epidemiology of CMA. The review also briefly discusses topics related to prevalence studies and specific implications of CMA, including severity, natural course, nutritional impact, and risk factors.

Keywords: cow's milk allergy; epidemiology; natural history; prevalence

1. Introduction

Cow's milk allergy (CMA) is defined as an immune-mediated response to proteins in cow's milk that occurs consistently with ingestion. It is one of the most common food allergies in early life [1–3] with an estimated prevalence in developed countries ranging from 0.5% to 3% at age 1 year (reviewed in [1,4–8]). In this review, we summarize prevalence estimates of cow's milk allergy worldwide and discuss the clinical and public health implications of understanding risk factors for development and persistence of CMA. Diagnosis, prevention, and treatment are briefly discussed in relation to their implications on prevalence estimates.

2. Subtypes of Immune-Mediated Reactions to Cow's Milk

CMA refers to immune-mediated reactions to cow's milk that are categorized as immunoglobulin E (IgE)-mediated, non-IgE mediated, and mixed (IgE combined with non-IgE) [4,8–13]. This review focuses on IgE-mediated cow's milk allergy (CMA), a type I hypersensitivity reaction in which symptoms usually occur within minutes to 1 to 2 hours of ingestion. IgE antibodies to proteins in cow's milk bind to mast cells, and subsequent exposure to the protein leads to mast cell degranulation and release of mediators, including histamine. This leads to symptoms including urticaria; angioedema; throat tightness; respiratory symptoms, including difficulty breathing, coughing, and wheezing; gastrointestinal symptoms, including abdominal pain, vomiting, and diarrhea; and cardiovascular symptoms, including dizziness, confusion, and hypotension [8,9,11]. Approximately 60% of those with CMA have the IgE-mediated form, although estimates vary by study population and age [8,12,14].

Mixed and non-IgE mediated forms of CMA have different underlying mechanisms, presentation, and implications, which complicate assessment of the epidemiology of IgE-mediated CMA (Table 1). Mixed forms of CMA (both IgE and non-IgE mediated) include atopic dermatitis, allergic eosinophilic esophagitis, and eosinophilic gastritis. Given the lack of validated testing, cow's milk as a trigger for these diseases is often identified through history and elimination diets. Non-IgE mediated forms of CMA include cow's milk enteropathy [12], food protein induced proctitis/proctocolitis [6], food protein induced enterocolitis syndrome (FPIES) [13], and Heiner syndrome [8]. FPIES generally manifests with severe vomiting at least 2 hours after ingestion, with negative skin and blood testing (reviewed in [4]) and no validated diagnostic testing [6]. One large Israeli study estimated cow's milk FPIES prevalence in infancy at 0.34% [15]. Non-immune mediated reactions, such as lactose intolerance, typically lead to overestimates of prevalence in population-based studies that rely on self-report [8,10,14,16].

3. Diagnosis of CMA

The remainder of this review focuses on IgE-mediated CMA. Methods available for diagnosis have limitations, which impact the ability to elucidate the underlying epidemiology (Table 1) [3,9,17,18]. The gold standard for diagnosis is the double blind, placebo-controlled oral food challenge (DBPCFC) [18]. The unblinded oral food challenge (OFC) is less rigorous, but well validated, especially in young children [19]. However, both tests are time- and resource-intensive and carry inherent risk of anaphylaxis. OFCs are therefore not always appropriate for use in clinical practice or especially in large epidemiologic studies. Studies that employ these methods may have incomplete assessments due to parental refusal and safety concerns in highly atopic children [20]. Objective measures used routinely in both epidemiologic studies and clinical practice include serum-specific IgE (sIgE) and skin prick tests (SPT). These two tests predict the likelihood of reaction, but, in isolation, are not sufficient for diagnosis (reviewed in [9]). Sensitization measured via SPT is often defined as a wheal at least 3 mm larger than the negative control. Cow's milk sIgE, measured by in vitro immunoassay, measures IgE binding to specific proteins; sensitization is defined as detectable sIgE (often sIgE ≥ 0.35 kU/L, sometimes ≥ 0.10 kU/L). Self-report of CMA [16] and reliance on sensitization based solely on serum sIgE and/or skin testing as a means to identify CMA tend to overestimate prevalence [18,21,22]. Further, variations in assays for sIgE can lead to conflicting interpretations and limit comparability between study populations [23].

Prevalence estimates are also impacted by variation in tolerance as defined by the types of milk products that are ingested. Approximately 70% of patients with IgE-mediated cow's milk allergy who would react to whole cow's milk products (e.g., milk, ice cream, yogurt) tolerate extensively heated, baked milk products (e.g., cookie, muffin) [24] because baking alters certain proteins in milk, leading to a loss of conformational epitopes and decreased allergenicity (reviewed in [6,25,26]). This further complicates our understanding of the epidemiology and natural history and it is rare that epidemiologic studies distinguish tolerance of baked cow's milk.

4. Prevalence of CMA

4.1. Prevalence of IgE-Mediated CMA: Meta-Analysis and Systematic Reviews

Despite the limitations in assessment, there are a large number of studies in the US and worldwide that attempt to estimate the incidence or prevalence of CMA. There are select meta-analyses and systematic reviews that summarize existing data on CMA in the US and worldwide [2]; however, there is heterogeneity between the studies, which complicates comparisons and summary estimates (Table 2A). Rona et al. [3] performed a meta-analysis of papers on food allergy published from January 1990 to December 2005. They reported a range in prevalence by study methodology, with estimates from studies relying on self-report ranging from 1.2% to 17%, those using SPT alone from 0.2% to 2.5% and sIgE alone from 2% to 9%, studies using symptoms and sensitization (SPT ≥ 3 mm or sIgE ≥ 0.35) ranging from 0% to 2.0%, and those relying on food challenges (OFC or DBPCFC) ranging from 0% to 3.0%.

Nwaru et al. performed a systematic review and meta-analysis of CMA prevalence in European studies published between 2000 and 2012 and included 42 primary articles on CMA [27]. They also observed a variation in prevalence by means of diagnosis. By self-report, the point prevalence of CMA was 2.3% (95% CI 2.1–2.5); by SPT alone, 0.3% (95% CI 0.03–0.6); and by sIgE alone, 4.7% (95% CI 4.2–5.1). The prevalence of CMA diagnosed by food challenge was 0.6% (95% CI 0.5–0.8) and by food challenge or reported history of CMA was 1.6% (95% CI 1.2–1.9). The authors also reported a higher prevalence among younger ages.

These studies demonstrate that appreciable variation is seen in estimates varying by factors, including geographic region, source population (high risk referral vs. general population), age and participation rates, and limitations of diagnosis [9,18].

4.2. Prevalence of IgE-Mediated CMA: Select Studies

Systematic reviews and meta-analyses are helpful in characterizing overall trends and observing differences in prevalence estimates. However, there are nuances in individual studies that provide a number of insights into the epidemiology of CMA as demonstrated by the studies reviewed below (Table 2B).

The EuroPrevall birth cohort included 12,000 children from nine countries in Europe enrolled at birth and followed through the age of 24 to 30 months [28]. Diagnosis was based on DBPCFC administered to all infants with no regular consumption of cow's milk and evidence of sensitization or a reported history of reaction or improvement in symptoms after dietary elimination. They observed an overall prevalence of CMA of 0.59% (adjusted for loss to follow-up and DBPCFC placebo reactors), with prevalence ranging from 0% to 1.3% in various countries. The strengths of the study include the large sample size, a source population representative of the general population, and use of DBPCFC for all participants with possible allergy at age 1 year. The study also carefully distinguished between IgE and non-IgE mediated CMA. However, it is limited to European countries and it is not clear if the results can be generalized.

In Israel, Katz et al. performed a prospective study of all births at a single medical center over 2 years (n = 13,234) and identified a cumulative incidence of 0.5% with a mean age of onset of 3.9 months [29]. The study was strengthened by its prospective design, high recruitment rate (98.4%), and use of OFC for almost 75% of diagnoses (with the exception of those with a history of life-threatening reaction or family refusal). In an unselected population of 3 year olds in Denmark (started in late 1990s), Osterballe et al. reported a CMA prevalence of 0.6% [30]. In a birth cohort of 1749 infants in Denmark, Host et al. observed an incidence of CMA in the first year of life of 2.2% (95% CI 1.5%–2.9%); 0.5% (95% CI 0.2%–0.9%) were confirmed by OFC [31].

The studies described above benefit from recruitment in early life and some have prospective follow-up enabling estimates of incidence. Additional studies, primarily in the US, provide population-based estimates, which are also important for informing public health. Liu et al. [32] used data from the National Health and Nutrition Examination Survey (NHANES) 2005–2006 to estimate the prevalence of food allergy to select foods. They measured cow's milk sIgE from stored serum samples from 8203 child and adult participants. The prevalence of milk sensitization (defined as sIgE $\geq$ 0.35) was 5.7% overall and the estimated clinical prevalence of CMA (based on a 95% positive predictive value for milk of 15 kU/L) was 0.40. The prevalence was appreciably higher among younger children (aged 1 to 5 years), with 22% sensitized and an estimated clinical food allergy rate of 1.8%. Among all other ages (aged 6 to >60 years), estimated rates ranged from 0.16% to 0.49%. This study is limited by the lack of data on clinical reactivity; however, the authors made use of a large, nationally representative sample and prevalence estimates are based upon a conservative estimate of sIgE. Gupta et al. [7] performed a cross-sectional study in a representative sample of 38,380 children from the US population using a telephone survey. Overall prevalence of CMA was 1.7% (95% CI

1.5–1.8) and peaked in children aged 0 to 5 years at 2.0%, with a decrease in prevalence over childhood that did not reach statistical significance.

There are fewer reports of the prevalence of CMA in adults. Based on what is currently understood about the natural history of CMA, prevalence in adults is expected to be lower and estimates are often around 0.5% [3]. However, higher estimates have been reported, with some misclassification likely due to inaccuracies in self-report. McGowan et al. [33] investigated the self-reported prevalence of CMA in the National Health and Nutrition Examination Survey (NHANES) 2007–2010 and, surprisingly, report a higher prevalence in adults (2.64%) than children (1.94%), with an overall sample average of 2.47%. When the authors excluded those with consumption in the previous month, the prevalence of CMA in the overall sample decreased to 1.62% [33].

Recent data from the US also reports a higher prevalence of CMA in adults. In a representative cross-sectional survey of 40,443 adults who completed an internet or telephone-based survey in 2015–2016, Gupta et al. [34] observed a prevalence of 1.9% (95% CI 1.8%–2.1%), which peaked at the ages of 18 to 29 years at 2.4%. Diagnosis of CMA was not based on standardized assessment across the population, with the authors using objective criteria to identify "convincing" allergy, defined as those with a history of reaction limited to a symptom list defined by an expert panel. They further used physician diagnosis and history of severe reactions to limit misclassification. Of those with CMA in adulthood, 47.1% (95% CI 43.0%,51.3%) were physician diagnosed and 39.3% had severe reactions (95% CI 35.2%,43.5%). If these estimates are accurate, they would support an increase in prevalence in adults. Interestingly, 22.7% (19.6%,26.3%) reported onset in adulthood. There is likely some misclassification by self-report, and if replicated or validated, further assessment of the underlying mechanisms of adult-onset cases would be warranted.

4.3. Patterns of CMA Prevalence Over Time

Several lines of evidence suggest an increase in sensitization and reported FA over time [9,35]; however, it is unclear if there has been a change in the prevalence of CMA. Challenges in estimating patterns of CMA over time are complicated by its natural history of resolution in a majority of individuals, as well as non-IgE mediated CMA, non-immune mediated reactions, and heterogeneity within CMA, with a subset able to tolerate baked cow's milk or having a high threshold of ingestion of unheated cow's milk prior to having symptoms. If an increase in prevalence is not due to increased self-report, it could be due to changes in sensitization, incidence, or persistence of CMA.

Studies demonstrate a relative stability of estimates of sensitization over time. In a population-based study in the US, McGowan et al. [5] compared the prevalence of sensitization to cow's milk using stored serum samples in children aged 6 to 19 years in NHANES III (1988–1994, $n = 4995$) and NHANES 2005–2006 ($n = 2901$). Sensitization was based on serum cow's milk specific IgE (≥ 0.35) and was remarkably similar over time with a sensitization of 8.3% in 1988–1994 and 8.1% in 2005–2006. There were no significant differences observed using alternate cutoffs for moderate- and high-level sensitization [36]. The study was limited to older children (there was no stored serum for younger NHANES III participants), but demonstrates fairly constant levels of sensitization over approximately 15 years.

Peters et al. also assessed changes in sensitization over time [37] in two high risk cohorts in Australia (high risk defined as at least one first degree relative with atopy) in the Melbourne Atopy Cohort Study (MACS) (born 1990–1994) and the high-risk subgroup of the HealthNuts study (born 2006–2010). In both studies, children had SPT to milk and other allergens at 12 months and the prevalence of sensitization was similar at 2.4% and 2.6%, respectively. These studies were careful to use similar definitions of CMA, which strengthens internal comparisons, although a high-risk population may not be representative of general population trends.

Prevalence may also vary over time if there are changes in the rate of resolution. There is some evidence that the rate of resolution is slowing for some foods, including milk, leading to resolution at older ages [9,38]. This is supported by the reports of higher than expected rates of CMA in US adults described above [34]. However, there are inherent methodologic challenges in comparing resolution over time between different studies for several reasons, including heterogeneity in the source population and the method of follow up.

5. Natural History of CMA

The natural history of CMA is unique in that resolution is common. However, this complicates prevalence estimates. CMA most frequently presents in infancy and early childhood, typically in the first 12 months of life, and tends to resolve with age (reviewed in [1,4,8,11,19]). For reasons previously described, there is heterogeneity in the estimated rates of resolution. Lack of recognition of resolution can lead to unnecessary exclusion of cow's milk with subsequent nutritional and growth implications. In a study in Swedish schoolchildren aged 12 years, of those with allergy to milk reporting complete avoidance (*n* = 87), only 3% had true IgE-mediated CMA on DBPCFC and 32% (*N* = 28) had resolved CMA [14]. This demonstrates the need to consider follow-up evaluations both for population patterns and to address management on an individual patient level.

Several studies of natural history have been conducted in high risk cohorts, which are better powered to identify predictors of resolution. In the Consortium for Food Allergy Research (CoFAR) [39], 512 individuals aged 3 to 15 months selected for 1) moderate-to-severe atopic dermatitis (AD) and positive SPT to egg or milk, or 2) clinical history of egg or milk allergy with confirmatory SPT were followed. At a mean age of 53 months, 53% of participants with CMA developed tolerance (defined using clinical history and OFC if needed). Similarly, in the EuroPrevall population, of those with DBPCFC-confirmed CMA, 57% developed tolerance within 1 year [28]. In the Isle of Wight cohort, CMA prevalence (defined as exposure to cow's milk with IgE-mediated symptoms within 4 hours) decreased from a maximum of 3.5% at age 1 year to 0.3% at age 18 years [40].

Sensitization has been shown to decline with age. In a high-risk cohort in Australia, there was an 8.7% prevalence of milk sensitization at 12 months, which decreased to <5% at age 18 years [41]. This study was biased by higher rates of loss to follow-up in participants with lower SES, younger parents, and lower rates of food sensitization at age 2 years, limiting the generalizability and possibly overestimating prevalence in older children. In large samples from the general population, it is often more feasible to use self-report; however, this can bias estimates. In a cross-sectional study of US children aged 0 to 18 years, no significant variation in self-reported CMA was observed by age although a non-significant higher prevalence was observed in children under 5 years (2.0%) than in older groups (1.4%–1.6%). The study was limited by a lack of standardized assessment and the cross-sectional study design, which does not provide for prospective follow-up over time [7].

Skripak et al. [42] reviewed the charts of 4117 patients seen at private and university-based practices. Of these, 1073 had a CMA diagnosis and 807 had complete data. Diagnostic criteria for CMA were (1) symptoms with exposure to cow's milk, (2) improvement in atopic dermatitis or other symptoms with avoidance, or (3) positive SPT or sIgE. They evaluated resolution within approximately 12 months (via home introduction with negative sIgE and no reaction or OFC) and identified rates of resolution at age 4, 8, 12, and 16 years of 19%, 42%, 64%, and 79%, respectively. This rate of resolution may be lower than other studies because the authors included a high risk cohort and may have missed resolved cases given that OFC was restricted to those with at least a 50% chance of passing.

Other factors impacting the determination of natural history include older age at recruitment, which may be associated with lower estimates of resolution [11], and type of CMA. Non-IgE mediated CMA is associated with a faster rate of resolution than IgE-mediated CMA and has different associated risk factors and outcomes [28,31,43,44]. For example, in EuroPrevall, 100% of patients with non-IgE mediated CMA developed tolerance within 1 year (compared with 57% of those with IgE-mediated CMA) [28].

Whereas there is clearly notable heterogeneity among studies to date, in a combined analysis of those with CMA in infancy, by age 5 years, 50% developed tolerance and by early adolescence, 75%

developed tolerance [45]. Understanding these overall trends and reasons for variation has important implications for management and treatment.

6. Factors Associated with Resolution

There is great interest in understanding factors associated with the development of tolerance because the prediction of resolution may impact decisions about treatment and the timing of diagnostic tests. In general, severe initial reactions and co-morbid atopic conditions are associated with lower rates of resolution [9,46]. Wood et al. [47] investigated predictors of development of tolerance in a subset of milk allergic patients in CoFAR and reported a significantly increased rate of resolution among those with lower baseline sIgE, SPT size, and less severe eczema [47] (reviewed in [39,48]). Specifically, among those with a baseline SPT < 5 mm, CMA resolved in 72%; among those with baseline SPT 5 to 10 mm, CMA resolved in 52%; and among those with a SPT > 10 mm, 37% developed tolerance. Thus, a baseline SPT < 5 mm was associated with a 3.7-fold higher chance of resolution compared to a baseline SPT > 10 mm. Baseline sIgE levels were similarly helpful in prediction. Among those with the lowest level of sensitization (sIgE < 2), 72% developed tolerance; among those with moderate levels (sIgE 2–10), 54% developed tolerance; and among those with higher levels (sIgE > 10), 23% developed tolerance [47].

The main proteins in cow's milk include casein (which accounts for approximately 80% of the total milk protein) and whey (which accounts for approximately 20%). The primary whey allergens are α-lactalbumin (Bosd4), β-lactoglobulin (Bosd5) and bovine serum albumin (Bosd6). Extensive heating decreases the allergenicity of whey proteins. Serum sIgE to casein can be particularly useful for both the prediction of resolution and management with regard to offering baked forms of milk. Higher levels of sIgE to casein have been associated with persistent CMA [20,48]. Lower levels of casein sIgE are associated with an increased likelihood of tolerating milk in a baked form. There is evidence that approximately 70% of patients with CMA can tolerate baked milk [25,26] and that exposure to baked milk may lead to faster development of tolerance to all forms of milk [1,49–51]. The ability to tolerate baked milk may also be a marker of a transient CMA phenotype with demonstrated immunologic changes, including increased IgG4 and decreased SPT, in patients ingesting baked milk [25]; however, the relative contributions of the phenotype and impact of intervention are still unknown [1,24,49].

7. Severity

Cow's milk is among the most common causes of food-induced anaphylaxis, along with peanuts and tree nuts [4,52]. In a prospective study of 512 children with likely milk or egg allergy, and many having multiple food allergies, including peanut (median 35 months, range 0–48 months), cow's milk was the most common cause of allergic reactions to food [53]. Cow's milk has also been implicated as a cause of severe reactions in several studies, including the European Anaphylaxis Registry, in which cow's milk was in particular an important cause of severe reactions in those less than 6 years old [54]. In another population, in 495 reactions to cow's milk, 9.1% were classified as severe [53]. Fatality is rare, but in a case series of fatalities in a European population with data on 1970 children in 10 countries with anaphylaxis to food, there were a total of 5 fatal anaphylactic reactions and 2 were attributed to cow's milk [54].

8. Nutritional and Growth Concerns

In addition to risk of severe and life-threatening anaphylaxis, CMA has important nutritional implications with impacts on growth that persist through adulthood [55]. Sinai et al. compared adult height in 87 patients with CMA compared with 36 individuals with no dietary limitations and found that patients with lifelong CMA had an average 3.8 cm lower adult height than controls [56]. The study was strengthened by the consideration of potential confounders, including chronic steroid exposure in the setting of asthma and use of attained height. Patients with CMA also have higher rates of vitamin D deficiency (reviewed in [57]).

9. Risk Factors for CMA

Risk factors and prevention are discussed briefly here in relation to epidemiology. Among children, those with CMA are more likely to be male (up to 2-fold higher risk; this reverses in adulthood with 80% of those with CMA being female). Those with CMA are also more likely to have other atopic diseases, with the prevalence of multiple food allergies identified in over 90% of CMA patients in a high risk population [29,42] and high rates of asthma, atopic dermatitis, and allergic rhinitis [31]. There is some evidence of variation by race/ethnicity, with studies suggesting non-Hispanic black and non-Hispanic white children are more likely to be sensitized to milk, based on serum IgE [32,58]. McGowan et al. compared the prevalence of sensitization by race in NHANES participants from 1991 to 1994 (aged 6 to 19 years) using sIgE to cow's milk measured in stored samples and found statistically significant differences, with the lowest rates of sensitization in white participants compared to black and Mexican-American participants (prevalence in non-Hispanic white, non-Hispanic black, and Mexican-American participants was 5.6%, 12.8%, and 12.2%, respectively) [59]. When estimates from NHANES III and 2005–2006 were compared, there were no changes in the prevalence of sensitization among any race/ethnicity included [36]. Several studies do not demonstrate differences in clinical CMA by race and ethnicity. A study that incorporated clinical and objective data in a New York City population from 1997 to 2007 found no differences in prevalence by race [52]. In another study, which compared prevalence based on self-report of reaction and physician diagnosis in a representative sample of the US population, non-Hispanic black and Asian participants had a statistically significant 50% lower risk of CMA than white participants, although limiting to those with a physician diagnosis decreased the precision and negated the significance of the estimate, and after restricting to those with positive OFC, no differences were seen by race [60]. It is unclear if disparities in access to care impacted these estimates. In the Wayne County Health, Environment, Allergy, and Asthma Longitudinal Study (WHEALS) birth cohort in Georgia, African American children were more likely to be sensitized based on sIgE to milk from birth through to age 3 to 6 years, but without increased clinically significant allergy [58].

There is evidence that genetic, epigenetic, and environmental factors play an important role in the development of CMA, although underlying mechanisms are still being elucidated [1,57]. CMA has been estimated to be 15% heritable [61], likely due to multiple genetic variants with small effect sizes. The prenatal and early childhood environment also plays a role, which is demonstrated by estimates of risk in immigrant populations. Among NHANES 2005–2006 participants aged 0 to 21 years, those who were US-born had a greater than 2-fold higher odds of sensitization to milk; among US-born children, those from immigrant families had a 1.7-fold higher risk of sensitization than children from non-immigrant households. There is evidence that children who immigrated in early life have a higher risk of sensitization to cow's milk than those who immigrated later in life. Among immigrant children, those who immigrated prior to age 2 years had non-significantly increased odds of sensitization to cow's milk (OR 3.47, $p = 0.09$) [62]. It will be important to further investigate the prenatal and early childhood exposures that underlie these differences and the implications on clinically relevant allergy.

10. Conclusions

Despite limitations in epidemiologic studies, the data support that CMA is an important problem worldwide with lifelong implications for health and, given its high prevalence in early life as well as natural history, may serve as a model for other food allergies. A better understanding of the epidemiology will guide clinician and public health efforts for prevention, diagnosis, and management of immediate hypersensitivity reactions, and treatment and prevention of growth restriction.

Table 1. Methodological issues that affect prevalence estimates of Cow's Milk Allergy (CMA).

Methodological Issue	
Variation in Reaction Types and Misclassification	- IgE-, mixed (IgE and non-IgE), and non-IgE mediated reactions - Non-immune mediated reactions (i.e., intolerance)
Study design	- Prospective cohort versus cross-sectional
Assessment of allergy	- Self-report - Physician diagnosis - Objective measures (sIgE, SPT) - Food challenge (DBPCFC—gold standard)
Diagnostic methods and distinguishing between sensitization and clinical allergy	- SPT and sIgE: Heterogeneity in method and assay used in study - Component-related diagnostics - Food challenge—double blind placebo controlled versus open
Study population	- High risk referral population versus average risk population based - Age at recruitment - Participation rates - Geographic region, genetic/environmental factors, differences in immigrant populations - Demographic factors (e.g., race/ethnicity, socioeconomic status)
Variations in phenotype	- Tolerate whole cow's milk versus extensively heated baked milk - Variation in thresholds resulting in symptoms
Natural history	- Incomplete identification of resolved cases

Table 2. (A) Estimates of prevalence from meta-analyses and individual studies. **Meta-analyses.** (B) Estimates of prevalence from meta-analyses and individual studies. **Individual Studies.**

(A)

Authors	Methods	Prevalence of IgE-Mediated Food Allergy
Nwaru et al. [27]	Systematic review and meta-analysis of European studies published between 2000 and 2012 (includes 42 primary articles on CMA)	Self-report: 2.3% (95% CI 2.1%,2.5%) SPT alone: 0.3% (95% CI 0.03%,0.6%) sIgE alone: 4.7% (95% CI 4.2%,5.1%) Food challenge: 0.6% (95% CI 0.5%,0.8%) Positive Food challenge or history of reaction: 1.6% (95% CI 1.2%,1.9%).
Rona et al. [3]	Meta-analysis of papers on food allergy published from January 1990 to December 2005	Prevalence Range Self-report: 1.2%–17% SPT alone: 0.2%–2.5% sIgE alone: 2%–9% Symptoms and sensitization: 0%–2.0% Food challenge: 0%–3.0%.

(B)

Cohort	n Age group	Prevalence of IgE-mediated CMA
EuroPrevall cohort [28]	n = 12049 enrolled n = 9336 followed to age 2Children	**Adjusted incidence:** 0.59% (adjusted for loss to follow-up/those not challenged/placebo reactors) **Natural History:** 57% developed tolerance within 1 year *Food allergy determined by (1) DBPCFC with sensitization on testing and no regular consumption of cow's milk or (2) history of reaction or improvement with elimination*
NHANES III (1988–1994) and NHANES 2005–2006 [36]	NHANES III n = 4995 NHANES 2005–2006 n = 2901 Children (age 6-19)	**Prevalence of sensitization to cow's milk** - **Overall (sIgE ≥ 0.35):** NHANESIII: 8.3% (95% CI 7.0%,9.8%); NHANES 2005–2006: 8.1% (95% CI 6.1%,10.2%) - **Moderate-level (IgE ≥ 2 kU/L):** NHANESIII: 0.4% (95% CI 0.1%,0.7%); NHANES 2005–2006: 0.5% (95% CI 0.1%,0.9%) - **High-level (IgE ≥ 15, 95% predictive probability cut-off):** NHANESIII: 0%; NHANES 2005–2006: 0.008% (95% CI −0.01%,0.03%)

Table 2. *Cont.*

Study	Sensitization	Estimated Clinical Food Allergy Rate
NHANES 2005–2006 [32] n = 8203 All ages	Overall = 5.7% By Age 1–5 year = 22.0% 6–19 year = 8.1% 20–39 year = 3.2% 40–59 year = 4.9% ≥60 = 3.8%	Overall = 0.40% By Age 1–5 year = 1.8% 6–19 year = 0.26% 20–39 year = 0.16% 40–59 year = 0.49% ≥60 = 0.33%
Melbourne, Australia High risk cohorts (first degree family history of atopy) [37] Melbourne Atopy Cohort Study (MACS) (born 1990–1994), n = 620 High-risk subset of HealthNuts Study (born 2006–2010), n = 3661 Children	**Prevalence of sensitization (SPT)** MACS 2.4% (95% CI 1.6%,3.1%) HealthNuts 2.6% (95% CI 2.0%,3.4%)	
US Cross-sectional Telephone Survey of children (2009–2010) [7] n = 38,380 Children		**Prevalence of self-reported CMA** 1.7% (95% CI = 1.5,1.8)
US Internet/Telephone Survey of adults (2015–2016) [34] n = 40,443 Adults		**Prevalence of self-reported CMA** 1.9% (95% CI = 1.8%,2.1%)
NHANES 2007–2010 [33] n = 20,686 All ages		**Prevalence of self-reported CMA** Children: 1.94% (95% CI = 1.43, 2.44) Adults: 2.64% (95% CI = 2.15, 3.13) *Prevalence in adults and children excluding those with ingestion* = 1.62% (95% CI = 1.32%, 1.92%)
New York City Urban Population (1997–2007) [52] Retrospective chart review n = 9184 Median age 7 years, range 0–21 years		**Prevalence of physician documented CMA** = 0.5% *(0.3% excluding those with no specific symptoms and no confirmatory testing)*
Israel, average-risk population (born 2004–2006) [29] n = 13,019 Enrolled at birth and followed through age 3–5 year		**Cumulative incidence of CMA diagnosis** = 0.5%
Denmark birth cohort (born 1985) [31] n = 1749 Enrolled at birth		**Incidence Age 1 y** = 2.2% (95% CI = 1.5%,2.9%) **Incidence Age 1 y confirmed by OFC** = 0.5% (95% CI = 0.2%,0.9%)
US Cross-sectional study of children (2009–2010) [60] n = 3218		**Prevalence** = 1.6% (95% CI = 1.4%,1.7%)

Author Contributions: Both authors made substantial contributions to the conception and design of the work, drafting and revising of the work and have approved the submitted version and agree to be personally accountable for ensuring that questions related to the accuracy or integrity of any part of the work are appropriately investigated, resolved, and documented in the literature.

Funding: No funding was received for this work.

Conflicts of Interest: J.D.F. has no conflicts of interest to report. S.H.S. reports royalty payments from UpToDate and from Johns Hopkins University Press; grants to his institution from the National Institute of Allergy and Infectious Diseases, from Food Allergy Research and Education, and from HAL Allergy; and personal fees from the American Academy of Allergy, Asthma, and Immunology, outside of the submitted work.

Abbreviations

95% CI	95% Confidence Interval
CoFAR	Consortium for Food Allergy Research
CMA	Cow's Milk Allergy
DBPCFC	Double Blind, Placebo-Controlled Oral Food Challenge
FPIES	Food Protein Induced Enterocolitis Syndrome
IgE	Immunoglobulin E
MACS	Melbourne Atopy Cohort Study
Mm	Millimeter
NHANES	National Health and Nutrition Examination Survey
OFC	Unblinded Oral Food Challenge
OR	Odds Ratio
sIgE	Serum-specific IgE (sIgE)
SPT	Skin Prick Test
US	United States
WHEALS	Wayne County Health, Environment, Allergy, and Asthma Longitudinal Study

References

1. Savage, J.; Johns, C.B. Food allergy: Epidemiology and natural history. *Immunol. Allergy Clin. N. Am.* **2015**, *35*, 45–59. [CrossRef] [PubMed]
2. Venter, C.; Arshad, S.H. Epidemiology of food allergy. *Pediatr. Clin. N. Am.* **2011**, *58*, 327–349. [CrossRef]
3. Rona, R.J.; Keil, T.; Summers, C.; Gislason, D.; Zuidmeer, L.; Sodergren, E.; Sigurdardottir, S.T.; Lindner, T.; Goldhahn, K.; Dahlstrom, J.; et al. The prevalence of food allergy: A meta-analysis. *J. Allergy Clin. Immunol.* **2007**, *120*, 638–646. [CrossRef] [PubMed]
4. Lifschitz, C.; Szajewska, H. Cow's milk allergy: Evidence-based diagnosis and management for the practitioner. *Eur. J. Pediatr.* **2015**, *174*, 141–150. [CrossRef] [PubMed]
5. Dunlop, J.H.; Keet, C.A. Epidemiology of food allergy. *Immunol. Allergy Clin. N. Am.* **2018**, *38*, 13–25. [CrossRef]
6. Kattan, J.D.; Cocco, R.R.; Jarvinen, K.M. Milk and soy allergy. *Pediatr. Clin. N. Am.* **2011**, *58*, 407–426. [CrossRef]
7. Gupta, R.S.; Springston, E.E.; Warrier, M.R.; Smith, B.; Kumar, R.; Pongracic, J.; Holl, J.L. The prevalence, severity, and distribution of childhood food allergy in the United States. *Pediatrics* **2011**, *128*, e9–e17. [CrossRef]
8. Boyce, J.A.; Assa'a, A.; Burks, A.W.; Jones, S.M.; Sampson, H.A.; Wood, R.A.; Plaut, M.; Cooper, S.F.; Fenton, M.J.; Arshad, S.H.; et al. Guidelines for the diagnosis and management of food allergy in the United States: Summary of the NIAID-sponsored expert panel report. *Nutrition* **2011**, *27*, 253–267. [CrossRef]
9. Sicherer, S.H.; Sampson, H.A. Food allergy: A review and update on epidemiology, pathogenesis, diagnosis, prevention, and management. *J. Allergy Clin. Immunol.* **2018**, *141*, 41–58. [CrossRef]
10. Sampson, H.A.; Aceves, S.; Bock, S.A.; James, J.; Jones, S.; Lang, D.; Nadeau, K.; Nowak-Wegrzyn, A.; Oppenheimer, J.; Perry, T.T.; et al. Food allergy: A practice parameter update-2014. *J. Allergy Clin. Immunol.* **2014**, *134*, 1016–1025. [CrossRef]
11. Host, A. Frequency of cow's milk allergy in childhood. *Ann. Allergy Asthma Immunol.* **2002**, *89*, 33–37. [CrossRef]

12. Sampson, H.A. Food allergy. Part 1: Immunopathogenesis and clinical disorders. *J. Allergy Clin. Immunol.* **1999**, *103*, 717–728. [CrossRef]

13. Agyemang, A.; Nowak-Wegrzyn, A. Food protein-induced enterocolitis syndrome: A comprehensive review. *Clin. Rev. Allergy Immunol.* **2019**. [CrossRef] [PubMed]

14. Winberg, A.; West, C.E.; Strinnholm, A.; Nordstrom, L.; Hedman, L.; Ronmark, E. Assessment of allergy to milk, egg, cod, and wheat in Swedish schoolchildren: A population based cohort study. *PLoS ONE* **2015**, *10*, e0131804. [CrossRef] [PubMed]

15. Katz, Y.; Goldberg, M.R.; Rajuan, N.; Cohen, A.; Leshno, M. The prevalence and natural course of food protein-induced enterocolitis syndrome to cow's milk: A large-scale, prospective population-based study. *J. Allergy Clin. Immunol.* **2011**, *127*, 647–653. [CrossRef]

16. Woods, R.K.; Stoney, R.M.; Raven, J.; Walters, E.H.; Abramson, M.; Thien, F.C. Reported adverse food reactions overestimate true food allergy in the community. *Eur. J. Clin. Nutr.* **2002**, *56*, 31–36. [CrossRef]

17. Sicherer, S.H. Epidemiology of food allergy. *J. Allergy Clin. Immunol.* **2011**, *127*, 594–602. [CrossRef]

18. Chafen, J.J.; Newberry, S.J.; Riedl, M.A.; Bravata, D.M.; Maglione, M.; Suttorp, M.J.; Sundaram, V.; Paige, N.M.; Towfigh, A.; Hulley, B.J.; et al. Diagnosing and managing common food allergies: A systematic review. *JAMA* **2010**, *303*, 1848–1856. [CrossRef] [PubMed]

19. Host, A. Cow's milk protein allergy and intolerance in infancy. Some clinical, epidemiological and immunological aspects. *Pediatr. Allergy Immunol.* **1994**, *5*, 1–36. [CrossRef]

20. Garcia-Ara, C.; Boyano-Martinez, T.; Diaz-Pena, J.M.; Martin-Munoz, F.; Reche-Frutos, M.; Martin-Esteban, M. Specific IgE levels in the diagnosis of immediate hypersensitivity to cow's milk protein in the infant. *J. Allergy Clin. Immunol.* **2001**, *107*, 185–190. [CrossRef]

21. Branum, A.M.; Lukacs, S.L. Food allergy among children in the United States. *Pediatrics* **2009**, *124*, 1549–1555. [CrossRef] [PubMed]

22. Keet, C.A.; Wood, R.A.; Matsui, E.C. Limitations of reliance on specific IgE for epidemiologic surveillance of food allergy. *J. Allergy Clin. Immunol.* **2012**, *130*, 1207–1209. [CrossRef] [PubMed]

23. Wang, J.; Godbold, J.H.; Sampson, H.A. Correlation of serum allergy (IgE) tests performed by different assay systems. *J. Allergy Clin. Immunol.* **2008**, *121*, 1219–1224. [CrossRef] [PubMed]

24. Leonard, S.A.; Caubet, J.C.; Kim, J.S.; Groetch, M.; Nowak-Wegrzyn, A. Baked milk- and egg-containing diet in the management of milk and egg allergy. *J. Allergy Clin. Immunol. Pract.* **2015**, *3*, 13–23. [CrossRef] [PubMed]

25. Wang, J.; Sampson, H.A. Food allergy. *J. Clin. Investig.* **2011**, *121*, 827–835. [CrossRef] [PubMed]

26. Nowak-Wegrzyn, A.; Bloom, K.A.; Sicherer, S.H.; Shreffler, W.G.; Noone, S.; Wanich, N.; Sampson, H.A. Tolerance to extensively heated milk in children with cow's milk allergy. *J. Allergy Clin. Immunol.* **2008**, *122*, 342–347. [CrossRef]

27. Nwaru, B.I.; Hickstein, L.; Panesar, S.S.; Muraro, A.; Werfel, T.; Cardona, V.; Dubois, A.E.; Halken, S.; Hoffmann-Sommergruber, K.; Poulsen, L.K.; et al. The epidemiology of food allergy in Europe: A systematic review and meta-analysis. *Allergy* **2014**, *69*, 62–75. [CrossRef] [PubMed]

28. Schoemaker, A.A.; Sprikkelman, A.B.; Grimshaw, K.E.; Roberts, G.; Grabenhenrich, L.; Rosenfeld, L.; Siegert, S.; Dubakiene, R.; Rudzeviciene, O.; Reche, M.; et al. Incidence and natural history of challenge-proven cow's milk allergy in European children–EuroPrevall birth cohort. *Allergy* **2015**, *70*, 963–972. [CrossRef] [PubMed]

29. Katz, Y.; Rajuan, N.; Goldberg, M.R.; Eisenberg, E.; Heyman, E.; Cohen, A.; Leshno, M. Early exposure to cow's milk protein is protective against IgE-mediated cow's milk protein allergy. *J. Allergy Clin. Immunol.* **2010**, *126*, 77–82. [CrossRef]

30. Osterballe, M.; Hansen, T.K.; Mortz, C.G.; Host, A.; Bindslev-Jensen, C. The prevalence of food hypersensitivity in an unselected population of children and adults. *Pediatr. Allergy Immunol.* **2005**, *16*, 567–573. [CrossRef]

31. Host, A.; Halken, S.; Jacobsen, H.P.; Christensen, A.E.; Herskind, A.M.; Plesner, K. Clinical course of cow's milk protein allergy/intolerance and atopic diseases in childhood. *Pediatr. Allergy Immunol.* **2002**, *13*, 23–28. [CrossRef]

32. Liu, A.H.; Jaramillo, R.; Sicherer, S.H.; Wood, R.A.; Bock, S.A.; Burks, A.W.; Massing, M.; Cohn, R.D.; Zeldin, D.C. National prevalence and risk factors for food allergy and relationship to asthma: Results from the National Health and Nutrition Examination Survey 2005–2006. *J. Allergy Clin. Immunol.* **2010**, *126*, 798–806. [CrossRef]

33. McGowan, E.C.; Keet, C.A. Prevalence of self-reported food allergy in the National Health and Nutrition Examination Survey (NHANES) 2007–2010. *J. Allergy Clin. Immunol.* **2013**, *132*, 1216–1219. [CrossRef] [PubMed]

34. Gupta, R.S.; Warren, C.M.; Smith, B.M.; Jiang, J.; Blumenstock, J.A.; Davis, M.M.; Schleimer, R.P.; Nadeau, K.C. Prevalence and severity of food allergies among US adults. *JAMA Netw. Open* **2019**, *2*, e185630. [CrossRef]

35. Keet, C.A.; Savage, J.H.; Seopaul, S.; Peng, R.D.; Wood, R.A.; Matsui, E.C. Temporal trends and racial/ethnic disparity in self-reported pediatric food allergy in the United States. *Ann. Allergy Asthma Immunol.* **2014**, *112*, 222–229. [CrossRef] [PubMed]

36. McGowan, E.C.; Peng, R.D.; Salo, P.M.; Zeldin, D.C.; Keet, C.A. Changes in food-specific IgE over time in the National Health and Nutrition Examination Survey (NHANES). *J. Allergy Clin. Immunol. Pract.* **2016**, *4*, 713–720. [CrossRef]

37. Peters, R.L.; Koplin, J.J.; Allen, K.J.; Lowe, A.J.; Lodge, C.J.; Tang, M.L.K.; Wake, M.; Ponsonby, A.L.; Erbas, B.; Abramson, M.J.; et al. The prevalence of food sensitization appears not to have changed between 2 Melbourne cohorts of high-risk infants recruited 15 years apart. *J. Allergy Clin. Immunol. Pract.* **2018**, *6*, 440–448. [CrossRef] [PubMed]

38. Kattan, J. The prevalence and natural history of food allergy. *Curr. Allergy Asthma Rep.* **2016**, *16*, 47. [CrossRef] [PubMed]

39. Sampson, H.A.; Berin, M.C.; Plaut, M.; Sicherer, S.H.; Jones, S.; Burks, A.W.; Lindblad, R.; Leung, D.Y.M.; Wood, R.A. The consortium for food allergy research (CoFAR): The first generation. *J. Allergy Clin. Immunol.* **2019**, *143*, 486–493. [CrossRef]

40. Venkataraman, D.; Erlewyn-Lajeunesse, M.; Kurukulaaratchy, R.J.; Potter, S.; Roberts, G.; Matthews, S.; Arshad, S.H. Prevalence and longitudinal trends of food allergy during childhood and adolescence: Results of the Isle of Wight Birth Cohort study. *Clin. Exp. Allergy* **2018**, *48*, 394–402. [CrossRef] [PubMed]

41. Alduraywish, S.A.; Lodge, C.J.; Vicendese, D.; Lowe, A.J.; Erbas, B.; Matheson, M.C.; Hopper, J.; Hill, D.J.; Axelrad, C.; Abramson, M.J.; et al. Sensitization to milk, egg and peanut from birth to 18 years: A longitudinal study of a cohort at risk of allergic disease. *Pediatr. Allergy Immunol.* **2016**, *27*, 83–91. [CrossRef]

42. Skripak, J.M.; Matsui, E.C.; Mudd, K.; Wood, R.A. The natural history of IgE-mediated cow's milk allergy. *J. Allergy Clin. Immunol.* **2007**, *120*, 1172–1177. [CrossRef]

43. Host, A.; Jacobsen, H.P.; Halken, S.; Holmenlund, D. The natural history of cow's milk protein allergy/intolerance. *Eur. J. Clin. Nutr.* **1995**, *49*, S13–S18. [PubMed]

44. Santos, A.; Dias, A.; Pinheiro, J.A. Predictive factors for the persistence of cow's milk allergy. *Pediatr. Allergy Immunol.* **2010**, *21*, 1127–1134. [CrossRef] [PubMed]

45. Spergel, J.M. Natural history of cow's milk allergy. *J. Allergy Clin. Immunol.* **2013**, *131*, 813–814. [CrossRef]

46. Hill, D.J.; Firer, M.A.; Ball, G.; Hosking, C.S. Natural history of cow's milk allergy in children: Immunological outcome over 2 years. *Clin. Exp. Allergy* **1993**, *23*, 124–131. [CrossRef] [PubMed]

47. Wood, R.A.; Sicherer, S.H.; Vickery, B.P.; Jones, S.M.; Liu, A.H.; Fleischer, D.M.; Henning, A.K.; Mayer, L.; Burks, A.W.; Grishin, A.; et al. The natural history of milk allergy in an observational cohort. *J. Allergy Clin. Immunol.* **2013**, *131*, 805–812. [CrossRef]

48. Host, A.; Halken, S. Cow's milk allergy: Where have we come from and where are we going? *Endocr. Metab. Immune Disord. Drug Targets* **2014**, *14*, 2–8. [CrossRef]

49. Kim, J.S.; Nowak-Wegrzyn, A.; Sicherer, S.H.; Noone, S.; Moshier, E.L.; Sampson, H.A. Dietary baked milk accelerates the resolution of cow's milk allergy in children. *J. Allergy Clin. Immunol.* **2011**, *128*, 125–131. [CrossRef]

50. Matsuo, H.; Yokooji, T.; Taogoshi, T. Common food allergens and their IgE-binding epitopes. *Allergol. Int.* **2015**, *64*, 332–343. [CrossRef]

51. Caubet, J.C.; Nowak-Wegrzyn, A.; Moshier, E.; Godbold, J.; Wang, J.; Sampson, H.A. Utility of casein-specific IgE levels in predicting reactivity to baked milk. *J. Allergy Clin. Immunol.* **2013**, *131*, 222–224. [CrossRef]

52. Taylor-Black, S.; Wang, J. The prevalence and characteristics of food allergy in urban minority children. *Ann. Allergy Asthma Immunol.* **2012**, *109*, 431–437. [CrossRef] [PubMed]

53. Fleischer, D.M.; Perry, T.T.; Atkins, D.; Wood, R.A.; Burks, A.W.; Jones, S.M.; Henning, A.K.; Stablein, D.; Sampson, H.A.; Sicherer, S.H. Allergic reactions to foods in preschool-aged children in a prospective observational food allergy study. *Pediatrics* **2012**, *130*, e25–e32. [CrossRef] [PubMed]

54. Grabenhenrich, L.B.; Dolle, S.; Moneret-Vautrin, A.; Kohli, A.; Lange, L.; Spindler, T.; Rueff, F.; Nemat, K.; Maris, I.; Roumpedaki, E.; et al. Anaphylaxis in children and adolescents: The European anaphylaxis registry. *J. Allergy Clin. Immunol.* **2016**, *137*, 1128–1137. [CrossRef] [PubMed]

55. Robbins, K.A.; Wood, R.A.; Keet, C.A. Milk allergy is associated with decreased growth in US children. *J. Allergy Clin. Immunol.* **2014**, *134*, 1466–1468. [CrossRef] [PubMed]

56. Sinai, T.; Goldberg, M.R.; Nachshon, L.; Amitzur-Levy, R.; Yichie, T.; Katz, Y.; Monsonego-Ornan, E.; Elizur, A. Reduced final height and inadequate nutritional intake in cow's milk-allergic young adults. *J. Allergy Clin. Immunol. Pract.* **2019**, *7*, 509–515. [CrossRef]

57. Jansen, P.R.; Petrus, N.C.M.; Venema, A.; Posthuma, D.; Mannens, M.; Sprikkelman, A.B.; Henneman, P. Higher polygenetic predisposition for asthma in cow's milk allergic children. *Nutrients* **2018**, *10*, 1582. [CrossRef] [PubMed]

58. Joseph, C.L.; Zoratti, E.M.; Ownby, D.R.; Havstad, S.; Nicholas, C.; Nageotte, C.; Misiak, R.; Enberg, R.; Ezell, J.; Johnson, C.C. Exploring racial differences in IgE-mediated food allergy in the WHEALS birth cohort. *Ann. Allergy Asthma Immunol.* **2016**, *116*, 219–224. [CrossRef]

59. McGowan, E.C.; Matsui, E.C.; Peng, R.; Salo, P.M.; Zeldin, D.C.; Keet, C.A. Racial/ethnic and socioeconomic differences in self-reported food allergy among food-sensitized children in National Health and Nutrition Examination Survey III. *Ann. Allergy Asthma Immunol.* **2016**, *117*, 570–572. [CrossRef]

60. Warren, C.M.; Jhaveri, S.; Warrier, M.R.; Smith, B.; Gupta, R.S. The epidemiology of milk allergy in US children. *Ann. Allergy Asthma Immunol.* **2013**, *110*, 370–374. [CrossRef]

61. Tsai, H.J.; Kumar, R.; Pongracic, J.; Liu, X.; Story, R.; Yu, Y.; Caruso, D.; Costello, J.; Schroeder, A.; Fang, Y.; et al. Familial aggregation of food allergy and sensitization to food allergens: A family-based study. *Clin. Exp. Allergy* **2009**, *39*, 101–109. [CrossRef] [PubMed]

62. Keet, C.A.; Wood, R.A.; Matsui, E.C. Personal and parental nativity as risk factors for food sensitization. *J. Allergy Clin. Immunol.* **2012**, *129*, 169–175. [CrossRef] [PubMed]

Review

Cow's Milk Allergy: Immunomodulation by Dietary Intervention

Enza D'Auria [1,*], Silvia Salvatore [2], Elena Pozzi [1], Cecilia Mantegazza [1], Marco Ugo Andrea Sartorio [1], Licia Pensabene [3], Maria Elisabetta Baldassarre [4], Massimo Agosti [2], Yvan Vandenplas [5] and GianVincenzo Zuccotti [1]

[1] Department of Pediatrics, Vittore Buzzi Children's Hospital-University of Milan, 20154 Milan, Italy; elena.pozzi@asst-fbf-sacco.it (E.P.); cecilia.mantegazza@asst-fbf-sacco.it (C.M.); marcoua.sartorio@gmail.com (M.U.A.S.); gianvincenzo.zuccotti@unimi.it (G.V.Z.)

[2] Department of Pediatrics, Ospedale "F. Del Ponte", University of Insubria, 21100 Varese, Italy; silvia.salvatore@uninsubria.it (S.S.); massimo.agosti@uninsubria.it (M.A.)

[3] Department of Medical and Surgical Sciences, Pediatric Unit, University "Magna Graecia" of Catanzaro, 88100 Catanzaro, Italy; pensabene@unicz.it

[4] Neonatology and Neonatal Intensive Care Unit, Department of Biomedical Science and Human Oncology, "Aldo Moro" University of Bari, P.zza Giulio Cesare 11, 70124 Bari, Italy; mariaelisabetta.baldassarre@uniba.it

[5] KidZ Health Castle, Universitair Ziekenhuis Brussel, Vrije Universiteit Brussel, 1090 Brussels, Belgium; yvan.vandenplas@uzbrussel.be

* Correspondence: enza.dauria@unimi.it

Received: 25 May 2019; Accepted: 17 June 2019; Published: 21 June 2019

Abstract: Cow's milk proteins cause allergic symptoms in 2% to 3% of all infants. In these individuals, the physiological mechanism of tolerance is broken with subsequent possible sensitization to antigens, which can lead eventually to allergic responses. The present review aims to provide an overview of different aspects of immune modulation by dietary intervention in cow's milk allergy (CMA). It focuses on pathogenetic mechanisms of different CMA related disorders, e.g., gastroesophageal reflux and eosinophilic esophagitis, highlighting the role of dietary management on innate and adaptive immune systems. The traditional dietary management of CMA has greatly changed in the last years, moving from a passive approach, consisting of an elimination diet to relieve symptoms, to a "proactive" one, meaning the possibility to actively modulate the immune system. Thus, new insights into the role of hydrolysates and baked milk in immunomodulation are addressed here. Additionally, nutritional components, such as pre- and probiotics, may target the immune system via microbiota, offering a possible road map for new CMA prevention and treatment strategies.

Keywords: cow's milk allergy; immune system; dietary intervention; bioactive peptides; gut microbiota; prebiotics; probiotics

1. Introduction

Cow's milk allergy (CMA) is one of the most common food allergies in early childhood, with an estimated prevalence of 2% to 3% [1]. A growing body of evidence suggests a close relationship between immunoinflammation and gastrointestinal (GI) motility triggered by dietary antigens [2]. Cow's milk (CM) free diets and in particular extensive hydrolyzed formulas may reduce gastrointestinal (GI) symptoms due to both immune mechanisms and motility alterations, such as reduced gastric emptying time. Food allergy plays a central role in driving the allergic reaction in eosinophilic esophagitis (EoE) and cow's milk is the single most common food allergen causing esophageal inflammation [3].

Dietary elimination therapy is thought to target the adaptive immune system, by suppressing antigen-driven T-cell response. Moreover, the role of milk lipids as potential triggers of milk-induced inflammation in EoE is emerging. These findings provide new insights into EoE pathogenetic mechanisms that might change the paradigm of allergy, as a protein antigen-driven response. In the last decade, much has changed in the treatment of food allergy, switching from a passive approach, consisting of a restrictive diet to relieve symptoms, to a "proactive" one, meaning the possibility to actively modulate the immune system.

Protein hydrolysates have been recognized as a potent source of bioactive peptides [4]. They may act locally, e.g., in the gut, by modulating the intestinal microbiota, thereby playing a role in inducing oral tolerance to milk proteins. Additionally, the role of baked milk as a possible form of oral immunotherapy has emerged [5,6]. Maintaining tolerance requires complex interactions between non-immune cells and cells that belong to the gut-associated lymphoid tissue (GALT). Regulatory T cells (Treg) play a crucial role in tolerance. Although different subtypes of Tregs have been identified, the pivotal roles of Foxp3+ in oral tolerance are not completely understood. Gut microbes induce the activation of Tregs, while the same cells are depleted in germ-free mice [7]. Gut microbiota dysbiosis induces alterations in gut function resulting in aberrant Th2 responses towards allergic, rather than tolerogenic response [8]. Therefore, the possibility to actively immunomodulate the immune system targeting microbiota by nutritional factors, e.g., prebiotics and probiotics, represents a novel research strategy. The present review aims to give an overview of the different aspects of immunomodulation by dietary intervention in CMA, based on the most recent evidence.

2. Cow's Milk Allergy and Allergic Dysmotility: A Pleiomorphic Disorder

CMA affects many organs with immediate and delayed reactions [9]. According to the Hill and Hosking classification, CMA may manifest in three different ways: (1) The IgE-sensitized group showing immediate cutaneous reactions and anaphylaxis; (2) the non-IgE-sensitized group with gastrointestinal (GI) symptoms, developing within hours after ingesting moderate amounts of CM; and (3) the group with GI disturbances with or without respiratory symptoms and/or eczema/urticaria, occurring after several hours or days [10].

Allergies may involve the GI tract from mouth to rectum, and may be characterized by an acute (anaphylaxis) or delayed onset [9], the latter including eosinophilic gastroenteropathy, allergic proctocolitis, food protein-induced enterocolitis syndrome, and enteropathy [11]. Allergic dysmotility encompasses different entities, including gastroesophageal reflux disease (GERD), dyspepsia, and constipation, where digestive motility is altered by the neuro-immune-muscle inflammatory interaction triggered by the cow's milk proteins in predisposed individuals [9,11]. Up to half of the cases of GERD in infants younger than 1 year have been related to CMA based on clinical presentation and improvement on CM [12–14]. However, many symptoms, such as weight loss, failure to thrive, food-refusal, irritability, excessive crying, regurgitation, vomiting, anemia, wheezing, and sleep disturbances, may be expressions of both entities [13,14]. By contrast, multiple organ involvement, mucous or bloody stools, an increase in eosinophils blood count, atopic dermatitis, or recurrent bronchitis are more suggestive of CMA [13,14].

However, the reasons for clinical improvement after being started on a milk free diet may vary: The physiological resolution of symptoms over time, an improvement in gastric emptying due to the use of hydrolyzed proteins, or a placebo effect on parental anxiety.

The strongest evidence that intestinal allergic responses can modulate enteric motility originates from a series of studies in animal models. Cytokines production by T helper (Th) 2 cells, and recruitment and activation of either mast cells or eosinophils have been suggested as the major mechanisms potentially linking allergic responses and dysmotility [15]. A Th 2 polarized response determines the release of interleukin (IL)-4 and -13, cytokines that alter motility by upregulating transforming growth factor-beta, with spontaneous contractility of smooth muscle [15]. In a murine model of luminal sensitization, the allergen exposure induced a skew towards a Th2 response with tissue infiltration of IgE degranulating mast cells in the mucosa and mesenteric lymph nodes, causing enteropathy with loose stools and poor weight gain [16]. Moreover, mast cells and their mediators may cause sensorimotor dysfunction of the gut through interactions with the enteric nervous system [17–19].

An increase in mast cell density and number in close proximity to submucosal nerve endings has been demonstrated in children with functional dyspepsia and allergies [2]. In allergic children, milk allergen exposure induces rapid degranulation of gastric antral lamina propria mast cells and eosinophils and the release of mast cell tryptase, which interacts with proteinase-activated receptors that colocalize with gastric mucosal nerve fibers. Subsequent electrogastrographic myoelectrical abnormalities occur, determining atopy-related dyspeptic symptoms [2].

In the esophagus, animal studies have shown the degranulation of mast cells and the release of histamine when the mucosa was exposed and injured by acid [20] or when the stress-induced corticotrophin-releasing factor (CRF) signaling system [21,22] was involved. A rise in mast cell numbers and released cytokines has also been demonstrated in humans with non-erosive reflux disease and chest pain syndromes [23,24]. Mast cells play an important role in the esophageal inflammatory reaction and nociception by increasing vagal nociceptive C fibers' excitability [25,26]. Proteinase activated receptors 2 (PAR2), a target receptor of mast cell derived tryptase, is expressed in epithelial cells, GI smooth muscle cells, and capsaicin-sensitive neurons and regulates GI mucosa barrier functioning and inflammation [27–29]. PAR2-mediated pathways have been demonstrated to be important in the pathogenesis of GERD-associated mucosal alterations, such as dilated intercellular spaces and a decrease of tight junction proteins [23,30].

The diagnosis of CMA in patients with GI symptoms is often challenging because of the delayed type of allergic reaction and the absence of specific diagnostic tests: Skin prick or serum specific IgE are usually negative, while atopy patch tests have shown conflicting data [31,32]. Hence, elimination diet followed by an oral open or double blind standardized challenge, in infants or older children is the recommended test to diagnose CMA [33,34]. In allergic patients, GI symptoms disappear in up to 2 to 4 weeks on a CM free diet and relapse when milk is reintroduced [33]. Extensive hydrolisate milk formulas are indicated as the first dietetic choice, whereas elemental formulas should be reserved for more severe cases or eosinophilic disorders [33,35]. A hypoallergenic diet has been proven effective in reducing mast cell mucosal infiltration, thus normalizing immune-nerve interactions and improving motor abnormalities [2,36]. At the same time, hydrolyzed proteins may be effective in these children due to accelerated gastric emptying [13,14]. In patients with persistent symptoms who are on a diet, esophageal pH-impedance may provide data on acid and non-acid reflux exposure and temporal reflux–symptoms association whereas esophagogastroduodenoscopy with esophageal and duodenal biopsies may reveal the presence and the type of esophagitis and/or enteropathy [13].

3. Eosinophilic Esophagitis: Insights on Pathogenetic Mechanisms and Dietary Immunomodulation

EoE is a chronic immune-mediated antigen-driven inflammatory disorder characterized by symptoms of esophageal dysfunction and histologic evidence of eosinophilic-predominant inflammation of the esophagus [37,38].

Clinical presentation varies according to age. In infants and younger children, the most common symptoms are food refusal, vomiting, irritability, and failure to thrive. Dysphagia, choking, and food impaction are the most common symptoms in school children and adolescents, as well as in adults [39,40].

Diagnostic criteria for EoE are: (a) Symptoms of esophageal dysfunction; (b) eosinophilic esofageal inflammation, ≥15 eosinophils (Eo)/per high power field (HPF); and (c) exclusion of other causes of esophageal eosinophilia [37,41].

EoE pathogenesis is closely related to atopy. About 70% of patients have a history of atopy, including asthma, IgE-mediated food allergy, allergic rhinitis, and atopic dermatitis. Similarly, about 2/3 of patients have at least one family member with an atopic condition [42]. Peripheral blood eosinophilia is observed in about 50% of patients, and elevated levels of IgE can be detected in 80% of patients. Moreover, up to 80% of patients have positive skin prick tests (SPTs) and/or specific IgE (sIgE) for food or aeroallergens.

Food allergy plays a central role in driving the allergic reaction in EoE, as demonstrated by clinical and histological remission on dietary restriction therapy and exacerbation after food reintroduction [43].

However, the lack of immediate symptoms after food ingestion, the low predictive value of SPTs or SIgE, and the poor response to anti-IgE therapy [44] disprove the hypothesis of a merely IgE- mediated food reaction. The pathogenesis of EoE is most likely a mixed IgE and non-IgE/cell mediated food reaction, in which Th2 cytokines, particularly thymic stromal lymphopoietin (TSLP), interleukin (IL)-4, IL5, IL13, and transforming growth factor-β (TGF-β), and eosinophilic chemokines (eotaxin 1-3/CCL11-CCL24-CCL26 and RANTES/CCL5) play a central role in eosinophilic recruitment, perpetuating local Th2-inflammation. Eosinophils cause tissue damage, remodeling, and fibrosis.

Antigens, primarily food ones, activate the innate and adaptive immune systems, priming the Th2 immune response [45,46].

The goal of therapy is to induce clinical and histological remission (defined as esophageal Eo < 15/HPF). Treatment strategies include drugs (e.g., proton pump inhibitors, corticosteroids) and elimination diets. These therapies both act on esophageal inflammation.

An avoidance diet is thought to target the adaptive immune system, by suppressing antigen-driven T-cell response; it requires elimination of food antigen/s, demonstration of remission, and subsequent sequential reintroduction of each single food in order to identify the causative agent [47–49].

Different elimination strategies are currently used in EoE: Elemental diet and empirical elimination diets, such as the six-food groups elimination diet (milk, wheat, soybean/legumes, egg, peanut/nuts, and fish/shellfish) (SFED), four-food elimination diet (milk, wheat, egg, legumes/soy) (FFED), or allergy testing-based food elimination diet (ATBD).

The efficacy of these different dietary treatments ranges from 90.8% for the elemental diet, to 72% for SFED, 55% for FFED, and 45.5% for ATBD [43,50].

After sequential food reintroduction, in the majority of patients (45–85%), one or two causative foods are identifiable. (2) Much evidence supports CM as a major trigger food for EoE. CM is the single most common food allergen causing esophageal inflammation. In both adults and children studies on empiric SFED-FFED or two-food elimination diet (TFED), CM was identified as the single trigger in 18% to 50% of adult patients [3,51,52] and from 30% up to 60% of pediatric patients, in prospective studies [3,52,53].

Moreover, CM elimination diet induced a significant reduction in the mean peak pre- and post-treatment eosinophil count in 68.2% of patients [54,55]. Sensitization to CM (serum sIgE and/or positive skin prick test) was detected in 45.9% of children with EoE, in a large cohort of European EoE children [42].

However, it is known that sIgE correlates poorly with food triggers. Furthermore, it has been observed that CM sIgE levels are paradoxically lower in responders to the CM elimination diet, than in non-responders [56]. These findings are in keeping with the evidence of a non-IgE-mediated reaction.

Nevertheless, even in patients with negative skin prick tests, sIgE to whey protein Bos d 4 (α-lactalbumin) and Bos d 5 (β-lactoglobulin) are frequently detectable by ImmunoCAP assay. Therefore, although IgE response is not the primary mechanism in EoE, Bos d 4 and Bos d 5, minor components of CM, can act as primary antigens for IgE response, triggering T cell-driven inflammation in EoE.

Another antibody isotype currently investigated in EoE is IgG4. IgG4 is an immunoglobulin involved in allergen tolerance and anti-inflammatory activity. High levels of serum and esophageal IgG4 have been found in active EoE in adults [57]. Recently, it has been demonstrated that levels of esophageal IgG4 in EoE patients correlate with the number of esophageal eosinophils, with basal zone hyperplasia, and with levels of IL4 to IL13, and especially IL10, providing evidence that IgG4 correlates with disease activity (i.e., eosinophils and basal zone hyperplasia) and Th 2 inflammation [58]. The highest titers of IgG4 in EoE are to CM and gluten. Levels of serum IgG4 to CM proteins (Bos d 4, Bos d 5 and casein, Bos d 8) are higher in active EoE than in controls. These data suggest a pathogenetic role for IgG4, especially to CM proteins, in EoE.

However, levels decrease on a CM elimination diet not only in subjects with histological remission, but also in subjects without remission, suggesting that IgG4 could be only an epiphenomenon in EoE [59].

Invariant natural killer T cells (iNKTs), a subset of T cells, play a key role in IgE-mediated CM allergy. They are activated by sphingolipids (SLs) rather than by protein antigens. Milk sphingomyelin (milk-SM) activate iNKTs, induce iNKTs' proliferation, and promote Th2 response [60]. In children with IgE mediated food allergy, especially to CM milk, iNKTs are reduced. Children with active EoE have lower peripheral blood iNKTs with greater Th2 response to milk-SM compared to children with controlled EoE and controls. Esophageal iNKTs are higher in active EoE than in controlled EoE and healthy children. Low peripheral iNKTs could reflect recruitment on site of esophageal inflammation, suggesting a pathogenetic role of iNKTs in EoE [61].

These findings could explain why some foods are more able to trigger EoE than others, and provide new insights into EoE pathogenetic mechanisms. Milk lipids as potential triggers of milk-induced inflammation may change the paradigm of allergy, as a protein antigen-driven response.

4. Immune Modulation by Hydrolysate Proteins

Great consideration has recently been given to hydrolysate proteins. Their capacity to reduce allergic symptoms due to the lack of IgE binding epitopes is common knowledge [62]. Therefore, infant formulas containing extensively hydrolyzed proteins are tolerated by allergic infants and are recommended for the management of children with CMA symptoms [63,64].

Furthermore, hydrolysates have been demonstrated as capable of reducing the gut intestinal permeability [65] in ex vivo models. The improved barrier function may decrease the antigen uptake and the antigen contact with the intestinal immune cells in the lamina propria, which may lead to a reduction in allergic symptoms [66].

More recent evidence, however, suggests that hydrolyzed peptides also have an active role in modulating the immune system through different mechanisms both in children with CMA and in those at risk of developing CMA [67,68]

In vitro and ex vivo studies have described hydrolysates as having local and systemic effects on the immune system, including their ability to strengthen the epithelial barrier, via many immunomodulatory mechanisms, such as increasing the regulatory cytokines (e.g., IL-10) or decreasing the inflammatory markers, including cyclo-oxygenase 2 (COX-2), NF-kB, and IL-8, and also by the expression of genes encoding for tight junction proteins [65].

Protein hydrolysates act on the intestinal mesenteric lymph nodes, increasing the number of Treg cells, which are crucial in inducing tolerance [69]. These effects have been demonstrated in murine models analyzing both peptides derived from casein and whey proteins [70–72]. Besides enhancing the Treg number in the mesenteric lymphonodes, other effects on the local immune system have been described. In particular, hydrolysates from bovine milk seem to have an anti-inflammatory effect in vivo that is dependent on the protein source (casein or whey). These effects have been observed in animal models after inducing experimental colitis. While pro-inflammatory cytokines, IL-1beta, IL-17, TNF-alpha, and IFN, decreased, an increase in the regulatory cytokine, IL-10, and reduced macroscopical and microscopical damage of the colon mucosa was observed after administration of casein hydrolysate or casein glycomacropeptide [71,73,74]. Based on this and other experimental findings, feeding with a casein eHF is actually recommended as a first-line choice for the management of food protein induced enterocolitis [75].

Feeding with a partially hydrolyzed whey protein diet reduced allergic skin response in a cow's milk allergy mouse model, by decreasing the levels of Th1, Th17, and enhancing regulatory T and B cells in Peyer plaques after whey challenge [76]. Interestingly, a sequenced peptide derived from whey has been demonstrated to reduce the whey antibody levels in animal models.

Protein hydrolysates also have an effect on the systemic immune system probably via small peptides that pass through the intestinal barrier and enter the systemic circulation. An increase in IL-10 producing regulatory B cells was observed after inducing oral tolerance by administration of intact casein in casein-allergic mice and in the spleen after whey hydrolysates' administration, respectively [76,77].

Peptides can exert their immune modulatory effects via different mechanisms, among which the direct stimulation of the receptors on immune cells via Toll-like receptors (TLR) is one of the most important [78]. Other mechanisms have been described, including cells' absorption via transporter or via endocytosis that leads to interactions with inflammatory signaling pathways or conversely to the inhibition of inflammatory signaling pathways [78] (Figure 1).

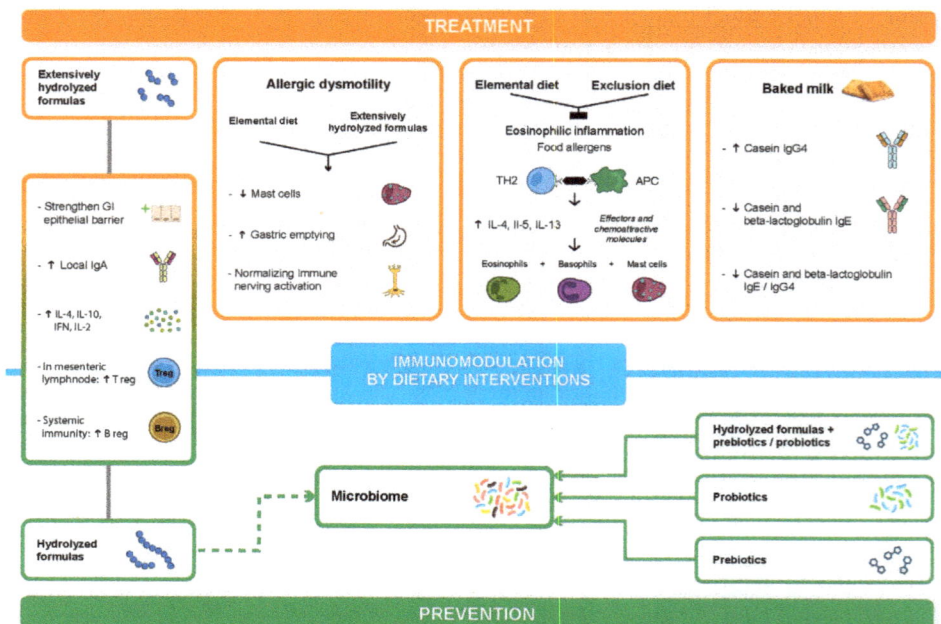

Figure 1. Immunomodulation by dietary interventions. TH2 = T cell helper 2; APC = Antigen Presenting Cell; IL-2 = Interleukin-2; IL-4 = Interleukin-4; IL-5 = Interleukin-5; IL-13 = Interleukin-13; IL-10 = Interleukin-10; IFN = Interferon; T-reg = Regulatory T Cell; B-reg = Regulatory B Cell; ↑ = increase; ↓ = decrease.

The majority of studies on the effects of immune modulation by hydrolyzed formula were conducted on ex vivo models; there are very few data regarding how to speed up an increase in tolerance in infants fed with hydrolyzed formulas. However, these studies mostly came from the same group of authors [79,80] and need confirmation before drawing firm conclusions.

The peptides with an immunomodulatory effect are mostly small in size (2 to 20 amino acids), although peptides with a molecular weight over 1000 daltons in whey and soy proteins hydrolysates also seem to have immunomodulatory properties [77,81]. While different protein hydrolysates seem able to directly modulate the local and systemic immune system, the final effect depends on the type of hydrolysate and on the protein source. Furthermore, only few immunomodulatory peptides have been identified up to now. Indeed, further research should be focused on identifying specific immunomodulatory peptides and investigating their immune effects in humans.

5. Baked Milk: A Possible Form of Oral Immunotherapy?

Most children with CMA can tolerate baked milk [82,83]. At baseline, children tolerant to baked milk differ from reactive children by having lower milk-specific, beta-lactoglobulin, and casein IgE concentrations [84] and a higher number of Treg cells [85] Although casein IgE levels have been shown to have the best accuracy in predicting the reactivity to a baked milk challenge [86], the test's performance relies on the decided cut-off points, which, in turn, depend on the sensibility and specificity of the test. Indeed, on an individual basis, an oral food challenge with baked milk should be performed to identify baked milk tolerant subjects as no laboratory testing can predict patient tolerance to baked milk in a reliable and conclusive way.

Many cohort and retrospective studies have hypothesized that CMA resolution occurs more rapidly in cases of regular baked milk assumption [82,84].

However, since cow's milk tolerance can spontaneously occur in the first years of life, studies without a control group could not explain whether the faster tolerance observed is due to real immune modulation via a baked product, or by a milder phenotype of those patients [82,87,88].

A recent systematic review [88], considering only published observational studies, found weak evidence that the ingestion of baked hens' eggs or cow's milk results in an acceleration of tolerance achievement. However, very recently, a controlled randomized clinical trial showed that introducing baked milk in cow's milk protein allergic patients accelerates the tolerance to fresh milk [5].

It is well known that oral immunotherapy (OIT) plays an immunological role by modulation of humoral and cell immunity. Humoral changes caused by OIT include a decrease in IgE levels and a rise in IgG levels, especially IgG4, which have a protective role on allergic reactions by blocking IgE-mediated basophil and mast cell activation. T cell response modifications include a reduction of Th2 cell line and Th2 cytokines' expression [89,90]. A study from Goldberg et al. [91] showed that baked-milk reactive patients, who underwent baked milk OIT and reached maintenance dose, present a decrease in IgE reactivity to casein and alpha-lactalbumin. Similar to what happens during OIT [6], studies on the immune profile have suggested that after regular ingestion of baked milk products in baked mild tolerant children, casein IgG4 levels increase [82,84], while casein and beta lactoglobulin-specific IgE levels and casein IgE/IgG4 and beta lactoglobulin IgE/IgG4 ratios decrease [82].

All these findings together suggest that the ingestion of baked milk products could drive a change in immune patterns, speeding up milk tolerance. However, further randomized studies are warranted to confirm this hypothesis.

Besides, the opportunity to reduce the child's dietary and label-related restrictions has been demonstrated to reduce the stress levels with a beneficial effect on the quality of life of food-allergic children and their families [92,93].

6. Gut Microbiota in Perinatal Period and Its Relationship with Immune Function and Allergy Development

Gut microbiota are influenced by several factors occurring during pregnancy and after birth [94,95].

Several studies have evaluated the relationship between bifidobacterial colonization and the development of allergic diseases, including cow's milk allergy [96–98].

Oral feeding determines the major modifications in the composition of intestinal microbiota. Breast milk contains important substances influencing the development of a newborn's immune system [99]. A recent study [100] suggests that lactoferrin passes throughout breast milk to the intestine of the newborn, promoting the growth of bifidobacteria, which in turn contributes to the regulation of postnatal intestinal development [101].

Human milk contains mainly Lactobacilli and Bifidobacteria with an estimated number of ingested bacteria of 1×10^5 to 1×10^7 per 800 mL of milk consumed daily [102]. These bacteria stimulate endogenous production of secretory IgA [103], activation of T regulatory cells [104], and anti-inflammation response [105,106]. Gut microbiota establishment in early life is crucial for the success of oral tolerance, mediated by Foxp3þ and Treg, known to inhibit immune activation [107,108].

Germ free mice showed a Th2-skewed response [109]. An early exposure to pro-and/or prebiotics during the prenatal period and in early life might be beneficial in preventing Th2- mediated allergic disease, including food allergy [110].

7. Prebiotics and CMA

An increasing body of evidence shows that the gut microbiota contributes to the maturation of the immune system [111]. An altered patterns of early colonization, e.g., "dysbiosis," predisposes people to allergic diseases. Prospective studies have demonstrated that a gut microbial imbalance due to reduced diversity in the early years of life is associated with an increased risk of developing food sensitization and atopic eczema [112–114]. Although the specific microbiota dysfunction in allergies remains unclear, both prebiotics and probiotics probably modulate immune development through a number of different pathways that can be modified by host and environmental factors. Prebiotic carbohydrates are a major substrate for bacterial growth, and stimulate selectively the growth and/or activity of beneficial species of the gut microbiota. The bifidogenic effect of human milk (a rich source of oligosaccharides) and of certain prebiotics (i.e., fructo- and galacto-oligosaccharides) added to infant milk formulas has long been reported [115,116].

Human milk oligosaccharides (HMOs) may both reduce the adhesion of pathogens and act as metabolic substrates for select species, contributing to the shaping of the infant gut microbiota and modulating the immune system [117] and health of infants [118]. However, there are hundreds of different HMOs, with specific properties and functions [97]. Up to now, only a small number of HMOs have been synthetized and added to infant formula, showing beneficial results [97]. According to a recent study, infants fed with human milk containing low Lacto-N-fucopentaose III (LNFP) concentrations were more likely to become affected by CMA compared to infants receiving high LNFP III-containing milk (odds ratio 6.7, 95% CI 2.0–22) [119].

A systematic review [120] on HMOs reported a protective effect against CMA by 18 months of age.

A beneficial effect of a special mixture of prebiotics (short-chain galacto- and long chain fructo-oligosaccharides) on the development of atopic dermatitis in a high risk population of infants was shown for the first time in 2006 [116].

Fewer infants in the intervention group (hydrolyzed protein formula + prebiotic mixture) developed atopic dermatitis compared to infants in the control group (hydrolyzed protein formula + maltodextrine) (9.8%; 95 CI 5.4–17.1% vs. 23.1%; 95 CI 16.0%–32.1%) In the intervention group, a significantly higher number of faecal bifidobacteria was detected compared to the controls [116]. A systematic review and meta-analysis found no effect on the onset of asthma whilst it did find a significant reduction in eczema (four studies, 1218 infants; risk ratio (RR) 0.68, 95% CI 0.48–0.97, with a number needed to treat 25 (95% CI 14 to >100)) [121]. Conversely, a more recent systematic review reported no difference in eczema (RR: 0.57, 95% CI: 0.30e1.08). Only one study evaluated the risk of food allergy and found a reduced risk (RR: 0.28, 95% CI 0.08e1.00) in prebiotic-treated infants [122].

A very recent study [111] assessed the effect of a partially hydrolyzed protein formula supplemented with non-digestible oligosaccharides on the prevention of eczema in 138 infants at high risk of allergy. Infants receiving the prebiotic formula had a fecal microbial composition, metabolites, and pH closer to that of breast-fed infants than that of infants receiving standard cow's milk formula. Infants with eczema by 18 months showed decreased acquisition of *Eubacterium* and *Anaerostipes* species with increased lactate and reduced butyrate levels [111].

A similar effect was also shown in non-at-risk infants [123]: In total, 414 infants received an intact protein formula containing a specific mixture of neutral oligosaccharides and pectin-derived acidic oligosaccharides compared to 416 infants fed with a control formula without oligosaccharides. Up to the first year of life, atopic dermatitis occurred in a significantly higher number of infants from the control group (9.7%) than the prebiotic group (5.7%) [123]. The addition of lactose to an extensively hydrolyzed formula significantly increased the total fecal counts of *Bifidobacteria* and lactic acid bacteria, and decreased that of *Bacteroides/Clostridia*.s. Moreover, lactose significantly increased the concentration of total short-chain fatty acids, especially acetic and butyric acids, as demonstrated by the metabolomic analysis [124].

A recent multicenter double-blind randomized controlled trial [125] investigated the effects of an amino acid-based formula (AAF), including fructo-oligosaccharides, and the probiotic strain, *Bifidobacterium* breve M-16V, in 35 infants with suspected non-IgE-mediated CMA. After 8 weeks of diet, the median percentage of *Bifidobacteria* was significantly ($p < 0.001$) higher in the test group than in the 36 control subjects fed non-supplemented AAF (35.4% vs. 9.7%), whereas *Eubacterium* rectale/*Clostridium coccoides* group in feces was lower (9.5% vs. 24.2%) and similar to that detected in breastfed infants (55% and 6.5%, respectively).

A subsequent double-blind randomized controlled multicenter trial with the same study groups and formulas confirmed the same fecal microbiota changes at 26 weeks [126]. Safety parameters were similar between groups.

Data from animal models have shown that in whey-sensitized mice, dietary supplementation with short chain galacto-oligosaccharides, long chain fructo-olgosaccharides, pectin-derived acidic oligosaccharides, and/or mixtures of the above prebiotics effectively reduced allergic symptoms but differentially affected mucosal immune activation. In whey-sensitized mice, mixtures of prebiotics increased the number of Foxp3+ cells in the proximal small intestine compared to sham-sensitized mice [127]. The increased expression of Th2 and Th17 mRNA markers in the small intestine of whey-sensitized mice was prevented by the mixture of galacto and fructo-oligosaccharides. Adding pectin-derived acidic oligosaccharides to this mixture enhanced Tbet (Th1), IL-10, and TGF-β mRNA expression, which was maintained in the distal small intestine and/or colon [127]. Interestingly, a more recent study [128] showed that co-administration of oligosaccharides and partially hydrolyzed whey protein can induce immunological tolerance in mice orally sensitized with whey and/or cholera toxin on day 35, particularly if the intake was on a daily basis. The oligosaccharide composition seems to influence the tolerance inducing mechanisms and was associated with the decrease of *Lactobacillus* species, being replaced by Bacteroidales family S24-7 members and with the relative abundance of Prevotella [128].

In 2011, the European Society for Paediatric Gastroenterology Hepatology and Nutrition (ESPGHAN) Committee on Nutrition concluded that there was insufficient evidence to recommend the use of prebiotics in infant formula to prevent atopic disease [129].

Conversely, based on the Grading of Recommendations Assessment, Development and Evaluation (GRADE) approach, in 2016, the World Allergy Organization guideline panel suggested the use of prebiotic supplementation in not-exclusively breastfed infants; however, both recommendations were based on a very low certainty of evidence [122].

At present, despite some promising results mainly related to the effect of specific prebiotics on the gut microbiota, clinical evidence of the beneficial effect of prebiotics in CMA is still inconclusive [128].

8. Probiotics and CM

Several studies have shown that probiotic supplementation given to women during pregnancy and lactation can modulate the microbial milk composition and immunity-modulating molecules, with health benefits ranging from gastrointestinal symptoms to allergies, transferred to the newborn [130]. Administration to mothers of a probiotic mixture (sold in continental Europe and the USA as Vivomixx® and Visbiome®-, -Danisco-Dupont, WI, USA,) resulted in an increase of Lactobacilli and Bifidobacteria in both colostrum and mature milk [131] in the "probiotic group" with respect to the "placebo group", and in breast milk concentrations of secretory IgA and TGF-β and IL-10 (anti-inflammatory and immunomodulatory cytokines) [132]. This increasing gut maturation influences a newborn's IgA production and seems to improve gastrointestinal functional symptoms in infants [132]. TGF-β ingested through breast milk restrains inflammatory responses in intestinal epithelial cells and T cells and exerts a modulation on the immune tolerance towards dietary antigens and indigenous intestinal microbes by induction of Treg cells [132]. It also increases the IgA production in newborns, improving the intestinal barrier function [133].

Maternal probiotic supplementation during pregnancy and breastfeeding seems to prevent atopic eczema in children [130]. The results of the main studies (RCT) are shown in Table 1 [134–141] (Table 1).

Table 1. Probiotics administration during pregnancy and breastfeeding for the prevention of allergic disorders.

Author, Year	Study	Subjects	Strain, Dose, Beginning of the Treatment (S), End of the Treatment (E)	Placebo	Outcomes	Follow-Up (Years)	Side Effects
Dotterud et al. [134]	RCT	415 pregnant women	LGG 5×10^{10} CFU, Bb-12 5×10^{10} CFU and La-5. 5×10^{9} (CFU) daily S: 4 weeks before expected delivery date E: 3 weeks after delivery (breastfeeding)	yes	Probiotic supplementation reduces incidence of atopic dermatitis (AD) in children	2	No
Enomoto et al. [135]	Open-trial	166 pregnant women and newborns	BB536 5×10^{9} CFU and BB M-16V 5×10^{9} CFU daily S: 4 weeks before expected delivery date E: 6 months after delivery (to infants)	no	Probiotic supplementation reduces incidence of AD in children	3	no
Wickens et al. [141]	RCT	423 pregnant women	LR HN001 6×10^{9} CFU daily S: from 14–16 weeks gestation E: 6 months after delivery (breastfeeding)	yes	Probiotic supplementation does not prevent AD in infants	1	no
Ou et al. [138]	RCT	191 pregnant women and related newborns	LGG ATCC 53103, 1×10^{10} CFU daily S: From the second trimester of pregnancy; E: 6 months after delivery (to mothers and infants) during breastfeeding	yes	Probiotic supplementation doesn't prevent infant allergic disease (AD, allergic rhinitis, asthma)	3	no
Rautava et al. [139]	RCT	241 pregnant women	LPR 1×10^{9} CFU BL999 1×10^{9} CFU ST11 1×10^{9} CFU daily S: 2 months before expected delivery E: 2 months after delivery (breast-feeding)	yes	Probiotic supplementation prevents infant eczema	2	Not observed
Kim et al. [136]	Randomized placebo-controlled trial	112 pregnant women and newborns	BGN4 1.6×10^{9} CFU, AD011 1.6×10^{9} CFU, and AD031 1.6×10^{9} CFU daily S: 4–8 weeks before expected delivery E: 6 months after delivery (to mothers during breastfeeding and to infants)	yes	Probiotics supplementation reduces incidence of AD in children	1	yes
Niers et al. [137]	Double-blind, randomized, placebo-controlled trial	136 pregnant women and newborns	BB: 1×10^{9} CFU; BL 1×10^{9} CFU; LL 1×10^{9} CFU S: last 6 weeks of pregnancy E: 12 months after delivery (to infants)	yes	Probiotics supplementation reduces the incidence of AD in children at 3 months of life	24 months after delivery	no
Simpson et al. [140]	Randomized placebo-controlled trial	415 pregnant women	Probiotic milk: LGG, 5×10^{10} CFU; La-5 5×10^{9} CFU and Bb-12 5×10^{10} CFU S: from 36 weeks gestation E: 3 months after delivery (breastfeeding)	yes	Probiotic supplementation reduces incidence of AD	6 years after delivery	no

LGG: *Lactobacillus rhamnosus* GG; Bb-12: *Bifidobacterium animalis* subsp. *Lactis* Bb-12; La-5: *L. acidophilus* La-5; CFU: colony-forming unit; BB536: *B. longum* BB536 [ATCC BAA-999]; BB M-16V: *B. breve* M-16V [LMG 23729]; LR HN001: *Lactobacillus Rhamnosus* HN001; LG LPR: *Lactobacillus rhamnosus* LPR; BL999: *Bifidobacterium longum* BL999. ST11: *L. paracasei* ST11; BGN4: *Bifidobacterium bifidum* BGN4; AD011: *Bifidobacterium lactis* AD011; AD031: *Lactobacillus acidophilus* AD031; BB: *Bifidobacterium bifidum*; BL: *Bifidobacterium lactis*; LL: *Lactococcus lactis*; AD: Atopic Dermatitis

Further studies are requested in order to confirm the possibility of preventing other allergic disorders with perinatal probiotic administration.

The World Allergy Organization (WAO) [142] recommends the use of probiotics in pregnant and breastfeeding women and in non-exclusively breastfed infants at high risk of allergic disease. On the other hand, the Academy of Allergy and Clinical Immunology [143] and European Society for Paediatric Gastroenterology, Hepatology, and Nutrition [129] do not recommend the use of probiotics and/or prebiotics for the prevention of allergic diseases. However, the WAO guideline panel recognizes that the recommendations of both probiotics and prebiotics are conditional and based on very low quality evidence.

In terms of the therapeutic property of probiotics, it has been demonstrated [144] that in infants with proctocolitis, the addition of *Lactobacillus rhamnosus* LGG to an extensively hydrolyzed cow's milk protein formula determines a greater decrease in fecal calprotectin [145] and a reduction in the number of infants with a persistence of occult blood in stools after 1 month. LGG could enhance the intestinal mucosa's barrier function and participate in the degradation of protein antigens, compete with pathogenic bacteria, and promote early immune system maturation towards non-allergy. A recent systematic review considered a randomized trial, involving 895 pediatric patients with CMA. The primary outcome of interest was relief of symptoms in terms of a reduction of the severity of atopic dermatitis (measured by the SCORing Atopic Dermatitis (SCORAD) index). Overall, a decrease of the SCORAD index was shown in subjects given probiotics, but the results were imprecise and do not permit firm conclusions to be drawn [146].

The results of Randomized Controlled Trials (RCT) on probiotics use in CMA treatment are shown in Table 2 (Table 2).

Table 2. Probiotics in cow's milk allergy CMA treatment.

Author, Year	Study Design	Subjects	Strain, Dose (D)	Placebo	Outcomes	Treatment Period (Months)	Side Effects
Baldassarre et al. [144]	RCT	30 infants	LGG 1×10^6 CFU/g	yes	Probiotic supplementation improves gastrointestinal symptoms (hematochezia and fecal calprotectin)	1	No
Berni Canani et al. [79]	RCT	80 infants	LGG, 1.4×10^7 CFU/100 mL	yes	Probiotic supplementation accelerates tolerance acquisition to cow's milk proteins	12	No
Berni Canani et al. [80]	RCT	260 infants	LGG (dose not specified)	yes	Probiotic supplementation accelerates tolerance acquisition to cow's milk proteins	12	No
Berni Canani et al. [147]	RCT	220 children	LGG (dose not specified)	yes	Probiotic supplementation reduces the incidence of other allergic manifestations and hastens the development of oral tolerance to cow's milk proteins	36	No
Dupont et al. [148]	RCT	119 infants	LC CRL431 and Bb-12 (dose not specified)	yes	Probiotic supplementation significantly improves the SCORAD index and growth indices	6	No
Hol et al. [149]	RCT	119 infants	LC CRL431 and Bb-12 1×10^7 CFU/g formula	yes	Probiotic supplementation does not accelerate tolerance acquisition to cow's milk proteins	6	No
Kirjavainen et al. [150]	RCT	35 infants	LGG ATCC 53103 1×10^9 CFU/g	yes	Supplementation with viable probiotics improves the SCORAD index	2	Diarrhea (with heat-inactivated LGG)
Majamaa et al. [151]	RCT	31 infants	LGG ATCC 53103- 5×10^8 CFU/g formula twice a day	yes	Probiotic supplementation improves the SCORAD index and reduces markers of intestinal inflammation	1	No
Viljanen et al. [152]	RCT	230 infants	LGG (ATCC 53103) 5×10^9 CFU vs. LGG 5×10^9 CFU, LR LC705- 5×10^9 CFU, Bbi99- 2×10^8 CFU, and PJS- 2×10^9 CFU twice a day	yes	Probiotic supplementation improves the SCORAD index in IgE-sensitized infants but not in non-IgE-sensitized infants	1	No

LGG: *Lactobacillus rhamnosus* LGG; CFU: colony-forming unit; LC CRL431: *L. casei* CRL431; Bb12: *B. lactis* Bb-12 (*B animalis* subspecies *lactis*); LR LC705: *L. Rhamnosus* LC705 Bbi99: *Bifidobacterium breve* Bbi99; PJS: *Propionibacterium* JS.

Great interest has recently arisen regarding the possible role of probiotics administration in fasting tolerance. Despite some evidence that a specific strain, such as *Lactobacillus rhamnosus*, LGG administration may induce tolerance among infants with CMA with a long-lasting effect [147]. Although, no general conclusions can be drawn, due to inconclusive evidence and imprecise results [146]. Further studies are required to investigate the effects of pre- and postnatal probiotic supplementation on the development of systemic and mucosal immunity. Similarly, the most effective strains, dosages, or optimal duration of treatment still need to be defined.

The use of probiotics is in general safe during pregnancy and in newborns (see Table 1). Kuitunen et al. [153] reported that newborns supplemented with probiotics before and after birth had significantly lower hemoglobin levels compared to the placebo group at six months of life. This effect was considered to be transient

Without proper identification of the strains the clinical evidence regarding one product could not be transferred from one product to another. This is the reason why the limiting of information to probiotic genera/species is not the best choice [154]. Without consideration of current regulatory and commercial loopholes, assessing harm will be difficult for researchers, physicians, and patients. More stringent regulations mandating full disclosure of the probiotic microorganisms at the strain level and the origin of the product and manufacturing changes are a prerequisite for proper safety and efficacy reporting [155].

9. Conclusions

Much has changed in recent years in food allergy management, moving from a one-size approach to a personalized one, associated with the specific food allergy phenotype. While different protein hydrolysates seem able to modulate the immune system, the few in vivo data, although promising, do not allow us to draw conclusions on their effect on tolerance achievement. Furthermore, the paucity and heterogeneity among the studies currently limit one's ability to compare the results and to recommend the routine use of prebiotics and probiotics for prevention and treatment of CMA.

Author Contributions: Conceptualization, E.D. and G.Z.; Methodology, E.D., S.S., M.A. and Y.V.; Supervision, Y.V. and G.Z.; Writing—original draft, E.D., S.S., E.P., M.U.A.S., C.M., L.P. and M.E.B.; Writing—review and editing, E.D., S.S., M.U.A.S., L.P., and M.E.B.

Funding: This research received no external funding.

Conflicts of Interest: The authors declare no conflict of interests related to this paper.

References

1. Sicherer, S.H. Epidemiology of food allergy. *J. Allergy Clin. Immunol.* **2011**, *127*, 594–602. [CrossRef] [PubMed]
2. Schappi, M.G.; Borrelli, O. Mast cell-nerve interactions in children with functional dyspepsia. *J. Pediatr. Gastroenterol. Nutr.* **2008**, *47*, 472–480. [CrossRef] [PubMed]
3. Molina-Infante, J.; Arias, A. Four-food group elimination diet for adult eosinophilic esophagitis: A prospective multicenter study. *J. Allergy Clin. Immunol.* **2014**, *134*, 1093–1099.e1. [CrossRef] [PubMed]
4. Bougle, D.; Bouhallab, S. Dietary bioactive peptides: Human studies. *Crit. Rev. Food Sci. Nutr.* **2017**, *57*, 335–343. [CrossRef] [PubMed]
5. Esmaeilzadeh, H.; Alyasin, S. The effect of baked milk on accelerating unheated cow's milk tolerance: A control randomized clinical trial. *Pediatr. Allergy Immunol.* **2018**, *29*, 747–753. [CrossRef]
6. Huang, F.; Nowak-Wegrzyn, A. Extensively heated milk and egg as oral immunotherapy. *Curr. Opin. Allergy Clin. Immunol.* **2012**, *12*, 283–292. [CrossRef]
7. Berni Canani, R.; Gilbert, J.A. The role of the commensal microbiota in the regulation of tolerance to dietary allergens. *Curr. Opin. Allergy Clin. Immunol.* **2015**, *15*, 243–249. [CrossRef]
8. Plunkett, C.H.; Nagler, C.R. The Influence of the Microbiome on Allergic Sensitization to Food. *J. Immunol.* **2017**, *198*, 581–589. [CrossRef]
9. Dupont, C. Diagnosis of cow's milk allergy in children: Determining the gold standard? *Expert Rev. Clin. Immunol.* **2014**, *10*, 257–267. [CrossRef]

10. Hill, D.J.; Hosking, C.S. The cow milk allergy complex: Overlapping disease profiles in infancy. *Eur. J. Clin. Nutr.* **1995**, *49* (Suppl. 1), S1–S12.

11. du Toit, G.; Meyer, R. Identifying and managing cow's milk protein allergy. *Arch. Dis. Child. Educ. Pract. Ed.* **2010**, *95*, 134–144. [CrossRef] [PubMed]

12. Heine, R.G. Gastroesophageal reflux disease, colic and constipation in infants with food allergy. *Curr. Opin. Allergy Clin. Immunol.* **2006**, *6*, 220–225. [CrossRef] [PubMed]

13. Rosen, R.; Vandenplas, Y. Pediatric Gastroesophageal Reflux Clinical Practice Guidelines: Joint Recommendations of the North American Society for Pediatric Gastroenterology, Hepatology, and Nutrition and the European Society for Pediatric Gastroenterology, Hepatology, and Nutrition. *J. Pediatr. Gastroenterol. Nutr.* **2018**, *66*, 516–554. [CrossRef] [PubMed]

14. Salvatore, S.; Vandenplas, Y. Gastroesophageal reflux and cow milk allergy: Is there a link? *Pediatrics* **2002**, *110*, 972–984. [CrossRef] [PubMed]

15. Murch, S. Allergy and intestinal dysmotility—Evidence of genuine causal linkage? *Curr. Opin. Gastroenterol.* **2006**, *22*, 664–668. [CrossRef] [PubMed]

16. Nakajima-Adachi, H.; Ebihara, A. Food antigen causes TH2-dependent enteropathy followed by tissue repair in T-cell receptor transgenic mice. *J. Allergy Clin. Immunol.* **2006**, *117*, 1125–1132. [CrossRef] [PubMed]

17. Barbara, G.; Stanghellini, V. Functional gastrointestinal disorders and mast cells: Implications for therapy. *Neurogastroenterol. Motil.* **2006**, *18*, 6–17. [CrossRef]

18. Jakate, S.; Demeo, M. Mastocytic enterocolitis: Increased mucosal mast cells in chronic intractable diarrhea. *Arch. Pathol. Lab. Med.* **2006**, *130*, 362–367.

19. Ramsay, D.B.; Stephen, S. Mast cells in gastrointestinal disease. *Gastroenterol. Hepatol.* **2010**, *6*, 772–777.

20. Feldman, M.J.; Morris, G.P. Mast cells mediate acid-induced augmentation of opossum esophageal blood flow via histamine and nitric oxide. *Gastroenterology* **1996**, *110*, 121–128. [CrossRef]

21. Wu, S.V.; Yuan, P.Q. Identification and characterization of multiple corticotropin-releasing factor type 2 receptor isoforms in the rat esophagus. *Endocrinology* **2007**, *148*, 1675–1687. [CrossRef] [PubMed]

22. Zhong, C.J.; Wang, K. Mast cell activation is involved in stress-induced epithelial barrier dysfunction in the esophagus. *J. Dig. Dis.* **2015**, *16*, 186–196. [CrossRef] [PubMed]

23. Yu, Y.; Ding, X. Alterations of Mast Cells in the Esophageal Mucosa of the Patients with Non-Erosive Reflux Disease. *Gastroenterol. Res.* **2011**, *4*, 70–75. [CrossRef] [PubMed]

24. Zhong, C.; Liu, K. Developing a diagnostic understanding of GERD phenotypes through the analysis of levels of mucosal injury, immune activation, and psychological comorbidity. *Dis. Esophagus* **2018**, *31*, doy039. [CrossRef] [PubMed]

25. Yu, S.; Gao, G. TRPA1 in mast cell activation-induced long-lasting mechanical hypersensitivity of vagal afferent C-fibers in guinea pig esophagus. *Am. J. Physiol. Gastrointest. Liver Physiol.* **2009**, *297*, G34–G42. [CrossRef] [PubMed]

26. Yu, S.; Kollarik, M. Mast cell-mediated long-lasting increases in excitability of vagal C fibers in guinea pig esophagus. *Am. J. Physiol. Gastrointest. Liver Physiol.* **2007**, *293*, G850–G856. [CrossRef]

27. Cenac, N.; Chin, A.C. PAR2 activation alters colonic paracellular permeability in mice via IFN-gamma-dependent and -independent pathways. *J. Physiol.* **2004**, *558*, 913–925. [CrossRef]

28. Itoh, Y.; Sendo, T. Physiology and pathophysiology of proteinase-activated receptors (PARs): Role of tryptase/PAR-2 in vascular endothelial barrier function. *J. Pharmacol. Sci.* **2005**, *97*, 14–19. [CrossRef]

29. Liu, H.; Miller, D.V. Proteinase-activated receptor-2 activation evokes oesophageal longitudinal smooth muscle contraction via a capsaicin-sensitive and neurokinin-2 receptor-dependent pathway. *Neurogastroenterol. Motil.* **2010**, *22*, 210–216, e67. [CrossRef]

30. Kandulski, A.; Wex, T. Proteinase-activated receptor-2 in the pathogenesis of gastroesophageal reflux disease. *Am. J. Gastroenterol.* **2010**, *105*, 1934–1943. [CrossRef]

31. De Boissieu, D.; Waguet, J.C. The atopy patch tests for detection of cow's milk allergy with digestive symptoms. *J. Pediatr.* **2003**, *142*, 203–205. [CrossRef] [PubMed]

32. Majamaa, H.; Moisio, P. Cow's milk allergy: Diagnostic accuracy of skin prick and patch tests and specific IgE. *Allergy* **1999**, *54*, 346–351. [CrossRef] [PubMed]

33. Koletzko, S.; Niggemann, B. Diagnostic approach and management of cow's-milk protein allergy in infants and children: ESPGHAN GI Committee practical guidelines. *J. Pediatr. Gastroenterol. Nutr.* **2012**, *55*, 221–229. [CrossRef] [PubMed]

34. Sampson, H.A.; Gerth van Wijk, R. Standardizing double-blind, placebo-controlled oral food challenges: American Academy of Allergy, Asthma & Immunology-European Academy of Allergy and Clinical Immunology PRACTALL consensus report. *J. Allergy Clin. Immunol.* **2012**, *130*, 1260–1274. [PubMed]

35. Giovannini, M.; D'Auria, E. Nutritional management and follow up of infants and children with food allergy: Italian Society of Pediatric Nutrition/Italian Society of Pediatric Allergy and Immunology Task Force Position Statement. *Ital. J. Pediatr.* **2014**, *40*, 1. [CrossRef]

36. Borrelli, O.; Barbara, G. Neuroimmune interaction and anorectal motility in children with food allergy-related chronic constipation. *Am. J. Gastroenterol.* **2009**, *104*, 454–463. [CrossRef]

37. Lucendo, A.J.; Molina-Infante, J. Guidelines on eosinophilic esophagitis: Evidence-based statements and recommendations for diagnosis and management in children and adults. *United Eur. Gastroenterol. J.* **2017**, *5*, 335–358. [CrossRef]

38. Papadopoulou, A.; Koletzko, S. Management guidelines of eosinophilic esophagitis in childhood. *J. Pediatr. Gastroenterol. Nutr.* **2014**, *58*, 107–118. [CrossRef]

39. Miehlke, S. Clinical features of eosinophilic esophagitis. *Dig. Dis.* **2014**, *32*, 61–67. [CrossRef]

40. Papadopoulou, A.; Dias, J.A. Eosinophilic esophagitis: An emerging disease in childhood—Review of diagnostic and management strategies. *Front. Pediatr.* **2014**, *2*, 129. [CrossRef]

41. Dellon, E.S.; Gonsalves, N. ACG clinical guideline: Evidenced based approach to the diagnosis and management of esophageal eosinophilia and eosinophilic esophagitis (EoE). *Am. J. Gastroenterol.* **2013**, *108*, 679–692. [CrossRef] [PubMed]

42. Hoofien, A.; Dias, J.A. Pediatric Eosinophilic Esophagitis: Results of the European Retrospective Pediatric Eosinophilic Esophagitis Registry (RetroPEER). *J. Pediatr. Gastroenterol. Nutr.* **2019**, *68*, 552–558. [CrossRef] [PubMed]

43. Arias, A.; Gonzalez-Cervera, J. Efficacy of dietary interventions for inducing histologic remission in patients with eosinophilic esophagitis: A systematic review and meta-analysis. *Gastroenterology* **2014**, *146*, 1639–1648. [CrossRef] [PubMed]

44. Loizou, D.; Enav, B. A pilot study of omalizumab in eosinophilic esophagitis. *PLoS ONE* **2015**, *10*, e0113483. [CrossRef] [PubMed]

45. Aceves, S.S.; Chen, D. Mast cells infiltrate the esophageal smooth muscle in patients with eosinophilic esophagitis, express TGF-beta1, and increase esophageal smooth muscle contraction. *J. Allergy Clin. Immunol.* **2010**, *126*, 1198–1204.e4. [CrossRef] [PubMed]

46. Leung, J.; Beukema, K.R. Allergic mechanisms of Eosinophilic oesophagitis. *Best Pract. Res. Clin. Gastroenterol.* **2015**, *29*, 709–720. [CrossRef] [PubMed]

47. Kagalwalla, A.F.; Sentongo, T.A. Effect of six-food elimination diet on clinical and histologic outcomes in eosinophilic esophagitis. *Clin. Gastroenterol. Hepatol.* **2006**, *4*, 1097–1102. [CrossRef] [PubMed]

48. Liacouras, C.A.; Spergel, J.M. Eosinophilic esophagitis: A 10-year experience in 381 children. *Clin. Gastroenterol. Hepatol.* **2005**, *3*, 1198–1206. [CrossRef]

49. Spergel, J.M.; Andrews, T. Treatment of eosinophilic esophagitis with specific food elimination diet directed by a combination of skin prick and patch tests. *Ann. Allergy Asthma Immunol.* **2005**, *95*, 336–343. [CrossRef]

50. Gomez Torrijos, E.; Gonzalez-Mendiola, R. Eosinophilic Esophagitis: Review and Update. *Front. Med. (Lausanne)* **2018**, *5*, 247. [CrossRef]

51. Gonsalves, N.; Yang, G.Y. Elimination diet effectively treats eosinophilic esophagitis in adults; food reintroduction identifies causative factors. *Gastroenterology* **2012**, *142*, 1451–1459.e1. [CrossRef] [PubMed]

52. Molina-Infante, J.; Arias, A. Step-up empiric elimination diet for pediatric and adult eosinophilic esophagitis: The 2-4-6 study. *J. Allergy Clin. Immunol.* **2018**, *141*, 1365–1372. [CrossRef] [PubMed]

53. Kagalwalla, A.F.; Wechsler, J.B. Efficacy of a 4-Food Elimination Diet for Children with Eosinophilic Esophagitis. *Clin. Gastroenterol. Hepatol.* **2017**, *15*, 1698–1707.e7. [CrossRef] [PubMed]

54. Kagalwalla, A.F.; Amsden, K. Cow's milk elimination: A novel dietary approach to treat eosinophilic esophagitis. *J. Pediatr. Gastroenterol. Nutr.* **2012**, *55*, 711–716. [CrossRef] [PubMed]

55. Kruszewski, P.G.; Russo, J.M. Prospective, comparative effectiveness trial of cow's milk elimination and swallowed fluticasone for pediatric eosinophilic esophagitis. *Dis. Esophagus* **2016**, *29*, 377–384. [CrossRef] [PubMed]

56. Erwin, E.A.; Tripathi, A. IgE Antibody Detection and Component Analysis in Patients with Eosinophilic Esophagitis. *J. Allergy Clin. Immunol. Pract.* **2015**, *3*, 896–904.e3. [CrossRef]

57. Clayton, F.; Fang, J.C. Eosinophilic esophagitis in adults is associated with IgG4 and not mediated by IgE. *Gastroenterology* **2014**, *147*, 602–609. [CrossRef] [PubMed]

58. Rosenberg, C.E.; Mingler, M.K. Esophageal IgG4 levels correlate with histopathologic and transcriptomic features in eosinophilic esophagitis. *Allergy* **2018**, *73*, 1892–1901. [CrossRef]

59. Schuyler, A.J.; Wilson, J.M. Specific IgG4 antibodies to cow's milk proteins in pediatric patients with eosinophilic esophagitis. *J. Allergy Clin. Immunol.* **2018**, *142*, 139–148.e12. [CrossRef]

60. Jyonouchi, S.; Abraham, V. Invariant natural killer T cells from children with versus without food allergy exhibit differential responsiveness to milk-derived sphingomyelin. *J. Allergy Clin. Immunol.* **2011**, *128*, 102–109.e13. [CrossRef]

61. Jyonouchi, S.; Smith, C.L. Invariant natural killer T cells in children with eosinophilic esophagitis. *Clin. Exp. Allergy* **2014**, *44*, 58–68. [CrossRef] [PubMed]

62. Tanabe, S. Analysis of food allergen structures and development of foods for allergic patients. *Biosci. Biotechnol. Biochem.* **2008**, *72*, 649–659. [CrossRef] [PubMed]

63. Kneepkens, C.M.; Meijer, Y. Clinical practice. Diagnosis and treatment of cow's milk allergy. *Eur. J. Pediatr.* **2009**, *168*, 891–896. [CrossRef] [PubMed]

64. Sicherer, S.H.; Sampson, H.A. Food allergy: A review and update on epidemiology, pathogenesis, diagnosis, prevention, and management. *J. Allergy Clin. Immunol.* **2018**, *141*, 41–58. [CrossRef] [PubMed]

65. Visser, J.T.; Lammers, K. Restoration of impaired intestinal barrier function by the hydrolysed casein diet contributes to the prevention of type 1 diabetes in the diabetes-prone BioBreeding rat. *Diabetologia* **2010**, *53*, 2621–2628. [CrossRef] [PubMed]

66. Korhonen, H.; Pihlanto, A. Food-derived bioactive peptides–opportunities for designing future foods. *Curr. Pharm. Des.* **2003**, *9*, 1297–1308. [CrossRef] [PubMed]

67. Kiewiet, M.B.G.; Gros, M. Immunomodulating properties of protein hydrolysates for application in cow's milk allergy. *Pediatr. Allergy Immunol.* **2015**, *26*, 206–217. [CrossRef]

68. Wichers, H. Immunomodulation by food: Promising concept for mitigating allergic disease? *Anal. Bioanal. Chem.* **2009**, *395*, 37–45. [CrossRef]

69. Sakaguchi, S.; Powrie, F. Emerging challenges in regulatory T cell function and biology. *Science* **2007**, *317*, 627–629. [CrossRef]

70. Meulenbroek, L.A.; van Esch, B.C. Oral treatment with beta-lactoglobulin peptides prevents clinical symptoms in a mouse model for cow's milk allergy. *Pediatr. Allergy Immunol.* **2013**, *24*, 656–664. [CrossRef]

71. Ortega-Gonzalez, M.; Capitan-Canadas, F. Validation of bovine glycomacropeptide as an intestinal anti-inflammatory nutraceutical in the lymphocyte-transfer model of colitis. *Br. J. Nutr.* **2014**, *111*, 1202–1212. [CrossRef] [PubMed]

72. van Esch, B.C.; Schouten, B. Oral tolerance induction by partially hydrolyzed whey protein in mice is associated with enhanced numbers of Foxp3+ regulatory T-cells in the mesenteric lymph nodes. *Pediatr. Allergy Immunol.* **2011**, *22*, 820–826. [CrossRef] [PubMed]

73. Espeche Turbay, M.B.; de Moreno de LeBlanc, A. Beta-Casein hydrolysate generated by the cell envelope-associated proteinase of Lactobacillus delbrueckii ssp. lactis CRL 581 protects against trinitrobenzene sulfonic acid-induced colitis in mice. *J. Dairy Sci.* **2012**, *95*, 1108–1118. [CrossRef] [PubMed]

74. Requena, P.; Daddaoua, A. Bovine glycomacropeptide induces cytokine production in human monocytes through the stimulation of the MAPK and the NF-kappaB signal transduction pathways. *Br. J. Pharmacol.* **2009**, *157*, 1232–1240. [CrossRef] [PubMed]

75. Nowak-Wegrzyn, A.; Chehade, M. International consensus guidelines for the diagnosis and management of food protein-induced enterocolitis syndrome: Executive summary-Workgroup Report of the Adverse Reactions to Foods Committee, American Academy of Allergy, Asthma & Immunology. *J. Allergy Clin. Immunol.* **2017**, *139*, 1111–1126.e4.

76. Kiewiet, M.B.G.; van Esch, B. Partially hydrolyzed whey proteins prevent clinical symptoms in a cow's milk allergy mouse model and enhance regulatory T and B cell frequencies. *Mol. Nutr. Food Res.* **2017**, *61*, 1700340. [CrossRef] [PubMed]

77. Kim, A.R.; Kim, H.S. Mesenteric IL-10-producing CD5+ regulatory B cells suppress cow's milk casein-induced allergic responses in mice. *Sci. Rep.* **2016**, *6*, 19685. [CrossRef] [PubMed]

78. Kiewiet, M.B.G.; Dekkers, R. Toll-like receptor mediated activation is possibly involved in immunoregulating properties of cow's milk hydrolysates. *PLoS ONE* **2017**, *12*, e0178191. [CrossRef]

79. Berni Canani, R.; Nocerino, R. Effect of Lactobacillus GG on tolerance acquisition in infants with cow's milk allergy: A randomized trial. *J. Allergy Clin. Immunol.* **2012**, *129*, 580–582. [CrossRef]

80. Berni Canani, R.; Nocerino, R. Formula selection for management of children with cow's milk allergy influences the rate of acquisition of tolerance: A prospective multicenter study. *J. Pediatr.* **2013**, *163*, 771–777.e1. [CrossRef]

81. Kiewiet, M.B.G.; Dekkers, R. Immunomodulating protein aggregates in soy and whey hydrolysates and their resistance to digestion in an in vitro infant gastrointestinal model: New insights in the mechanism of immunomodulatory hydrolysates. *Food Funct.* **2018**, *9*, 604–613. [CrossRef] [PubMed]

82. Kim, J.S.; Nowak-Wegrzyn, A. Dietary baked milk accelerates the resolution of cow's milk allergy in children. *J. Allergy Clin. Immunol.* **2011**, *128*, 125–131 e2. [CrossRef]

83. Leonard, S.A.; Nowak-Wegrzyn, A.H. Baked Milk and Egg Diets for Milk and Egg Allergy Management. *Immunol. Allergy Clin. North Am.* **2016**, *36*, 147–159. [CrossRef] [PubMed]

84. Nowak-Wegrzyn, A.; Bloom, K.A. Tolerance to extensively heated milk in children with cow's milk allergy. *J. Allergy Clin. Immunol.* **2008**, *122*, 342–347. [CrossRef] [PubMed]

85. Shreffler, W.G.; Wanich, N. Association of allergen-specific regulatory T cells with the onset of clinical tolerance to milk protein. *J. Allergy Clin. Immunol.* **2009**, *123*, 43–52.e7. [CrossRef]

86. Caubet, J.C.; Nowak-Wegrzyn, A. Utility of casein-specific IgE levels in predicting reactivity to baked milk. *J. Allergy Clin. Immunol.* **2013**, *131*, 222–224. [CrossRef] [PubMed]

87. Dang, T.D.; Peters, R.L. Debates in allergy medicine: Baked egg and milk do not accelerate tolerance to egg and milk. *World Allergy Organ. J.* **2016**, *9*, 2. [CrossRef]

88. Lambert, R.; Grimshaw, K.E.C. Evidence that eating baked egg or milk influences egg or milk allergy resolution: A systematic review. *Clin. Exp. Allergy* **2017**, *47*, 829–837. [CrossRef]

89. Tordesillas, L.; Berin, M.C. Immunology of Food Allergy. *Immunity* **2017**, *47*, 32–50. [CrossRef]

90. Upton, J.; Nowak-Wegrzyn, A. The Impact of Baked Egg and Baked Milk Diets on IgE- and Non-IgE-Mediated Allergy. *Clin. Rev. Allergy Immunol.* **2018**, *55*, 118–138. [CrossRef]

91. Goldberg, M.R.; Nachshon, L. Efficacy of baked milk oral immunotherapy in baked milk-reactive allergic patients. *J. Allergy Clin. Immunol.* **2015**, *136*, 1601–1606. [CrossRef] [PubMed]

92. D'Auria, E.; Abrahams, M. Personalized Nutrition Approach in Food Allergy: Is It Prime Time Yet? *Nutrients* **2019**, *11*, 359. [CrossRef] [PubMed]

93. Lee, E.; Mehr, S. Adherence to extensively heated egg and cow's milk after successful oral food challenge. *J. Allergy Clin. Immunol. Pract.* **2015**, *3*, 125–127.e4. [CrossRef] [PubMed]

94. Collado, M.C.; Rautava, S. Human gut colonisation may be initiated in utero by distinct microbial communities in the placenta and amniotic fluid. *Sci. Rep.* **2016**, *6*, 23129. [CrossRef] [PubMed]

95. Pickard, J.M.; Zeng, M.Y. Gut microbiota: Role in pathogen colonization, immune responses, and inflammatory disease. *Immunol. Rev.* **2017**, *279*, 70–89. [CrossRef]

96. Van Zwol, A.; Van Den Berg, A. Intestinal microbiota in allergic and nonallergic 1-year-old very low birth weight infants after neonatal glutamine supplementation. *Acta Paediatr.* **2010**, *99*, 1868–1874. [CrossRef] [PubMed]

97. Vandenplas, Y. Prevention and Management of Cow's Milk Allergy in Non-Exclusively Breastfed Infants. *Nutrients* **2017**, *9*, 731. [CrossRef]

98. Wopereis, H.; Oozeer, R. The first thousand days—Intestinal microbiology of early life: Establishing a symbiosis. *Pediatr. Allergy Immunol.* **2014**, *25*, 428–438. [CrossRef]

99. Pratico, G.; Capuani, G. Exploring human breast milk composition by NMR-based metabolomics. *Nat. Prod. Res.* **2014**, *28*, 95–101. [CrossRef]

100. Mastromarino, P.; Capobianco, D. Correlation between lactoferrin and beneficial microbiota in breast milk and infant's feces. *Biometals* **2014**, *27*, 1077–1086. [CrossRef]

101. Buccigrossi, V.; de Marco, G. Lactoferrin induces concentration-dependent functional modulation of intestinal proliferation and differentiation. *Pediatr. Res.* **2007**, *61*, 410–414. [CrossRef] [PubMed]

102. Donnet-Hughes, A.; Perez, P.F. Potential role of the intestinal microbiota of the mother in neonatal immune education. *Proc. Nutr. Soc.* **2010**, *69*, 407–415. [CrossRef] [PubMed]

103. Jost, T.; Lacroix, C. New insights in gut microbiota establishment in healthy breast fed neonates. *PLoS ONE* **2012**, *7*, e44595. [CrossRef] [PubMed]

104. Schwartz, S.; Friedberg, I. A metagenomic study of diet-dependent interaction between gut microbiota and host in infants reveals differences in immune response. *Genome Biol.* **2012**, *13*, r32. [CrossRef] [PubMed]

105. Campeotto, F.; Baldassarre, M. Fecal expression of human beta-defensin-2 following birth. *Neonatology* **2010**, *98*, 365–369. [CrossRef] [PubMed]

106. Furuta, S.; Toyama, S. Disposition of polaprezinc (zinc L-carnosine complex) in rat gastrointestinal tract and effect of cimetidine on its adhesion to gastric tissues. *J. Pharm. Pharmacol.* **1995**, *47*, 632–636. [CrossRef]

107. Berin, M.C.; Shreffler, W.G. Mechanisms Underlying Induction of Tolerance to Foods. *Immunol. Allergy Clin. N. Am.* **2016**, *36*, 87–102. [CrossRef]

108. Weissler, K.A.; Caton, A.J. The role of T-cell receptor recognition of peptide:MHC complexes in the formation and activity of Foxp3(+) regulatory T cells. *Immunol. Rev.* **2014**, *259*, 11–22. [CrossRef]

109. West, C.E.; Jenmalm, M.C. The gut microbiota and its role in the development of allergic disease: A wider perspective. *Clin. Exp. Allergy* **2015**, *45*, 43–53. [CrossRef]

110. West, C.E.; Jenmalm, M.C. Probiotics for treatment and primary prevention of allergic diseases and asthma: Looking back and moving forward. *Expert Rev. Clin. Immunol.* **2016**, *12*, 625–639. [CrossRef]

111. Wopereis, H.; Sim, K. Intestinal microbiota in infants at high risk for allergy: Effects of prebiotics and role in eczema development. *J. Allergy Clin. Immunol.* **2018**, *141*, 1334–1342.e5. [CrossRef] [PubMed]

112. Abrahamsson, T.R.; Jakobsson, H.E. Low diversity of the gut microbiota in infants with atopic eczema. *J. Allergy Clin. Immunol.* **2012**, *129*, 434–440. [CrossRef] [PubMed]

113. Azad, M.B.; Konya, T. Infant gut microbiota and food sensitization: Associations in the first year of life. *Clin. Exp. Allergy* **2015**, *45*, 632–643. [CrossRef] [PubMed]

114. Ismail, I.H.; Oppedisano, F. Reduced gut microbial diversity in early life is associated with later development of eczema but not atopy in high-risk infants. *Pediatr. Allergy Immunol.* **2012**, *23*, 674–681. [CrossRef] [PubMed]

115. Boehm, G.; Lidestri, M. Supplementation of a bovine milk formula with an oligosaccharide mixture increases counts of faecal bifidobacteria in preterm infants. *Arch. Dis. Child. Fetal Neonatal Ed.* **2002**, *86*, F178–F181. [CrossRef]

116. Moro, G.; Arslanoglu, S. A mixture of prebiotic oligosaccharides reduces the incidence of atopic dermatitis during the first six months of age. *Arch. Dis. Child.* **2006**, *91*, 814–819. [CrossRef] [PubMed]

117. Alderete, T.L.; Autran, C. Associations between human milk oligosaccharides and infant body composition in the first 6 mo of life. *Am. J. Clin. Nutr.* **2015**, *102*, 1381–1388. [CrossRef]

118. Bode, L.; Contractor, N. Overcoming the limited availability of human milk oligosaccharides: Challenges and opportunities for research and application. *Nutr. Rev.* **2016**, *74*, 635–644. [CrossRef]

119. Seppo, A.E.; Autran, C.A. Human milk oligosaccharides and development of cow's milk allergy in infants. *J. Allergy Clin. Immunol.* **2017**, *139*, 708–711.e5. [CrossRef]

120. Doherty, A.M.; Lodge, C.J. Human Milk Oligosaccharides and Associations with Immune-Mediated Disease and Infection in Childhood: A Systematic Review. *Front. Pediatr.* **2018**, *6*, 91. [CrossRef]

121. Osborn, D.A.; Sinn, J.K. Prebiotics in infants for prevention of allergy. *Cochrane Database Syst. Rev.* **2013**, CD006474. [CrossRef] [PubMed]

122. Cuello-Garcia, C.A.; Fiocchi, A. World Allergy Organization-McMaster University Guidelines for Allergic Disease Prevention (GLAD-P): Prebiotics. *World Allergy Organ. J.* **2016**, *9*, 10. [CrossRef] [PubMed]

123. Gruber, C.; van Stuijvenberg, M. Reduced occurrence of early atopic dermatitis because of immunoactive prebiotics among low-atopy-risk infants. *J. Allergy Clin. Immunol.* **2010**, *126*, 791–797. [CrossRef]

124. Francavilla, R.; Calasso, M. Effect of lactose on gut microbiota and metabolome of infants with cow's milk allergy. *Pediatr. Allergy Immunol.* **2012**, *23*, 420–427. [CrossRef] [PubMed]

125. Candy, D.C.A.; Van Ampting, M.T.J. A synbiotic-containing amino-acid-based formula improves microbiota in non-IgE-mediated allergic infants. *Pediatr. Res.* **2018**, *83*, 677–686. [CrossRef]

126. Fox, A.T.; Wopereis, H. A specific synbiotic-containing amino acid-based formula in dietary management of cow's milk allergy: A randomized controlled trial. *Clin. Transl. Allergy* **2019**, *9*, 5. [CrossRef] [PubMed]

127. Kerperien, J.; Jeurink, P.V. Non-digestible oligosaccharides modulate intestinal immune activation and suppress cow's milk allergic symptoms. *Pediatr. Allergy Immunol.* **2014**, *25*, 747–754. [CrossRef]

128. Kleinjans, L.; Veening-Griffioen, D.H. Mice co-administrated with partially hydrolysed whey proteins and prebiotic fibre mixtures show allergen-specific tolerance and a modulated gut microbiota. *Benef. Microbes* **2019**, *10*, 165–178. [CrossRef]

129. Braegger, C.; Chmielewska, A. Supplementation of infant formula with probiotics and/or prebiotics: A systematic review and comment by the ESPGHAN committee on nutrition. *J. Pediatr. Gastroenterol. Nutr.* **2011**, *52*, 238–250. [CrossRef]

130. Baldassarre, M.E.; Palladino, V. Rationale of Probiotic Supplementation during Pregnancy and Neonatal Period. *Nutrients* **2018**, *10*, 1693. [CrossRef]

131. Mastromarino, P.; Capobianco, D. Administration of a multistrain probiotic product (VSL#3) to women in the perinatal period differentially affects breast milk beneficial microbiota in relation to mode of delivery. *Pharmacol. Res.* **2015**, *95–96*, 63–70.

132. Baldassarre, M.E.; Di Mauro, A. Administration of a Multi-Strain Probiotic Product to Women in the Perinatal Period Differentially Affects the Breast Milk Cytokine Profile and May Have Beneficial Effects on Neonatal Gastrointestinal Functional Symptoms. A Randomized Clinical Trial. *Nutrients* **2016**, *8*, 677. [CrossRef] [PubMed]

133. Rautava, S.; Walker, W.A. Academy of Breastfeeding Medicine founder's lecture 2008: Breastfeeding–an extrauterine link between mother and child. *Breastfeed. Med.* **2009**, *4*, 3–10. [CrossRef] [PubMed]

134. Dotterud, C.K.; Storro, O. Probiotics in pregnant women to prevent allergic disease: A randomized, double-blind trial. *Br. J. Dermatol.* **2010**, *163*, 616–623. [CrossRef] [PubMed]

135. Enomoto, T.; Sowa, M. Effects of bifidobacterial supplementation to pregnant women and infants in the prevention of allergy development in infants and on fecal microbiota. *Allergol. Int.* **2014**, *63*, 575–585. [CrossRef] [PubMed]

136. Kim, J.Y.; Kwon, J.H. Effect of probiotic mix (Bifidobacterium bifidum, Bifidobacterium lactis, Lactobacillus acidophilus) in the primary prevention of eczema: A double-blind, randomized, placebo-controlled trial. *Pediatr. Allergy Immunol.* **2010**, *21*, e386–e393. [CrossRef]

137. Niers, L.; Martin, R. The effects of selected probiotic strains on the development of eczema (the PandA study). *Allergy* **2009**, *64*, 1349–1358. [CrossRef]

138. Ou, C.Y.; Kuo, H.C. Prenatal and postnatal probiotics reduces maternal but not childhood allergic diseases: A randomized, double-blind, placebo-controlled trial. *Clin. Exp. Allergy* **2012**, *42*, 1386–1396. [CrossRef]

139. Rautava, S.; Kainonen, E. Maternal probiotic supplementation during pregnancy and breast-feeding reduces the risk of eczema in the infant. *J. Allergy Clin. Immunol.* **2012**, *130*, 1355–1360. [CrossRef]

140. Simpson, M.R.; Dotterud, C.K. Perinatal probiotic supplementation in the prevention of allergy related disease: 6 year follow up of a randomised controlled trial. *BMC Dermatol.* **2015**, *15*, 13. [CrossRef]

141. Wickens, K.; Barthow, C. Maternal supplementation alone with Lactobacillus rhamnosus HN001 during pregnancy and breastfeeding does not reduce infant eczema. *Pediatr. Allergy Immunol.* **2018**, *29*, 296–302. [CrossRef] [PubMed]

142. Fiocchi, A.; Pawankar, R. World Allergy Organization-McMaster University Guidelines for Allergic Disease Prevention (GLAD-P): Probiotics. *World Allergy Organ. J.* **2015**, *8*, 4. [CrossRef] [PubMed]

143. Muraro, A.; Agache, I. EAACI food allergy and anaphylaxis guidelines: Managing patients with food allergy in the community. *Allergy* **2014**, *69*, 1046–1057. [CrossRef] [PubMed]

144. Baldassarre, M.E.; Laforgia, N. Lactobacillus GG improves recovery in infants with blood in the stools and presumptive allergic colitis compared with extensively hydrolyzed formula alone. *J. Pediatr.* **2010**, *156*, 397–401. [CrossRef] [PubMed]

145. Kapel, N.; Campeotto, F. Faecal calprotectin in term and preterm neonates. *J. Pediatr. Gastroenterol. Nutr.* **2010**, *51*, 542–547. [CrossRef]

146. Tan-Lim, C.S.C.; Esteban-Ipac, N.A.R. Probiotics as treatment for food allergies among pediatric patients: A meta-analysis. *World Allergy Organ. J.* **2018**, *11*, 25. [CrossRef] [PubMed]

147. Berni Canani, R.; Di Costanzo, M. Extensively hydrolyzed casein formula containing Lactobacillus rhamnosus GG reduces the occurrence of other allergic manifestations in children with cow's milk allergy: 3-year randomized controlled trial. *J. Allergy Clin. Immunol.* **2017**, *139*, 1906–1913.e4. [CrossRef]

148. Dupont, C.; Hol, J.; Nieuwenhuis, E.E. An extensively hydrolysed casein-based formula for infants with cows' milk protein allergy: tolerance/hypo-allergenicity and growth catch-up. *Br. J. Nutr.* **2015**, *113*, 1102–1112. [CrossRef]

149. Hol, J.; Van leer, E.H.; Elink schuurman, B.E.; de Ruiter, L.F.; Samsom, J.N.; Hop, W.; Neijens, H.J.; de Jongste, J.C.; Nieuwenhuis, E.E.; Cow's Milk Allergy Modified by Elimination and Lactobacilli study group. The acquisition of tolerance toward cow's milk through probiotic supplementation: a randomized, controlled trial. *J. Allergy Clin. Immunol.* **2008**, *121*, 1448–1454. [CrossRef]

150. Kirjavainen, P.V.; Salminen, S.J.; Isolauri, E. Probiotic bacteria in the management of atopic disease: underscoring the importance of viability. *J. Pediatr. Gastroenterol. Nutr.* **2003**, *36*, 223–227. [CrossRef]

151. Majamaa, H.; Isolauri, E. Probiotics: a novel approach in the management of food allergy. *J. Allergy Clin. Immunol.* **1997**, *99*, 179–185. [CrossRef]

152. Viljanen, M.; Savilahti, E.; Haahtela, T.; Juntunen-Backman, K.; Korpela, R.; Poussa, T.; Tuure, T.; Kuitunen, M. Probiotics in the treatment of atopic eczema/dermatitis syndrome in infants: a double-blind placebo-controlled trial. *Allergy* **2005**, *60*, 494–500. [CrossRef] [PubMed]

153. Kuitunen, M.; Kukkonen, K. Pro- and prebiotic supplementation induces a transient reduction in hemoglobin concentration in infants. *J. Pediatr. Gastroenterol. Nutr.* **2009**, *49*, 626–630. [CrossRef] [PubMed]

154. Baldassarre, M.E. Probiotic Genera/Species Identification Is Insufficient for Evidence-Based Medicine. *Am. J. Gastroenterol.* **2018**, *113*, 1561. [CrossRef] [PubMed]

155. Baldassarre, M.E. Harms Reporting in Randomized Controlled Trials of Interventions Aimed at Modifying Microbiota. *Ann. Intern. Med.* **2019**, *170*, 143. [CrossRef] [PubMed]

nutrients

MDPI

Review

Molecular Approaches for Diagnosis, Therapy and Prevention of Cow's Milk Allergy

Birgit Linhart [1,*], Raphaela Freidl [1], Olga Elisyutina [2], Musa Khaitov [2], Alexander Karaulov [3] and Rudolf Valenta [1,2,3]

1   Department of Pathophysiology and Allergy Research, Medical University of Vienna, 1090 Vienna, Austria
2   NRC Institute of Immunology FMBA of Russia, 115478 Moscow, Russia
3   Laboratory of Immunopathology, Department of Clinical Immunology and Allergy, Sechenov First Moscow State Medical University, 119435 Moscow, Russia
*   Correspondence: birgit.linhart@meduniwien.ac.at; Tel.: +43-1-40400-51150

Received: 3 June 2019; Accepted: 25 June 2019; Published: 29 June 2019

Abstract: Cow's milk is one of the most important and basic nutrients introduced early in life in our diet but can induce IgE-associated allergy. IgE-associated allergy to cow's milk can cause severe allergic manifestations in the gut, skin and even in the respiratory tract and may lead to life-threatening anaphylactic shock due to the stability of certain cow's milk allergens. Here, we provide an overview about the allergen molecules in cow's milk and the advantages of the molecular diagnosis of IgE sensitization to cow's milk by serology. In addition, we review current strategies for prevention and treatment of cow's milk allergy and discuss how they could be improved in the future by innovative molecular approaches that are based on defined recombinant allergens, recombinant hypoallergenic allergen derivatives and synthetic peptides.

Keywords: cow's milk allergy; cow's milk allergens; diagnosis of cow's milk allergy; molecular diagnosis; treatment of cow's milk allergy; prevention of cow's milk allergy

1. Introduction

IgE-associated food allergy is much less common than respiratory allergy [1]. It usually affects only a small percent of the population, whereas more than 25% of the population suffers from respiratory allergy. Nevertheless, food allergy is a very important topic because the number of subjects with a perception of having food allergy is much higher than the number of patients with confirmed IgE-associated food allergy [2]. This can be attributed to the fact that there are different forms of food intolerance, of which immunologically mediated food allergy represents only a relatively small portion. IgE-associated allergy to milk and in particular to cow's milk is one of the most important forms of food allergy because it can cause severe and life-threatening symptoms and affects children early in life [2]. In the European EuroPrevall birth cohort, 12,049 children were enrolled, of whom 9336 (77.5%) were followed up to 2 years of age. Cow's milk allergy CMA was confirmed in 55 children by oral food challenge, resulting in an incidence of challenge-proven CMA of 0.54% [3]. The results from the EuroPrevall study thus confirm that CMA is rare and indicate that robust and simple diagnostic tests are needed to confirm and, perhaps even more important, to exclude IgE sensitization to cow's milk allergens early in life. We have suggested the use of micro-arrayed allergens in the format of allergen chips as a possibility to test for IgE sensitization to multiple allergen molecules with small amounts of blood in children and demonstrated the feasibility of this approach in European birth cohorts in the FP7-funded European project 'Mechanisms of the Development of Allergy' MeDALL [4,5]. This technology is based on purified single allergen molecules which are produced by recombinant expression or purification from the natural allergen sources. Major advantages of molecular testing are that the culprit allergen molecules can be precisely identified. The latter is of particular interest for

CMA because it was found that clinical phenotypes of patients with CMA may vary depending on the molecular sensitization pattern. Therefore, in the next section, we provide a short overview of the cow's milk allergen molecules, their characteristics and how IgE recognition of these molecules may be linked to clinical phenotypes.

2. Cow's Milk Allergen Molecules

In principle, it is possible to obtain large amounts of natural cow's milk allergens by biochemical purification from milk but the recombinant expression of allergen-encoding DNA allows obtaining highly pure allergens without contaminations with other allergens from the same allergen source (Figure 1). Accordingly, recombinant cow's milk allergens can be obtained by expression in different host systems and used in multiallergen tests. According to the amino acid sequences of the individual cow's milk allergens, it is possible to construct different recombinant hypoallergenic derivatives and to prepare synthetic peptides for innovative forms of treatment and prevention (Figure 1).

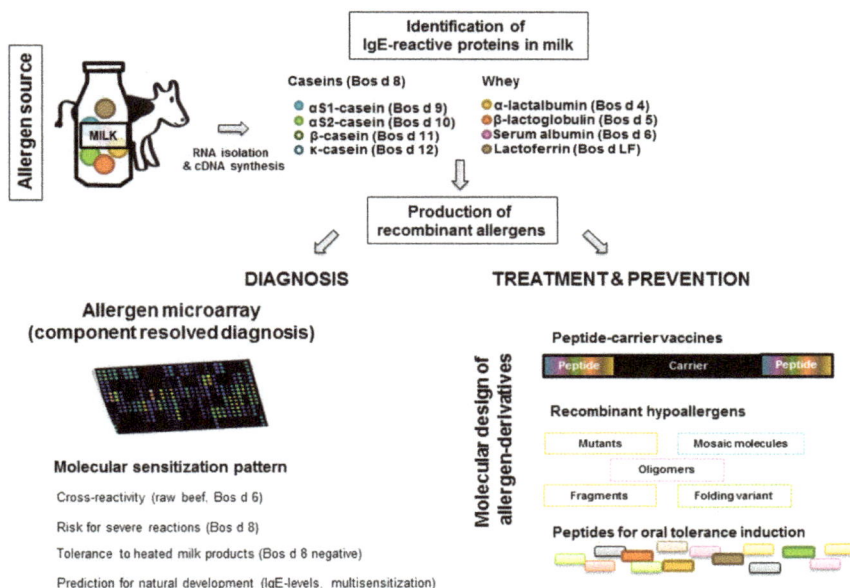

Figure 1. Component-resolved diagnosis, treatment, and prevention of cow's milk allergy. Cow's milk contains several allergenic molecules (components), which can be produced as recombinant proteins. Microarray technology allows determining reactivity profiles of patients and their sensitization to cross-reactive allergens and identifying individual allergens that cause disease. The severity of reactions and natural tolerance development can also be predicted. Based on the DNA sequences of cow's milk allergens, molecules for treatment and prevention can be designed.

2.1. Whey Proteins

2.1.1. Alpha-Lactalbumin (Bos d 4)

Alpha-lactalbumin is a small, monomeric (14.19 kDa) Ca^{2+}-binding protein (Figure 1, Table 1). Besides calcium binding, four disulfide bridges stabilize the structure of the molecule [6–8].

Bovine and human alpha-lactalbumin show a high (>70%) amino acid sequence similarity [7,8]. Despite this fact, alpha-lactalbumin contains genuine, milk-specific IgE epitopes which can be explained by the observation that the IgE epitopes are clustered at the N- and the C-terminal end of the protein, which differ greatest between human and bovine alpha-lactalbumin [8,9]. Human IgE-antibodies are mainly directed to conformational epitopes, which can be possibly attributed to its high stability and structural refolding capacity to heat [6]. However, IgE-reactivity fragments generated from alpha-lactalbumin by tryptic digestion has also been demonstrated [8].

2.1.2. Beta-Lactoglobulin (Bos d 5)

Beta-lactoglobulin is a small (18.3 kDa) lipocalin family protein (Figure 1, Table 1). The protein contains two disulfide bridges and one free cysteine that may cause dimerization of lactoglobulin [6,8]. The presence of disulfide bridges in the molecule is also associated with its high stability to proteolytic cleavage [7,10]. Even after digestion in the gastrointestinal tract, bovine beta-lactoglobulin is found to be secreted in human breast milk [8]. Peptides generated by the tryptic *in vitro* digestion of beta-lactoglobulin, as well as synthetic beta-lactoglobulin derived peptides, showed IgE reactivity and thus confirmed the presence of linear IgE binding epitopes in the beta-lactoglobulin amino acid sequence [11–13]. Therefore, the secretion of beta-lactoglobulin in breast milk could potentially cause symptoms in cow's milk allergic infants or sensitization. No human homologous protein to bovine beta-lactoglobulin is found in breast milk. It is thought that heating from 50–90 °C increases the allergenicity of beta-lactoglobulin, as linear, usually hidden epitopes become available due to the alteration of the native conformation of the protein. In contrast heating above 90 °C is thought to mask linear, as well as conformational epitopes, as beta-lactoglobulin forms aggregates [6].

2.1.3. Serum Albumin (Bos d 6)

Bovine serum albumin (BSA) only accounts for 5% of total whey protein content but is recognized by up to 50% of cow' milk allergic patients [6,11] (Figure 1, Table 1). Similar to alpha-lactalbumin, BSA (67 kDa) shows a high amino acid sequence homology to its human counterpart [11]. BSA not only plays a role in cow's milk allergy but seems to be also important in beef allergy. There is, however, evidence that cooking destroys the allergenic activity of BSA [14]. The identification of sensitization to BSA by component resolved diagnosis may therefore potentially predict whether a cow's milk allergic infant will also suffer from allergic symptoms when beef is introduced into the diet. Since BSA shows sequence and structural similarity with albumins from respiratory allergen sources such as cats, dogs, horses and small furry animals [15–17], it may be important to test for cross-reactivity to confirm whether the genuinely sensitizing allergen source is milk or a respiratory allergen source.

Table 1. Characteristics of cow's milk allergens

Allergen (UNIPROT)	Whey					Caseins				
	Bos d 4 (B6V3I5)	Bos d 5 (G5E5H7)	Bos d 6 (B0JYQ0)	Bos d 7	Bos d LF (B9VPZ5)	Bos d 8	Bos d 9 (B5B3R8)	Bos d 10 (P02663)	Bos d 11 (P02666)	Bos d 12 (P02668)
Isoallergen (Accession number)	Bos d 4.0101 (P00711)	Bos d 5.0101 (P02754), Bos d 5.0102 (B5B0D4)	Bos d 6 (P02769)	Bos d 7.0101	Bos d LF (P24627)		Bos d 9.0101 (P02662)	Bos d 10.0101 (P02663)	Bos d 11.0101 (P02666)	Bos d 12.0101 (P02668)
Protein family	Albumins	Globulins, Lipocalins	Albumins	Immuno-globulins	Transferrins	Caseins	Caseins	Caseins	Caseins	Caseins
Protein name	Alpha-lactalbumin	Beta-lactoglobulin	Serum albumin	IgG	Lactoferrin	AlphaS1-casein, alphaS2-casein, beta-casein, kappa-casein	AlphaS1-casein	AlphaS2-casein	Beta-casein	Kappa-casein
Molecular weight [kiloDalton]	14.19	18.31	67.20	160	76.14		22.89	24.35	23.58	18.97
Isoelectric point	4.80	4.83	5.60	n.a.	8.67		4.95	8.34	5.13	5.93
Number of amino acids (AA)	123 (AA 1-19 signal; AA 20-142 chain)	162 (AA 1-16 signal; AA 17-162 chain)	589 (AA 1-18 signal; AA 19-607)	n.a.	689 (AA v1-19 signal; AA 20-708 chain)		199 (AA 1-15 signal; AA 16-214 chain)	207 (AA 1-15 signal; AA 16-222 chain)	209 (AA 1-15 signal; AA 16-224 chain)	169 (AA 1-21 signal; AA 22-190 chain)
Number of patients (n), %IgE-positive	n = 58, 27.6% [18]; n = 140, 25% [19]; n = 78, 62.8% [20]; n = 51, 19.6% [21]; n = 45, 19% [22]	n = 58, 38.7% [18]; 10% [19]; n = 78, 43.6% [20]; n = 140, n = 51, 19.6% [21]; n = 45, 23% [22]	n = 58, 12.9% [18]; 25% [19]; n = 78, 3.8% [20]; n = 140; n = 51, 21.56% [21]; n = 45, 23% [22]	n = 58, 10.3% [18]; n = 140, 40% [19]	n = 58, 10.3% [18]; n = 140, 10% [19]; n = 51, 66.67% [21]	n = 58, 46.5% [18]; n = 140, 40% [19]; n = 51, 49.02% [21]; n = 45, 40% [22]	n = 58, 58% [18]; n = 140, 25% [19]		n = 58, 71.0% [18]; n = 140, 20% [19]	n = 58, 58.1% [18]; n = 140, 10% [19]; n = 78, 29.5% [20]
Cross reactivity		n = 6,beta-lactoglobulin from buffalo's, ewe's milk, goat's milk [23]	raw meat (beef, lamb, deer, pork) [24]; Cap h 6 (goat), Ovi a 6 (Sheep), Equ c 3 (horse), Equ as 6 (Donkey), Sus s 1 (Pig) [6]			Casein from buffalo's, ewe's milk, goat's milk [23]; 30S component from soy (IgE-reactivity) [25]				
Heat stability	Aggregation [26]	Aggregation [26]	Heat-sensitive Bos d 6-sensitized beef allergic children react to CM [27]; patients tolerated heated meat products [14,24]	?	?	Heat-stable Bos d 8-sensitized CM allergic patients react to heated products, Bos d 8-sIgE correlates with symptom severity [28]				
Associated symptoms										

2.1.4. Bos d 7 (Immunoglobulin)

Immunoglobulin G (IgG) has been reported to be recognized by IgE from a varying percentage (i.e.; 10–40%) of cow's milk allergic patients [18,19] (Table 1). The allergenic activity of IgG has not yet been evaluated and there is no information regarding its clinical relevance. It is possible that carbohydrate epitopes of IgG represent IgE epitopes similarly as has been reported for the cat allergens Fel d 5 (cat IgA) and Fel d 6 (cat IgM), both of which seem to have no clinical importance and allergenic activity [29,30]. A recent study revealed that the dominant IgE epitope on cat IgA2 is the carbohydrate α-gal, which exists on a large number of other mammalian proteins [31]. However, the role of carbohydrate epitopes on Bos d 7 for IgE recognition still has to be determined.

2.1.5. Lactoferrin (Bos d LF)

Lactoferrin is an iron-binding glycoprotein belonging to the transferrin protein family. By chelating iron, it deprives bacteria from iron uptake and thus acts as natural antimicrobial protein in milk [7,32]. It has been reported that lactoferrin is recognized by IgE antibodies from cow's milk allergic patients at widely varying percentages (i.e.; 5–66%) (Table 1) [18–21]. The clinical relevance of lactoferrin is not known. Lactoferrin-specific IgE antibodies have been identified in sera from cow's milk allergic patients but its allergenic activity and the impact of lactoferrin sensitization on severity of clinical symptoms have not yet been investigated [11].

2.2. Caseins (Bos d 8)

AlphaS1-Casein, AlphaS2-Casein, Beta-Casein, Kappa-Casein (Bos d 9–12)

Caseins are calcium-binding phosphoproteins that constitute 80% of total milk protein. The high calcium content in milk enables the formation of casein micelles. Four distinct casein proteins are recognized as individual allergens (alphaS1-casein, alphaS2-casein, beta-casein, kappa-casein, Bos d 9–12) [6] (Figure 1, Table 1). Bos d 9–12 contain cross-reactive as well as non-cross-reactive IgE epitopes among each other [20]. In contrast to whey proteins, caseins are heat stable [6]. However, caseins are highly susceptible to enzymatic degradation [6]. Tryptic digest fragments of aS1-casein exerted no immunoreactivity [8,33] It might be reasonable to assume that the sensitivity to digestion is the reason why casein-specific IgE antibodies are directed towards linear epitopes [11].

The cross-reactivity of caseins is potentially explained by the conserved region containing the phosphorylation site for alpha- and beta-caseins [8]. However, the measurement of IgE reactivity to purified recombinant caseins indicated that patients' IgE antibodies could discriminate the different casein allergens [20,33]. Moreover, natural milk allergen preparations may be contaminated with allergens from the same source, and residual casein fragments were detected in the whey fraction of cow's milk [34].

3. Diagnosis of Cow's Milk Allergy: From Classical Procedures Towards Molecular Diagnosis

Several guidelines for the diagnosis of cow's milk allergy have been published during the recent years [2,35,36]. Recording the medical history and physical examination of the patient is one cornerstone in the diagnosis. CMA can show a variety of clinical manifestations and it is important to discriminate immune-mediated reactions from non-immune-mediated adverse reactions, like toxic, and pharmacologic reactions, or lactose intolerance the most common form of milk intolerance which results from a reduced availability of enzymes to digest lactose and causes symptoms only in the bowel [1,37]. Immune-mediated symptoms can be classified according to Coombs and Gell in type I hypersensitivity reactions mediated by IgE antibodies due to sensitization to cow's milk allergens and type II-type IV reactions [1]. IgE-mediated cow's milk allergy can be clearly diagnosed by demonstration of the presence of allergen-specific IgE antibodies while for non-IgE-mediated food allergy no unambiguous diagnostic tests are available. Symptoms of IgE-associated cow's milk allergy can appear as immediate (early) reactions and delayed (late) reactions. While immediate reactions are observed within the first minutes up to 2 hours after milk consumption and represent IgE-mediated effects, delayed reactions may appear up to 48 hours or even 1 week after ingestion and may be attributed to cell-mediated immune mechanisms [36]. IgE-mediated allergen presentation to T cells is one possible mechanism involved in the late response to allergens [38] but it has also been shown that T cell-mediated allergen-specific inflammation can occur via a non-IgE-mediated mechanism [39,40]. It has to be considered that both early and late symptoms can occur in the same patient and may affect different organs. This mainly includes the gastrointestinal tract, leading to abdominal pain, vomiting, diarrhea, dysphagia, constipation, occult blood loss, iron-deficiency anemia, and the skin by developing urticaria, atopic eczema, or angioedema, and also respiratory symptoms like runny nose, wheezing, and allergic asthma, and more general and severe reactions as systemic anaphylaxis may occur, but are fortunately rare. Even acute coronary syndromes associated with anaphylactic reactions to milk have been observed [41–43]. IgE-mediated symptoms occur in approximately 50% of children with CMA, while they are rarer in adults, as CMA resolves in more than 85% of cases [2].

In contrast, non-IgE-mediated immune reactions associated with cow's milk ingestion symptoms are often restricted to the gastrointestinal tract and the skin, they are often of the delayed type and develop up to several days after milk intake. In these patients, no circulating cow's milk allergen-specific IgE antibodies can be detected and they have a negative skin reaction to cow's milk proteins. These reactions include food protein-induced enterocolitis syndrome, cow's milk protein-induced enteropathy, and cow's milk-induced proctitis and proctocolitis [2], and also conditions where milk protein-specific IgE antibodies are present may occur, like in eosinophilic esophagitis or gastroesophageal reflux disease.

Once the medical history suggests the presence of cow's milk allergy, a number of *in vitro* and *in vivo* diagnostic tests are performed. IgE sensitization to milk proteins is evaluated by the determination of specific IgE antibodies in serum. Skin testing includes the skin prick test (SPT) for IgE-mediated immediate allergic responses and the atopy patch test (APT) measuring T cell-mediated late responses to allergens. The APT can be useful to demonstrate the involvement of T cells in the pathogenesis. Still the double blind, placebo-controlled food challenge (DBPCFC) is considered as a gold standard for the diagnosis of milk allergy [2]. However, many clinicians avoid it because they fear severe and life-threatening side effects and the test procedure is very cumbersome, involving hospitalization and the availability of emergency care.

3.1. Cow's Milk-Specific IgE Measurement

The *in vitro* determination of allergen-specific IgE levels represents an important tool for the diagnosis of IgE-mediated milk allergy because it allows the unambiguous demonstration of IgE sensitization. It is performed in serum samples obtained from patients, whereby several test systems are available like the ImmunoCAP system, the Immulite assay system, or multiplex allergy tests [44,45]. It was recently proposed to measure specific IgE in the saliva instead of blood samples, because it is difficult to obtain blood samples from young children [46]. However, IgE concentrations in saliva and other body fluids are usually lower than in blood and may vary strongly due to secretion. Therefore, multiallergen testing using chips containing micro-arrayed allergens should be preferred because it requires on few microliters of blood and serum [47].

The sensitivity and specificity of IgE measurements is very high. However, the presence of allergen-specific IgE in the blood as well as a positive skin test result does not necessarily correlate with the development of allergic symptoms [2]. Therefore, specific IgE levels have to be evaluated in the context of the clinical history of the patient. High milk allergen-specific IgE levels were shown to be predictive for the development of clinical symptoms during an oral food challenge. A number of studies analyzed the cut-off levels for milk-specific IgE for predicting the onset of milk-induced allergic symptoms [48]. However, different values were obtained due to the various study populations, and the criteria to evaluate the severity of symptoms.

In the context of serological testing, it is important to state clearly that non-IgE-mediated cow's milk allergy cannot be diagnosed by determination of milk allergen-specific IgG and IgA antibody levels, which was shown to be inappropriate for diagnosing cow's milk allergy [49] and is also not recommended by national and international guidelines [50,51].

3.2. Skin Testing

3.2.1. Skin Prick Test (SPT)

The SPT represents a commonly used diagnostic method in clinical practice, as it is cheap and easy to perform. However, its diagnostic value is limited, as the specificity of this test is rather low. Several studies tried to define a cut-off value to predict CMA in the tested patients. Using commercial cow's milk extracts, wide variations of cut-offs ranging from 4.3–20 mm were found [48,52–57]. Studies comparing cow's milk extract to purified natural α-lactalbumin, ß-lactoglobulin, and casein indicated that fresh cow's milk was the most sensitive SPT reagent [57]. Having a positive SPT for all three cow's milk allergens increased the likelihood of a positive response to challenge [55]. Therefore, reagents containing defined amounts of pure allergen components would be needed for the *in vivo* diagnosis of CMA [58,59].

3.2.2. Atopy Patch Test (APT)

The APT is most often used for the diagnosis of atopic dermatitis but has also been proposed for the diagnosis of non-IgE-mediated CMA and eosinophilic esophagitis [2,60]. As the APT measures cow's milk allergen-induced T cell activation, it can only detect allergic reactions of the delayed type which may be IgE-dependent or IgE independent [39,40]. It is performed by the application of cow's milk allergens to the skin via a patch, which is fixed for 24–48 h. After the removal of the patch skin reactions are recorded, and results from different studies show great variations regarding sensitivity, specificity, and positive predictive value [61–64]. However, there might be variations in reagents, procedure and interpretation of skin signs, limiting the accuracy of the test. One study suggested that development of erythema plus papules and/or vesicles in the skin was of high diagnostic value in children with CMA-related gastrointestinal symptoms [61]. APTs were also suggested to predict tolerance induction in non-IgE-mediated cow's milk allergic children [65].

3.3. Food Challenges

Oral food challenges (OFC), especially the double-blind, placebo-controlled food challenge (DBPCFC) are recommended as the gold standard to confirm CMA, to determine the threshold dose of cow's milk allergens in individuals, and to evaluate acquired tolerance to cow's milk proteins [2,66]. Patients receive increasing doses of cow's milk until adverse reactions appear (positive challenge test) or after a certain amount has been administered without reactions (negative challenge test). However, oral food challenges are time consuming, expensive, and can induce severe anaphylactic reactions that require the availability of intensive care units. Patients at high risk for anaphylaxis should receive oral food challenges only in hospitals by trained specialists [67]. Adverse symptoms involve the skin, the gastrointestinal tract, the respiratory tract, cardiovascular, or neurological symptoms, which are graded according to a severity score [68]. A recent study found that the high incidence of anaphylaxis during an OFC was related to higher sIgE levels. In particular, there was a significant association observed between anaphylaxis during the OFC with cow's milk and the sIgE levels for caseins [28]. In addition, a history of anaphylaxis to the causative food and age were reported as risk factors for having severe anaphylactic reactions [68]. The fact that the classical path of diagnosis starting with case history and stepwise provocation testing is very time consuming is the reason why alternative approaches for diagnosis are currently emerging. These alternative approaches are based on the determination of allergen-specific IgE to a comprehensive panel of allergens in combination with medical history using provocation testing for verification of clinical relevance [69].

3.4. Molecular Diagnosis

In principle, pure allergen molecules can be produced, either by purification from the natural allergen source, or as recombinant proteins. However, it might be difficult to obtain pure preparations of natural allergens which are free of contaminating allergens from the same allergen source by biochemical purification procedures. Nowadays, the availability of cDNAs of allergens by molecular cloning techniques allows the production of recombinant proteins with high purity in big amounts (Figure 1). They represent defined isoforms which is difficult, if not impossible to achieve for natural proteins like in case of the cow's milk allergens αS1 casein and αS2 casein [7]. When used for molecular diagnosis, recombinant allergens allow the identification of the sensitizing allergen source and IgE-reactivities to cross-reactive allergens. Dependent on the expression system they can be obtained lacking cross-reactive carbohydrates, to avoid clinically irrelevant test results. Moreover, the biological activity of recombinant allergens can be studied in *in vitro* effector cell activation assays and by skin testing in allergic patients in order to identify their clinical relevance for the induction of allergic symptoms in patients.

So far, α-lactalbumin, ß-lactoglobulin, αS1-casein, αS2-casein, and κ-casein from cow's milk were produced as recombinant proteins in *Escherichia coli*. They have been used to study their frequencies of IgE recognition and allergenic activity in cow's milk allergic patients [20]. Furthermore, they represent the basis for the development of therapeutic and prophylactic vaccines for cow's milk allergy.

Since diagnostic *in vitro* tests for the measurement of specific IgE, which are based on single allergens like the ImmunoCAP system, require large amounts of serum, which is often difficult to obtain from small children, multiplex allergy tests requiring only few microliters of serum have been developed. In the recent years, these protein microarrays and other multiplex techniques have been applied to the field of allergy diagnosis. This has led to the development of allergen microarrays which are either commercially available or were produced for research purpose [18–20,70]. Only tiny amounts of serum are required for the determination of the IgE reactivity profiles to cow's milk allergens and a comprehensive set of other food and respiratory allergens. A study using an experimental allergen microarray containing purified natural and recombinant cow's milk allergens suggested that increased IgE levels to cow's milk allergens were associated with oligo-sensitization to several cow's milk allergens [20]. Besides specific IgE levels, the number of recognized allergens seemed to contribute to a positive food challenge result suggesting that allergen microarrays could provide useful tools to diagnose symptomatic cow's milk allergy [18,19]. Combining the allergen microarray results with basophil activation studies using recombinant cow's milk proteins was shown to have an additional diagnostic value, as basophil activation was mainly observed in patients with severe symptoms [20].

Cow's milk allergen-specific IgE was shown to be a prognostic marker for persistence of cow's milk allergy [71,72]. The development of multi-allergen assays including besides cow's milk allergens also allergen-derived peptides allowed the mapping of sequential IgE and IgG4 epitopes, indicating that recovery from CMA was associated with decreasing IgE and increasing IgG4 to cow's milk epitopes [73]. Furthermore, greater diversity of IgE epitope recognition and higher affinity were associated with clinical phenotypes and severity of milk allergy [74,75]. The chip-based diagnosis of comprehensive IgE reactivity profiles in conjunction with medical history has been shown to be extremely efficient for the analysis of complicated cases in children with complex IgE sensitization profiles and to facilitate the selection of personalized treatment strategies [76].

4. Current Treatment Strategies for Cow's Milk Allergy

4.1. Avoidance and Hypoallergenic Milk Formulas

Exclusive breastfeeding is currently recommended for all infants for the first 4–6 month due to many beneficial effects of breast milk. There is also evidence that breastfeeding may reduce the development of food allergies [66]. No recommendation is currently given for pregnant and breast-feeding women to avoid cow's milk, and maternal cow's milk avoidance was associated with the development of cow's milk allergy in infants [66,77,78]. In this context, the presence of allergen-specific IgA and IgG antibodies in human milk were suggested as protective factors [7,77]. A recent study highlighted the role of maternal allergen-specific IgG antibodies for the prevention of allergic sensitization to these allergens, but also peptides originating from cow's milk allergens have been detected in breast milk, however, a possible tolerogenic effect remains to be determined [79,80]. Sensitization to human milk proteins occurs rarely, and only a few cases of suspected allergy to breast milk are reported [81,82]. While standard cow's milk formulas are fed after the age of 4 month, hydrolysates of cow's milk protein are recommended for children, who cannot be breastfed, children at high risk, and those who have been diagnosed with cow's milk allergy [66] (Table 2).

Depending on the degree of hydrolysis, partially or extensively hydrolyzed milk formulas are available. Partially hydrolyzed milk formulas (phMF) usually contain whey protein derived-peptides smaller than 5 kDa, they are not considered to be completely hypoallergenic but are more easily digested than whole milk proteins. In contrast, hypoallergenic products are either extensively hydrolyzed milk formulas (ehMF), composed of small peptides <1.5 kDa, or amino acid formulas, consisting of essential and nonessential amino acids, the latter are recommended for infants who do not tolerate ehMF, but are significantly more expensive [83,84]. Milk protein hydrolysates have not only been shown to avoid allergic symptoms in cow's milk allergic children due to the destruction of IgE epitopes but might also have immunomodulating properties like the induction of T cell tolerance and the prevention of sensitization [85]. There is evidence that hydrolyzed infant formulas may have a long-lasting preventive effect on the development of allergic symptoms [86,87], though other studies do not support the effectiveness of hydrolyzed milk formulas [88]. However, the degree and method of milk protein hydrolysis may influence the preventive effect of infant formulas on the development of milk allergy. Indeed, a recent study compared different commercially available milk formulas regarding protein and peptide content, allergenic activity, and T cell responses in terms of proliferation and cytokine production [89]. Besides their varying allergenic activity, there were striking differences of the CM formulas to induce Th1, Th2, and pro-inflammatory cytokines. Interestingly, some formulas seemed to lack immuno-stimulatory peptides, as they failed to induce any T cell proliferation or the production of proinflammatory cytokines [89].

4.2. Substitution of CM by Non-Bovine Milk Sources

As the elimination of cow's milk may lead to malnutrition [90,91], milk from other sources has been proposed as a substitute for cow's milk (Table 2). However, allergens from cow, sheep, and goat show high cross-reactivity due to their sequence homology, while mare's, donkey's and camel milk are better tolerated by children with CMA. [6,23,92,93]. Selective hypersensitivity to goat milk and sheep milk is rare but several cases have been reported. The degree of phosphorylation of caseins was suggested as an explanation for the differences in allergenicity among milk proteins from cow, sheep, and goat [94,95]. The use of soy-based formulas is not recommended due to the presence of phytoestrogens [96]. In addition, cross-reactivity between soy and cow's milk proteins was reported in some patients [25,97]. Rice protein formulas represent another plant-based alternative to cow's milk, as they lack cross-reactivity and do not contain phytoestrogens [98]. Interestingly, a gene knockout cow was recently generated which produced ß-lactoglobulin-free milk. [99]

4.3. Pre- and Probiotics

Prebiotics are nondigestible substances providing a beneficial physiologic effect for the host by stimulation of growth or activity of a limited number of indigenous bacteria, while probiotics represent live microorganisms, which should confer a health benefit to the host upon administration [66] (Table 2). They are currently not recommended as a dietary supplement in cow's milk allergic patients as there is little evidence for a beneficial effect [66]. A recent meta-analysis of clinical studies investigating probiotics as treatment for food allergies in children provided some evidence that probiotics can improve symptoms of CMA but there was no evidence that they induce tolerance to cow's milk allergens. The effects seem to depend on the administered bacterial strains, showing a beneficial effect of *Lactobacillus rhamnosus* GG [100,101]. In a recent study performed in 10-month old children the daily application of *Lactobacillus rhamnosus* and *Bifidobacterium animalis* subsp *lactis* reduced the incidence of eczema in the treated group, though the development of allergic sensitization did not differ from the placebo group [102]. Both bacterial strains reduced the risk of eczema in children with a genetic predisposition to eczema due to single nucleotide polymorphisms to TLRs. Perinatal consumption of *L. rhamnosus* GG was investigated in 303 mothers and significantly reduced the development of allergic disease in infants [103], however, maternal supplementation with *Lactobacillus rhamnosus* HN001 was shown to have no effect on infant eczema development in another study [104]. Likewise, the addition of prebiotics to a partially hydrolyzed whey formula did not prevent eczema in the first year of life, although the supplementation of the milk formula with nondigestible oligosaccharides modulated the gut microbiota closer to that of breast-fed infants. [105,106].

4.4. Current Strategies for Allergen-Specific Immunotherapy of Cow's Milk Allergy

4.4.1. Oral Immunotherapy (OIT)

Oral immunotherapy of cow's milk allergic patients comprises the repeated consumption of milk allergens at regular intervals aiming to reduce the sensitivity to cow's milk allergens during treatment, referred to as desensitization, and to induce a state of sustained unresponsiveness to the allergen after discontinuation of therapy, also known as clinical tolerance [107] (Table 2). Although standardized protocols have not been established, OIT is generally performed in 3 steps, including an escalation phase, a build-up phase, and a maintenance phase. It is currently recommended for persistent cow's milk allergy for children from around 4–5 years of age but has recently been performed even in children less than 12 month [108]. However, it is not suited for a broad application, as it should be undertaken in clinical centers by experienced clinicians due to the risk of adverse reactions [107].

Several controlled clinical trials demonstrated the efficacy of CM OIT in cow's milk allergic children [109,110]. They found an increase of the tolerated threshold dose, and decreased cow's milk allergen-specific IgE levels and skin reactions in the treated groups [111–115]. One study reported a detailed analysis of IgE and IgG4 antibody responses to cow's milk allergens and allergen-derived peptides before and after milk OIT. Results indicated that patients who achieved clinical tolerance had a lower and less diverse epitope-specific IgE response. After milk OIT the amount of epitope-specific IgG4 increased in contrast to epitope-specific IgE [116]. This is in line with results from a recent study suggesting that baseline sIgE levels to CM allergens were predictive for the success and safety of milk OIT [117]. However, little is known about the long-term outcomes of milk OIT, and it has been suggested that the protective effect of milk OIT requires a continuous consumption of milk [118]. Though, follow-up studies over an extended period of time for larger numbers of patients would be required [119]

A major limitation of milk OIT is the frequent occurrence of adverse reactions. A recent study focusing on adverse events during milk OIT highlighted the inherent risks of OIT compared to allergen avoidance and identified higher IgE levels for α-lactalbumin and casein at baseline as risk factors for anaphylactic reactions during OIT [120].

Several approaches have been developed recently to reduce the occurrence of adverse reactions during milk OIT. Combining OIT and treatment with the monoclonal anti-IgE antibody omalizumab allowed an acceleration of the build-up phase and the reduction of the frequency and severity of side-effects in cow's milk allergic children, but varying outcomes of efficacy were observed [121–126]. However, the high costs of this treatment have to be considered. The administration of baked milk, which has been heated at 180 °C for 30 min, has been suggested as a hypoallergenic alternative in OIT, and the consumption of baked milk products may have a beneficial effect on tolerance induction to raw cow's milk [127]. However, severe adverse events were also observed in response to baked milk [128,129]. Hydrolyzed milk formulas, which possess lower allergenic activity than cow's milk proteins, were also tested in OIT of cow's milk allergic children to reduce adverse reactions during the treatment [130].

Several immunological changes that occur in the course of OIT have been described, as a decrease of allergen-specific IgE and the induction of allergen-specific IgG4 antibodies, a decline in antigen-specific Th2 cells, and the induction of regulatory T cells [126].

4.4.2. Sublingual Immunotherapy (SLIT)

During OIT, the allergen is immediately swallowed, and during SLIT, it is held under the tongue for a period of time (Table 2). Sublingual immunotherapy for the treatment of CM allergic children was so far reported in only two studies. Eight children with CMA received SLIT with increasing doses of CM for 6 months. Six children who completed the study showed an increase in the eliciting dose upon challenge [131]. Keet et al. compared a SLIT protocol to OIT and combined SLIT/OIT treatment in 30 cow's milk allergic children. Though OIT induced more severe adverse reactions, it was more efficacious for desensitization to CM than SLIT alone [132].

Table 2. Current strategies for the treatment of cow's milk allergy.

Intervention	Administration Route	Procedure	Aim of Intervention	Ref.
Avoidance of CM proteins	Oral	Introduction of ehMF or aaMF into diet; Substitution of CM by non-bovine milk	Reduction of allergic symptoms due to the lack of CM-specific epitopes	[23,86–95]
OIT	Oral	Application of increasing amounts of CM	Desensitization; Clinical tolerance to CM allergens	[108–130]
SLIT	Sublingual	Application of increasing amounts of CM	Desensitization; Clinical tolerance to CM allergens	[131,132]
EPIT	Epicutaneous	Delivery of CM proteins via patch application	Desensitization; Clinical tolerance to CM allergens	[133–138]
Pre/ Probiotics	Oral	Given alone or in combination with hydrolyzed milk formulas	Immunomodulation	[100–106]

CM: cow's milk; OIT: oral immunotherapy; SLIT: sublingual immunotherapy; EPIT: epicutaneous immunotherapy; ehMF: extensively hydrolyzed milk formulas; aaMF: amino acid milk formulas;

4.4.3. Epicutaneous Immunotherapy (EPIT)

As an alternative application route of allergens for allergen-specific immunotherapy AIT, epicutaneous immunotherapy (EPIT) has been investigated for the treatment of food allergy [133,134] (Table 2). The first EPIT study in children with CMA was reported in 2010 [135]. Delivery of skimmed cow's milk powder to the epidermis via a patch was tested in 19 children with CMA in a DBPC study. The treatment was well tolerated, adverse events were most frequently local skin reactions, and there was no incidence of systemic anaphylaxis. Though, no statistically significant improvement of the cumulative tolerated dose between the placebo and actively treated group was detected. A phase 1/2 DBPC trial (ClinicalTrials.gov identifier NCT02223182), which evaluated the safety and efficacy of a cow's milk protein containing patch at a dose of 150, 300, or 500 µg in a larger number of milk allergic children, was recently completed, but results have not been published yet. An abstract reports that, besides mild or moderate local skin reactions, no serious adverse events occurred [136]. EPIT was also tested for the treatment of peanut allergy and for inhalant allergens using different doses and duration of application [133,137]. In addition, allergens were applied to either intact, tape-stripped, or scarified skin, which might affect the penetration of the allergens into the dermis and the activation of keratinocytes [138], indicating that more development work will be necessary for a broad application. It should be also considered that epicutaneous allergen application was found not to induce relevant production of allergen-specific IgG which is important for the success of allergen-specific immunotherapy [139,140].

5. Emerging Strategies for the Treatment and Prevention of Cow's Milk Allergy

5.1. IgE-Targeting Therapies

Cow's milk allergic patients are often polysensitized and exhibit elevated total IgE levels. It therefore does not come as a surprise that IgE-targeting therapies have been considered for cow's milk allergy. In this context, a recent case report should be mentioned which described the use of IgE immunoadsorption preceding treatment with anti-IgE antibody due to extremely high IgE levels in a child with milk allergy, resulting in an increase of tolerance threshold to milk [141]. In fact, columns for selective IgE immunoadsorption have been developed and were shown to selectively deplete IgE by a few apheresis cycles thus preparing patients with very high total IgE levels for Omalizumab treatment [142,143]. In fact, beneficial effects of Omalizumab treatment on symptoms of food allergy have been recently reported for patients suffering from asthma and food allergy in a real-life study [144].

5.2. Subcutaneous Immunotherapy with Recombinant Hypoallergens

Although subcutaneous immunotherapy has been performed in patients suffering from respiratory allergies for more than 100 years and represents the most effective form of AIT, there is still no approved SCIT vaccine for the treatment of food allergy [1,145]. However, currently new innovative technologies for SCIT are emerging which are based on recombinant allergens and recombinant hypoallergenic allergen derivatives [140,146]. Based on the sequences of the major allergens from different allergen sources, recombinant and synthetic vaccines have been developed and already tested in clinical trials. More recently, B cell epitope-derived peptides lacking allergen-specific T cell epitopes were fused to a non-allergenic carrier and produced for the treatment of grass pollen allergy [147,148]. The latter two approaches aim at the induction of a blocking IgG antibody responses and eliminating immediate and late phase side effects. The grass pollen allergy vaccine BM32 which is based on recombinant hypoallergenic derivatives of the major grass pollen allergens has been tested now in several phase 2 clinical trials [149–152]. The studies conducted with the B cell epitope-based grass pollen allergy vaccine BM32 as well as a recent study in which cat allergic patients were treated by passive vaccination with allergen-specific monoclonal IgG antibodies highlight the important role of allergen-specific IgG-blocking antibodies for successful AIT [153]. In principle, both strategies could be also applied for treatment of food allergy [1]. A hypoallergenic derivative of the major fish allergen, parvalbumin, was developed by mutating the calcium-binding sites resulting in a loss of IgE reactivity but retained parvalbumin-specific T cell epitopes in the recombinant mutant [154]. The molecule was formulated as an aluminum hydroxide-adjuvanted vaccine for subcutaneous injection immunotherapy. It was evaluated in two phase 2 clinical trials and showed an excellent safety profile as well as the induction of parvalbumin-specific IgG antibody responses in vaccinated fish allergic patients [155,156]. Likewise, it should be feasible, that hypoallergens or B cell epitope-based recombinant vaccines derived from the major cow's milk allergens can be developed for vaccination against cow's milk allergy [7].

5.3. Strategies for the Prevention of Milk Allergy

In recent years, the analysis of sensitization profiles in large birth cohorts by molecular diagnostic tests as well as prevention trials in children at risk for developing food allergy have increased our knowledge regarding the development of IgE sensitization in childhood and provided directions for early prevention measures. As depicted in Figure 2, various intervention strategies could be applied and/or developed. While primary prevention strategies aim preventing the IgE sensitization to cow's milk allergens very early in life, secondary prevention intends to stop the transition from clinically silent IgE sensitization to symptoms of allergy and/or the progression of mild to severe symptoms. Several early treatment strategies are suggested for children with established milk allergy.

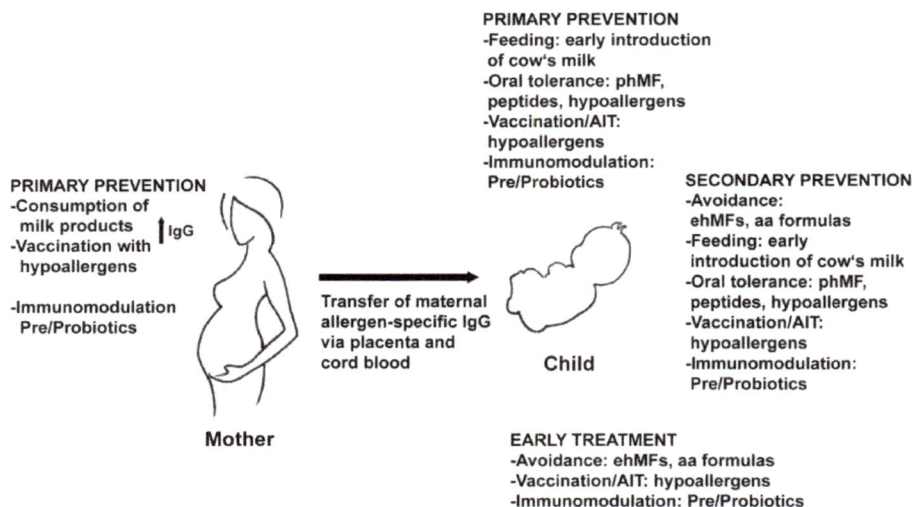

PRIMARY PREVENTION
-Feeding: early introduction
 of cow's milk
-Oral tolerance: phMF,
 peptides, hypoallergens
-Vaccination/AIT:
 hypoallergens
-Immunomodulation:
 Pre/Probiotics

PRIMARY PREVENTION
-Consumption of
 milk products
-Vaccination with
 hypoallergens

-Immunomodulation
 Pre/Probiotics

IgG

Transfer of maternal
allergen-specific IgG
via placenta and
cord blood

Child

SECONDARY PREVENTION
-Avoidance:
 ehMFs, aa formulas
-Feeding: early
 introduction of cow's milk
-Oral tolerance: phMF,
 peptides, hypoallergens
-Vaccination/AIT:
 hypoallergens
-Immunomodulation:
 Pre/Probiotics

Mother

EARLY TREATMENT
-Avoidance: ehMFs, aa formulas
-Vaccination/AIT: hypoallergens
-Immunomodulation: Pre/Probiotics

Figure 2. Primary, secondary prevention, and early treatment of cow's milk allergy. Primary prevention strategies can be applied to the mother and the child to avoid allergic sensitization to cow's milk allergens. The progression of the disease in sensitized children, or those who have already developed allergic symptoms, is inhibited by secondary prevention strategies or early treatment.

Within the EU-funded research project MeDALL, allergen-specific IgE responses in infancy were investigated based on an allergen microarray containing >160 allergens in large European birth cohorts [4,5]. It was shown that the evolution of IgE sensitization to food and respiratory allergens starts during the first few years of life with the recognition of major allergens [157–161]. Several factors, like allergen-specific IgE levels, the number of recognized allergens from an allergen source and cross-reactive allergens detected early in life seem to be predictive for the onset of allergic symptoms [157,159]. IgE sensitization profiles were shown to remain stable in adulthood, suggesting that the first years of life represent a window of opportunity for early interventions [162,163]. A very recent study investigating the effects of allergen-specific IgG responses in mothers on the development of IgE sensitization in the off-springs indicated that maternal allergen-specific IgG transmitted from the mothers to the children may protect against allergic sensitization [79]. IgG reactivity to microarrayed allergens was determined during pregnancy, in cord blood samples, in breastmilk, and in children in the first years of life. It was shown that allergen-specific IgG reactivity profiles were highly correlated in mothers, cord blood, and breast milk and that maternal IgG persisted up to 6 months in children. Most importantly, children from mothers with specific plasma IgG levels >30 ISU against an allergen did not develop IgE sensitization against that allergen at 5 years of age, suggesting a highly protective effect of maternal IgG on allergic sensitization. The finding that allergen-specific IgG from mothers might prevent allergic sensitization in the offspring has already been shown in several murine studies. Protection could either occur by neutralization of the antigen by specific IgG or by uptake of allergen-IgG immune complexes via Fcgamma receptors on immune cells mediating tolerogenic effects [164]. Accordingly, primary prevention of sensitization to cow's milk allergens in children might be initiated either by milk consumption of the mother to keep allergen-specific IgG levels high or by vaccinating the mother with hypoallergenic derivatives of cow's milk allergens (Figure 2). The feasibility of vaccinating healthy individuals with hypoallergenic molecules was recently demonstrated in a clinical study, showing that no allergic sensitization, but the induction of an allergen-specific IgG response developed in the treated individuals [165]. These results are in agreement with previous observations suggesting that AIT treatment of pregnant women could inhibit allergic sensitization in the offspring [166]. Transfer of maternal IgG could either occur via the placenta, cord blood, and less likely by breast milk. In principle, children could receive preventive vaccines also as a prophylactic treatment in the sense of primary prevention if given before sensitization took place (Figure 2). B cell epitope-based vaccines derived from the major cow's milk allergens might represent good candidates, as these vaccines are designed to lack allergen-specific T cell and IgE epitopes and direct the induced allergen-specific IgG responses to IgE binding sites without inducing IgE responses [151]. Likewise, passive immunization may be considered for primary prevention (Figure 2). It is known from mouse studies that the administration of allergen-specific IgG antibodies protected from subsequent allergic sensitization to this allergen [167]. However, the time window for this intervention would have to be determined in humans.

The importance of starting a preventive treatment of food allergy at a very early timepoint in children was recently demonstrated in a clinical study investigating the prevention of peanut allergy. It was shown that the early introduction of peanut in high risk infants between 4–11 months of age significantly decreased the frequency of developing peanut allergy [168]. However, this effect could not be seen in another trial introducing six allergenic foods including cow's milk in infants recruited from a general population [169]. At present it is not completely clear whether this intervention is an early form of OIT inducing blocking IgG or based on the induction of immunological tolerance. One also has to bear in mind that the early consumption of cow's milk allergens as a primary or secondary prevention strategy may induce allergic sensitization to cow's milk (Figure 2).

The feeding of hydrolyzed milk formulas may be an alternative because they are hypoallergenic and those formulas which contain intact cow's milk allergen peptides (i.e.; partially hydrolyzed formulas) may have the capacity to induce T cell tolerance (Figures 1 and 2). In fact, some cohort studies in children have indicated that hydrolyzed milk formulas may indeed suppress the development of allergic symptoms, but other studies did not confirm these results [86,87,170,171]. This might be explained by the heterogeneity of different hydrolyzed formulas [89] and the presence or absence of peptides critical for tolerance induction. As a consequence, the tolerogenic potential in hydrolyzed infant formulas is currently under investigation [172–174]. However, the presence of relevant T cell epitopes in these preparations depends on the use of enzymes for degradation of cow's milk proteins. Instead, it may be preferable to produce mixtures of defined synthetic peptides comprising the cow's milk allergen-derived T cell epitopes which could be administered for oral tolerance induction. In experimental animal models, oral tolerance induction has been shown to potently suppress allergic sensitization [175]. With the availability of recombinant allergens and synthetic allergen peptides, the doors seem to be open now for safe and robust oral tolerance induction for food as well as respiratory allergies as primary prevention [176].

6. Conclusions

Cow's milk allergy is rare but important because it represents one of the first forms of allergy affecting children and may induce severe and life-threatening allergic reactions. Therefore, there is a need for safe forms of diagnosis, treatment and prevention. Molecular allergy diagnosis performed with recombinant allergen molecules offers an attractive alternative possibility for diagnosis in addition to oral provocation testing which may be hazardous and time consuming. Based on the detailed knowledge of the disease-eliciting allergen sources, new forms of innovative molecular AIT treatments and strategies for primary and secondary prevention of cow's milk allergy appear on the horizon. They are based on AIT with recombinant hypoallergenic allergen molecules and tolerance induction with synthetic allergen derived peptides.

Author Contributions: B.L.; R.F.; O.E.; M.K.; A.K.; and R.V. contributed to the writing and editing of the manuscript.

Acknowledgments: This study was supported by grant F4605 of the Austrian Science Fund. Rudolf Valenta is recipient of a Megagrant of the Government of the Russian Federation, grant number 14.W03.31.0024. Alexander Karaulov is supported by the Russian Excellence Project "5-100". The authors are grateful to the Medical University of Vienna which has provided a strong and sustained basis as well as environment for Allergy Research in Vienna.

Conflicts of Interest: R.V. has received research grants from Biomay AG, Vienna, Austria and Viravaxx, Vienna, Austria and serves as consultant for Viravaxx. All other authors have no conflict of interest to declare.

References

1. Valenta, R.; Hochwallner, H.; Linhart, B.; Pahr, S. Food allergies: The basics. *Gastroenterology* **2015**, *148*, 1120–1131. [CrossRef] [PubMed]
2. Fiocchi, A.; Brozek, J.; Schünemann, H.; Bahna, S.L.; von Berg, A.; Beyer, K.; Bozzola, M.; Bradsher, J.; Compalati, E.; Ebisawa, M.; et al. World Allergy Organization (WAO) Diagnosis and Rationale for Action against Cow's Milk Allergy (DRACMA) Guidelines. *World Allergy Organ. J.* **2010**, *3*, 57–161. [CrossRef] [PubMed]
3. Schoemaker, A.A.; Sprikkelman, A.B.; Grimshaw, K.E.; Roberts, G.; Grabenhenrich, L.; Rosenfeld, L.; Siegert, S.; Dubakiene, R.; Rudzeviciene, O.; Reche, M.; et al. Incidence and natural history of challenge-proven cow's milk allergy in European children-EuroPrevall birth cohort. *Allergy* **2015**, *70*, 963–972. [CrossRef] [PubMed]
4. Lupinek, C.; Wollmann, E.; Baar, A.; Banerjee, S.; Breiteneder, H.; Broecker, B.M.; Bublin, M.; Curin, M.; Flicker, S.; Garmatiuk, T.; et al. Advances in allergen-microarray technology for diagnosis and monitoring of allergy: The MeDALL allergen-chip. *Methods* **2014**, *66*, 106–119. [CrossRef] [PubMed]
5. Anto, J.M.; Bousquet, J.; Akdis, M.; Auffray, C.; Keil, T.; Momas, I.; Postma, D.S.; Valenta, R.; Wickman, M.; Cambon-Thomsen, A.; et al. Mechanisms of the Development of Allergy (MeDALL): Introducing novel concepts in allergy phenotypes. *J. Allergy Clin. Immunol.* **2017**, *139*, 388–399. [CrossRef] [PubMed]
6. Villa, C.; Costa, J.; Oliveira, M.B.P.; Mafra, I. Bovine Milk Allergens: A Comprehensive Review. *Compr. Rev. Food Sci. Food Saf.* **2018**, *17*, 137–164. [CrossRef]
7. Hochwallner, H.; Schulmeister, U.; Swoboda, I.; Spitzauer, S.; Valenta, R. Cow's milk allergy: From allergens to new forms of diagnosis, therapy and prevention. *Methods* **2014**, *66*, 22–23. [CrossRef]
8. Wal, J.M. Cow's milk allergens. *Allergy* **1998**, *53*, 1013–1022. [CrossRef] [PubMed]
9. Hochwallner, H.; Schulmeister, U.; Swoboda, I.; Focke-Tejkl, M.; Civaj, V.; Balic, N.; Nystrand, M.; Härlin, A.; Thalhamer, J.; Scheiblhofer, S.; et al. Visualization of clustered IgE epitopes on alpha-lactalbumin. *J. Allergy Clin. Immunol.* **2010**, *125*, 1279–1285. [CrossRef] [PubMed]
10. Wal, J.M. Bovine milk allergenicity. *Ann. Allergy Asthma Immunol.* **2004**, *93* (Suppl. 5), S2–S11. [CrossRef]
11. Monaci, L.; Tregoat, V.; van Hengel, A.J.; Anklam, E. Milk allergens, their characteristics and their detection in food: A review. *Eur. Food Res. Technol.* **2006**, *223*, 149–179. [CrossRef]
12. Järvinen, K.M.; Chatchatee, P.; Bardina, L.; Beyer, K.; Sampson, H.A. IgE and IgG binding epitopes on alpha-lactalbumin and beta-lactoglobulin in cow's milk allergy. *Int. Arch. Allergy Immunol.* **2001**, *126*, 111–118. [CrossRef] [PubMed]
13. Sélo, I.; Clément, G.; Bernard, H.; Chatel, J.; Créminon, C.; Peltre, G.; Wal, J. Allergy to bovine beta-lactoglobulin: Specificity of human IgE to tryptic peptides. *Clin. Exp. Allergy* **1999**, *29*, 1055–1063. [CrossRef] [PubMed]
14. Vicente-Serrano, J.; Caballero, M.L.; Rodríguez-Pérez, R.; Carretero, P.; Pérez, R.; Blanco, J.G.; Juste, S.; Moneo, I. Sensitization to serum albumins in children allergic to cow's milk and epithelia. *Pediatr. Allergy Immunol.* **2007**, *18*, 503–507. [CrossRef] [PubMed]
15. Spitzauer, S.; Pandjaitan, B.; Söregi, G.; Mühl, S.; Ebner, C.; Kraft, D.; Valenta, R.; Rumpold, H. IgE cross-reactivities against albumins in patients allergic to animals. *J. Allergy Clin. Immunol.* **1995**, *96*, 951–959. [CrossRef]
16. Pandjaitan, B.; Swoboda, I.; Brandejsky-Pichler, F.; Rumpold, H.; Valenta, R.; Spitzauer, S. Escherichia coli expression and purification of recombinant dog albumin, a cross-reactive animal allergen. *J. Allergy Clin. Immunol.* **2000**, *105*, 279–285. [CrossRef]
17. Reininger, R.; Swoboda, I.; Bohle, B.; Hauswirth, A.W.; Valent, P.; Rumpold, H.; Valenta, R.; Spitzauer, S. Characterization of recombinant cat albumin. *Clin. Exp. Allergy* **2003**, *33*, 1695–1702. [CrossRef] [PubMed]
18. D'Urbano, L.E.; Pellegrino, K.; Artesani, M.C.; Donnanno, S.; Luciano, R.; Riccardi, C.; Tozzi, A.E.; Ravà, L.; De Benedetti, F.; Cavagni, G. Performance of a component-based allergen-microarray in the diagnosis of cow's milk and hen's egg allergy. *Clin. Exp. Allergy* **2010**, *40*, 1561–1570. [CrossRef]

19. Ott, H.; Baron, J.M.; Heise, R.; Ocklenburg, C.; Stanzel, S.; Merk, H.F.; Niggemann, B.; Beyer, K. Clinical usefulness of microarray-based IgE detection in children with suspected food allergy. *Allergy* **2008**, *63*, 1521–1528. [CrossRef]

20. Hochwallner, H.; Schulmeister, U.; Swoboda, I.; Balic, N.; Geller, B.; Nystrand, M.; Härlin, A.; Thalhamer, J.; Scheiblhofer, S.; Niggemann, B.; et al. Microarray and allergenic activity assessment of milk allergens. *Clin. Exp. Allergy* **2010**, *40*, 1809–1818. [CrossRef]

21. Röckmann, H.; van Geel, M.J.; Knulst, A.C.; Huiskes, J.; Bruijnzeel-Koomen, C.A.; de Bruin-Weller, MS. Food allergen sensitization pattern in adults in relation to severity of atopic dermatitis. *Clin. Transl. Allergy* **2014**, *4*, 9. [CrossRef]

22. Erwin, E.A.; Tripathi, A.; Ogbogu, P.U.; Commins, S.P.; Slack, M.A.; Cho, C.B.; Hamilton, R.G.; Workman, L.J.; Platts-Mills, T.A. IgE Antibody Detection and Component Analysis in Patients with Eosinophilic Esophagitis. *J. Allergy Clin. Immunol. Pract.* **2015**, *3*, 896–904. [CrossRef] [PubMed]

23. Restani, P.; Gaiaschi, A.; Plebani, A.; Beretta, B.; Cavagni, G.; Fiocchi, A.; Poiesi, C.; Velonà, T.; Ugazio, A.G.; Galli, C.L. Cross-reactivity between milk proteins from different animal species. *Clin. Exp. Allergy* **1999**, *29*, 997–1004. [CrossRef] [PubMed]

24. Kattan, J.D.; Cocco, R.R.; Järvinen, K.M. Milk and soy allergy. *Pediatr. Clin. North Am.* **2011**, *58*, 407–426. [CrossRef]

25. Rozenfeld, P.; Docena, G.H.; Añón, M.C.; Fossati, C.A. Detection and identification of a soy protein component that cross-reacts with caseins from cow's milk. *Clin. Exp. Immunol.* **2002**, *130*, 49–58. [CrossRef] [PubMed]

26. Bloom, K.A.; Huang, F.R.; Bencharitiwong, R.; Bardina, L.; Ross, A.; Sampson, H.A.; Nowak-Węgrzyn, A. Effect of heat treatment on milk and egg proteins allergenicity. *Pediatr. Allergy Immunol.* **2014**, *25*, 740–746. [CrossRef]

27. Martelli, A.; De Chiara, A.; Corvo, M.; Restani, P.; Fiocchi, A. Beef allergy in children with cow's milk allergy; cow's milk allergy in children with beef allergy. *Ann. Allergy Asthma Immunol.* **2002**, *89* (Suppl. 6), 38–43. [CrossRef]

28. Yanagida, N.; Sato, S.; Takahashi, K.; Nagakura, K.I.; Asaumi, T.; Ogura, K.; Ebisawa, M. Increasing specific immunoglobulin E levels correlate with the risk of anaphylaxis during an oral food challenge. *Pediatr. Allergy Immunol.* **2018**, *29*, 417–424. [CrossRef]

29. Adédoyin, J.; Johansson, S.G.; Grönlund, H.; van Hage, M. Interference in immunoassays by human IgM with specificity for the carbohydrate moiety of animal proteins. *J. Immunol. Methods* **2006**, *310*, 117–125. [CrossRef]

30. Adédoyin, J.; Grönlund, H.; Oman, H.; Johansson, S.G.; van Hage, M. Cat IgA, representative of new carbohydrate cross-reactive allergens. *J. Allergy Clin. Immunol.* **2007**, *119*, 640–645. [CrossRef]

31. Grönlund, H.; Adédoyin, J.; Commins, S.P.; Platts-Mills, T.A.; van Hage, M. The carbohydrate galactose-alpha-1,3-galactose is a major IgE-binding epitope on cat IgA. *J. Allergy Clin. Immunol.* **2009**, *123*, 1189–1191. [CrossRef] [PubMed]

32. Brock, J.H. Lactoferrin-50 years on. *Biochem. Cell Biol.* **2012**, *90*, 245–251. [CrossRef] [PubMed]

33. Schulmeister, U.; Hochwallner, H.; Swoboda, I.; Focke-Tejkl, M.; Geller, B.; Nystrand, M.; Härlin, A.; Thalhamer, J.; Scheiblhofer, S.; Keller, W.; et al. Cloning, expression, and mapping of allergenic determinants of alphaS1-casein, a major cow's milk allergen. *J. Immunol.* **2009**, *182*, 7019–7029. [CrossRef] [PubMed]

34. Egan, M.; Lee, T.; Andrade, J.; Grishina, G.; Mishoe, M.; Gimenez, G.; Sampson, H.A.; Bunyavanich, S. Partially hydrolyzed whey formula intolerance in cow's milk allergic patients. *Pediatr. Allergy Immunol.* **2017**, *28*, 401–405. [CrossRef] [PubMed]

35. Walsh, J.; O'Flynn, N. Diagnosis and assessment of food allergy in children and young people in primary care and community settings: NICE clinical guideline. *Br. J. Gen. Pract.* **2011**, *61*, 473–475. [CrossRef] [PubMed]

36. Koletzko, S.; Niggemann, B.; Arato, A.; Dias, J.A.; Heuschkel, R.; Husby, S.; Mearin, M.L.; Papadopoulou, A.; Ruemmele, F.M.; Staiano, A.; et al. Diagnostic approach and management of cow's-milk protein allergy in infants and children: ESPGHAN GI Committee practical guidelines. European Society of Pediatric Gastroenterology, Hepatology, and Nutrition. *J. Pediatr. Gastroenterol. Nutr.* **2012**, *55*, 221–229. [CrossRef]

37. Walsh, J.; Meyer, R.; Shah, N.; Quekett, J.; Fox, A.T. Differentiating milk allergy (IgE and non-IgE mediated) from lactose intolerance: Understanding the underlying mechanisms and presentations. *Br. J. Gen. Pract.* **2016**, *66*, e609–e611. [CrossRef] [PubMed]

38. Van der Heijden, F.L.; van Neerven, R.J.; van Katwijk, M.; Bos, J.D.; Kapsenberg, M.L. Serum-IgE-facilitated allergen presentation in atopic disease. *J. Immunol.* **1993**, *150*, 3643–3650.

39. Campana, R.; Mothes, N.; Rauter, I.; Vrtala, S.; Reininger, R.; Focke-Tejkl, M.; Lupinek, C.; Balic, N.; Spitzauer, S.; Valenta, R. Non-IgE-mediated chronic allergic skin inflammation revealed with rBet v 1 fragments. *J. Allergy Clin. Immunol.* **2008**, *121*, 528–530. [CrossRef]

40. Campana, R.; Moritz, K.; Marth, K.; Neubauer, A.; Huber, H.; Henning, R.; Blatt, K.; Hoermann, G.; Brodie, T.M.; Kaider, A.; et al. Frequent occurrence of T cell-mediated late reactions revealed by atopy patch testing with hypoallergenic rBet v 1 fragments. *J. Allergy Clin. Immunol.* **2016**, *137*, 601–609. [CrossRef]

41. Kounis, N.G.; Zavras, G.M. Histamine-induced coronary artery spasm: The concept of allergic angina. *Br. J. Clin. Pract.* **1991**, *45*, 121–128. [PubMed]

42. Tzanis, G.; Bonou, M.; Mikos, N.; Biliou, S.; Koniara, I.; Kounis, N.G.; Barbetseas, J. Early stent thrombosis secondary to food allergic reaction: Kounis syndrome following rice pudding ingestion. *World J. Cardiol.* **2017**, *9*, 283–288. [CrossRef] [PubMed]

43. Abdelghany, M.; Subedi, R.; Shah, S.; Kozman, H. Kounis syndrome: A review article on epidemiology, diagnostic findings, mamagement and complications of allergic acute coronary syndrome. *Int. J. Cardiol.* **2017**, *232*, 1–4. [CrossRef] [PubMed]

44. Yang, J.; Lee, H.; Choi, A.R.; Park, K.H.; Ryu, J.H.; Oh, E.J. Comparison of allergen-specific IgE levels between Immulite 2000 and ImmunoCAP systems against six inhalant allergens and ten food allergens. *Scand. J. Clin. Lab. Investig.* **2018**, *78*, 606–612. [CrossRef] [PubMed]

45. van Hage, M.; Hamsten, C.; Valenta, R. ImmunoCAP assays: Pros and cons in allergology. *J. Allergy Clin. Immunol.* **2017**, *140*, 974–977. [CrossRef] [PubMed]

46. Nunes, M.P.O.; van Tilburg, M.F.; Tramontina Florean, E.O.P.; Guedes, M.I.F. Detection of serum and salivary IgE and IgG1 immunoglobulins specific for diagnosis of food allergy. *PLoS ONE* **2019**, *14*, e0214745. [CrossRef] [PubMed]

47. Garib, V.; Rigler, E.; Gastager, F.; Campana, R.; Dorofeeva, Y.; Gattinger, P.; Zhernov, Y.; Khaitov, M.; Valenta, R. Determination of IgE and IgG reactivity to more than 170 allergen molecules in paper-dried blood spots. *J. Allergy Clin. Immunol.* **2019**, *143*, 437–440. [CrossRef] [PubMed]

48. Cuomo, B.; Indirli, G.C.; Bianchi, A.; Arasi, S.; Caimmi, D.; Dondi, A.; La Grutta, S.; Panetta, V.; Verga, M.C.; Calvani, M. Specific IgE and skin prick tests to diagnose allergy to fresh and baked cow's milk according to age: A systematic review. *Ital. J. Pediatr.* **2017**, *43*, 93. [CrossRef]

49. Hochwallner, H.; Schulmeister, U.; Swoboda, I.; Twaroch, T.E.; Vogelsang, H.; Kazemi-Shirazi, L.; Kundi, M.; Balic, N.; Quirce, S.; Rumpold, H.; et al. Patients suffering from non-IgE-mediated cow's milk protein intolerance cannot be diagnosed based on IgG subclass or IgA responses to milk allergens. *Allergy* **2011**, *66*, 1201–1207. [CrossRef]

50. Stapel, S.O.; Asero, R.; Ballmer-Weber, B.K.; Knol, E.F.; Strobel, S.; Vieths, S.; Kleine-Tebbe, J. Testing for IgG4 against foods is not recommended as a diagnostic. *Allergy* **2008**, *63*, 793–796. [CrossRef]

51. Bock, S.A. AAAAI support of the EAACI Position Paper on IgG4. *J. Allergy Clin. Immunol.* **2010**, *125*, 1410. [CrossRef] [PubMed]

52. Sporik, R.; Hill, D.J.; Hosking, C.S. Specificity of allergen skin testing in predicting positive open food challenges to milk, egg and peanut in children. *Clin. Exp. Allergy* **2000**, *30*, 1540–1546. [CrossRef] [PubMed]

53. Saarinen, K.M.; Suomalainen, H.; Savilahti, E. Diagnostic value of skin-prick and patch tests and serum eosinophil cationic protein and cow's milk-specific IgE in infants with cow's milk allergy. *Clin. Exp. Allergy* **2001**, *31*, 423–429. [CrossRef] [PubMed]

54. Verstege, A.; Mehl, A.; Rolinck-Werninghaus, C.; Staden, U.; Nocon, M.; Beyer, K.; Niggemann, B. The predictive value of the skin prick test weal size for the outcome of oral food challenges. *Clin. Exp. Allergy* **2005**, *35*, 1220–1226. [CrossRef] [PubMed]

55. Calvani, M.; Alessandri, C.; Frediani, T.; Lucarelli, S.; Miceli Sopo, S.; Panetta, V.; Zappalà, D.; Zicari, A.M. Correlation between skin prick test using commercial extract of cow's milk protein and fresh milk and food challenges. *Pediatr. Allergy Immunol.* **2007**, *18*, 583–588.

56. Calvani, M.; Berti, I.; Fiocchi, A.; Galli, E.; Giorgio, V.; Martelli, A.; Miceli Sopo, S.; Panetta, V. Oral food challenge: Safety, adherence to guidelines and predictive value of skin prick testing. *Pediatr. Allergy Immunol.* **2012**, *23*, 755–761. [CrossRef] [PubMed]

57. Onesimo, R.; Monaco, S.; Greco, M.; Caffarelli, C.; Calvani, M.; Tripodi, S.; Sopo, S.M. Predictive value of MP4 (Milk Prick Four), a panel of skin prick test for the diagnosis of pediatric immediate cow's milk allergy. *Eur. Ann. Allergy Clin. Immunol.* **2013**, *45*, 201–208. [PubMed]

58. Niederberger, V.; Eckl-Dorna, J.; Pauli, G. Recombinant allergen-based provocation testing. *Methods* **2014**, *66*, 96–105. [CrossRef] [PubMed]

59. Valenta, R.; Karaulov, A.; Niederberger, V.; Zhernov, Y.; Elisyutina, O.; Campana, R.; Focke-Tejkl, M.; Curin, M.; Namazova-Baranova, L.; Wang, J.Y.; et al. Allergen Extracts for In Vivo Diagnosis and Treatment of Allergy: Is There a Future? *J. Allergy Clin. Immunol. Pract.* **2018**, *6*, 1845–1855. [CrossRef]

60. Furuta, G.T.; Liacouras, C.A.; Collins, M.H.; Gupta, S.K.; Justinich, C.; Putnam, P.E.; Bonis, P.; Hassall, E.; Straumann, A.; Rothenberg, M.E. First International Gastrointestinal Eosinophil Research Symposium (FIGERS) Subcommittees. Eosinophilic esophagitis in children and adults: A systematic review and consensus recommendations for diagnosis and treatment. *Gastroenterology* **2007**, *133*, 1342–1363. [CrossRef]

61. Canani, R.B.; Buongiovanni, A.; Nocerino, R.; Cosenza, L.; Troncone, R. Toward a standardized reading of the atopy patch test in children with suspected cow's milk allergy-related gastrointestinal symptoms. *Allergy* **2011**, *66*, 1499–1500. [CrossRef] [PubMed]

62. Mowszet, K.; Matusiewicz, K.; Iwańczak, B. Value of the atopy patch test in the diagnosis of food allergy in children with gastrointestinal symptoms. *Adv. Clin. Exp. Med.* **2014**, *23*, 403–409. [CrossRef] [PubMed]

63. Caglayan Sozmen, S.; Povesi Dascola, C.; Gioia, E.; Mastrorilli, C.; Rizzuti, L.; Caffarelli, C. Diagnostic accuracy of patch test in children with food allergy. *Pediatr. Allergy Immunol.* **2015**, *26*, 416–422. [CrossRef] [PubMed]

64. Mansouri, M.; Rafiee, E.; Darougar, S.; Mesdaghi, M.; Chavoshzadeh, Z. Is the Atopy Patch Test Reliable in the Evaluation of Food Allergy-Related Atopic Dermatitis? *Int. Arch. Allergy Immunol.* **2018**, *175*, 85–90. [CrossRef] [PubMed]

65. Nocerino, R.; Granata, V.; Di Costanzo, M.; Pezzella, V.; Leone, L.; Passariello, A.; Terrin, G.; Troncone, R.; Berni Canani, R. Atopy patch tests are useful to predict oral tolerance in children with gastrointestinal symptoms related to non-IgE-mediated cow's milk allergy. *Allergy* **2013**, *68*, 246–248. [CrossRef] [PubMed]

66. Muraro, A.; Werfel, T.; Hoffmann-Sommergruber, K.; Roberts, G.; Beyer, K.; Bindslev-Jensen, C.; Cardona, V.; Dubois, A.; Dutoit, G.; Eigenmann, P.; et al. EAACI food allergy and anaphylaxis guidelines: Diagnosis and management of food allergy. *Allergy* **2014**, *69*, 1008–1025. [CrossRef] [PubMed]

67. Perry, T.T.; Matsui, E.C.; Conover-Walker, M.K.; Wood, R.A. Risk of oral food challenges. *J. Allergy Clin. Immunol.* **2004**, *114*, 1164–1168. [CrossRef]

68. Yanagida, N.; Sato, S.; Asaumi, T.; Ogura, K.; Ebisawa, M. Risk Factors for Severe Reactions during Double-Blind Placebo-Controlled Food Challenges. *Int. Arch. Allergy Immunol.* **2017**, *172*, 173–182. [CrossRef]

69. Matricardi, P.M.; Kleine-Tebbe, J.; Hoffmann, H.J.; Valenta, R.; Hilger, C.; Hofmaier, S.; Aalbrse, R.C.; Agache, I.; Asero, R.; Ballmer-Weber, B. EAACI Molecular Allergology User's Guide. *Pediatr. Allergy Immunol.* **2016**, *27* (Suppl. 23), 1–250. [CrossRef]

70. Sievers, S.; Cretich, M.; Gagni, P.; Ahrens, B.; Grishina, G.; Sampson, H.A.; Niggemann, B.; Chiari, M.; Beyer, K. Performance of a polymer coated silicon microarray for simultaneous detection of food allergen-specific IgE and IgG4. *Clin. Exp. Allergy* **2017**, *47*, 1057–1068. [CrossRef]

71. Ahrens, B.; Lopes de Oliveira, L.C.; Grabenhenrich, L.; Schulz, G.; Niggemann, B.; Wahn, U.; Beyer, K. Individual cow's milk allergens as prognostic markers for tolerance development? *Clin. Exp. Allergy* **2012**, *42*, 1630–1637. [CrossRef] [PubMed]

72. Petersen, T.H.; Mortz, C.G.; Bindslev-Jensen, C.; Eller, E. Cow's milk allergic children-Can component-resolved diagnostics predict duration and severity? *Pediatr. Allergy Immunol.* **2018**, *29*, 194–199. [CrossRef] [PubMed]

73. Savilahti, E.M.; Rantanen, V.; Lin, J.S.; Karinen, S.; Saarinen, K.M.; Goldis, M.; Mäkelä, M.J.; Hautaniemi, S.; Savilahti, E.; Sampson, H.A. Early recovery from cow's milk allergy is associated with decreasing IgE and increasing IgG4 binding to cow's milk epitopes. *J. Allergy Clin. Immunol.* **2010**, *125*, 1315–1321. [CrossRef] [PubMed]

74. Cerecedo, I.; Zamora, J.; Shreffler, W.G.; Lin, J.; Bardina, L.; Dieguez, M.C.; Wang, J.; Muriel, A.; de la Hoz, B.; Sampson, H.A. Mapping of the IgE and IgG4 sequential epitopes of milk allergens with a peptide microarray-based immunoassay. *J. Allergy Clin. Immunol.* **2008**, *122*, 589–594. [CrossRef] [PubMed]

75. Wang, J.; Lin, J.; Bardina, L.; Goldis, M.; Nowak-Wegrzyn, A.; Shreffler, W.G.; Sampson, H.A. Correlation of IgE/IgG4 milk epitopes and affinity of milk-specific IgE antibodies with different phenotypes of clinical milk allergy. *J. Allergy Clin. Immunol.* **2010**, *125*, 695–702. [CrossRef] [PubMed]

76. Fedenko, E.; Elisyutina, O.; Shtyrbul, O.; Pampura, A.; Valenta, R.; Lupinek, C.; Khaitov, M. Microarray-based IgE serology improves management of severe atopic dermatitis in two children. *Pediatr. Allergy Immunol.* **2016**, *27*, 645–649. [CrossRef] [PubMed]

77. Järvinen, K.M.; Westfall, J.E.; Seppo, M.S.; James, A.K.; Tsuang, A.J.; Feustel, P.J.; Sampson, H.A.; Berin, C. Role of maternal elimination diets and human milk IgA in the development of cow's milk allergy in the infants. *Clin. Exp. Allergy* **2014**, *44*, 69–78. [CrossRef]

78. Tuokkola, J.; Luukkainen, P.; Tapanainen, H.; Kaila, M.; Vaarala, O.; Kenward, M.G.; Virta, L.; Veijola, R.; Simell, O.; Ilonen, J.; et al. Maternal diet during pregnancy and lactation and cow's milk allergy in offspring. *Eur. J. Clin. Nutr.* **2016**, *70*, 554–559. [CrossRef]

79. Lupinek, C.; Hochwallner, H.; Johansson, C.; Mie, A.; Rigler, E.; Scheynius, A.; Alm, J.; Valenta, R. Maternal allergen-specific IgG might protect the child against allergic sensitization. *J. Allergy Clin. Immunol*, **2019**, in press.

80. Picariello, G.; De Cicco, M.; Nocerino, R.; Paparo, L.; Mamone, G.; Addeo, F.; Berni Canani, R. Excretion of Dietary Cow's Milk Derived Peptides into Breast Milk. *Front. Nutr.* **2019**, *12*, 6–25. [CrossRef]

81. Schulmeister, U.; Swoboda, I.; Quirce, S.; de la Hoz, B.; Ollert, M.; Pauli, G.; Valenta, R.; Spitzauer, S. Sensitization to human milk. *Clin. Exp. Allergy* **2008**, *38*, 60–68. [CrossRef]

82. Schocker, F.; Recke, A.; Kull, S.; Worm, M.; Jappe, U. Persistent cow's milk anaphylaxis from early childhood monitored by IgE and BAT to cow's and human milk under therapy. *Pediatr. Allergy Immunol.* **2018**, *29*, 210–214. [CrossRef] [PubMed]

83. Meyer, R.; Groetch, M.; Venter, C. When Should Infants with Cow's Milk Protein Allergy Use an Amino Acid Formula? *A Practical Guide. J. Allergy Clin. Immunol. Pract.* **2018**, *6*, 383–399. [CrossRef] [PubMed]

84. Dipasquale, V.; Serra, G.; Corsello, G.; Romano, C. Standard and Specialized Infant Formulas in Europe: Making, Marketing, and Health Outcomes. *Nutr. Clin. Pract.* **2019**. [CrossRef] [PubMed]

85. Kiewiet, M.B.G.; Gros, M.; van Neerven, R.J.J.; Faas, M.M.; de Vos, P. Immunomodulating properties of protein hydrolysates for application in cow's milk allergy. *Pediatr. Allergy Immunol.* **2015**, *26*, 206–217. [CrossRef] [PubMed]

86. von Berg, A.; Filipiak-Pittroff, B.; Krämer, U.; Link, E.; Bollrath, C.; Brockow, I.; Koletzko, S.; Grübl, A.; Heinrich, J.; Wichmann, HE.; et al. GINIplus study group. Preventive effect of hydrolyzed infant formulas persists until age 6 years: Long-term results from the German Infant Nutritional Intervention Study (GINI). *J. Allergy Clin. Immunol.* **2008**, *121*, 1442–1447. [CrossRef]

87. von Berg, A.; Filipiak-Pittroff, B.; Krämer, U.; Hoffmann, B.; Link, E.; Beckmann, C.; Hoffmann, U.; Reinhardt, D.; Grübl, A.; Heinrich, J.; et al. GINIplus study group. Allergies in high-risk schoolchildren after early intervention with cow's milk protein hydrolysates: 10-year results from the German Infant Nutritional Intervention (GINI) study. *J. Allergy Clin. Immunol.* **2013**, *131*, 1565–1573. [CrossRef] [PubMed]

88. Cabana, M.D. The Role of Hydrolyzed Formula in Allergy Prevention. *Ann. Nutr. Metab.* **2017**, *70* (Suppl. 2), 38–45. [CrossRef]

89. Hochwallner, H.; Schulmeister, U.; Swoboda, I.; Focke-Tejkl, M.; Reininger, R.; Civaj, V.; Campana, R.; Thalhamer, J.; Scheiblhofer, S.; Balic, N.; et al. Infant milk formulas differ regarding their allergenic activity and induction of T-cell and cytokine responses. *Allergy* **2017**, *72*, 416–424. [CrossRef]

90. Dupont, C.; Chouraqui, J.P.; Linglart, A.; Bocquet, A.; Darmaun, D.; Feillet, F.; Frelut, M.L.; Girardet, J.P.; Hankard, R.; Rozé, J.C.; et al. Committee on Nutrition of the French Society of Pediatrics. Nutritional management of cow's milk allergy in children: An update. *Arch. Pediatr.* **2018**, *25*, 236–243. [CrossRef]

91. Doulgeraki, A.E.; Manousakis, E.M.; Papadopoulos, N.G. Bone health assessment of food allergic children on restrictive diets: A practical guide. *J. Pediatr. Endocrinol. Metab.* **2017**, *30*, 133–139. [CrossRef]

92. Souroullas, K.; Aspri, M.; Papademas, P. Donkey milk as a supplement in infant formula: Benefits and technological challenges. *Food Res. Int.* **2018**, *109*, 416–425. [CrossRef] [PubMed]

93. Barni, S.; Sarti, L.; Mori, F.; Muscas, G.; Belli, F.; Pucci, N.; Novembre, E. Tolerability and palatability of donkey's milk in children with cow's milk allergy. *Pediatr. Allergy Immunol.* **2018**, *29*, 329–331. [CrossRef] [PubMed]

94. Umpiérrez, A.; Quirce, S.; Marañón, F.; Cuesta, J.; García-Villamuza, Y.; Lahoz, C.; Sastre, J. Allergy to goat and sheep cheese with good tolerance to cow cheese. *Clin. Exp. Allergy* **1999**, *29*, 1064–1068. [CrossRef] [PubMed]

95. Cases, B.; García-Ara, C.; Boyano, M.T.; Pérez-Gordo, M.; Pedrosa, M.; Vivanco, F.; Quirce, S.; Pastor-Vargas, C. Phosphorylation reduces the allergenicity of cow casein in children with selective allergy to goat and sheep milk. *J. Investig. Allergol. Clin. Immunol.* **2011**, *21*, 398–400. [PubMed]

96. Martin, C.R.; Ling, P.R.; Blackburn, G.L. Review of Infant Feeding: Key Features of Breast Milk and Infant Formula. *Nutrients* **2016**, *8*, 279. [CrossRef] [PubMed]

97. Candreva, Á.M.; Ferrer-Navarro, M.; Bronsoms, S.; Quiroga, A.; Curciarello, R.; Cauerhff, A.; Petruccelli, S.; Docena, G.H.; Trejo, S.A. Identification of cross-reactive B-cell epitopes between Bos d 9.0101(Bos Taurus) and Gly m 5.0101 (Glycine max) by epitope mapping MALDI-TOF MS. *Proteomics* **2017**, *17*, 15–16. [CrossRef] [PubMed]

98. Bocquet, A.; Dupont, C.; Chouraqui, J.-P.; Darmaun, D.; Feillet, F.; Frelut, M.-L.; Girardet, J.-P.; Hankard, R.; Lapillonne, A.; Rozé, J.-C.; et al. Efficacy and safety of hydrolyzed rice-protein formulas for the treatment of cow's milk protein allergy. *Archives de Pédiatrie* **2019**, *26*, 238–246. [CrossRef]

99. Sun, Z.; Wang, M.; Han, S.; Ma, S.; Zou, Z.; Ding, F.; Li, X.; Li, L.; Tang, B.; Wang, H.; et al. Production of hypoallergenic milk from DNA-free beta-lactoglobulin (BLG) gene knockout cow using zinc-finger nucleases mRNA. *Sci. Rep.* **2018**, *8*, 15430. [CrossRef]

100. Tan-Lim, C.S.C.; Esteban-Ipac, N.A.R. Probiotics as treatment for food allergies among pediatric patients: A meta-analysis. *World Allergy Organ. J.* **2018**, *11*, 25. [CrossRef]

101. Berni Canani, R.; Di Costanzo, M.; Bedogni, G.; Amoroso, A.; Cosenza, L.; Di Scala, C.; Granata, V.; Nocerino, R. Extensively hydrolyzed casein formula containing Lactobacillus rhamnosus GG reduces the occurrence of other allergic manifestations in children with cow's milk allergy: 3-year randomized controlled trial. *J. Allergy Clin. Immunol.* **2017**, *139*, 1906–1913. [CrossRef]

102. Schmidt, R.M.; Pilmann Laursen, R.; Bruun, S.; Larnkjaer, A.; Mølgaard, C.; Michaelsen, K.F.; Høst, A. Probiotics in late infancy reduce the incidence of eczema: A randomized controlled trial. *Pediatr. Allergy Immunol.* **2019**, *30*, 335–340. [CrossRef] [PubMed]

103. Lundelin, K.; Poussa, T.; Salminen, S.; Isolauri, E. Long-term safety and efficacy of perinatal probiotic intervention: Evidence from a follow-up study of four randomized, double-blind, placebo-controlled trials. *Pediatr. Allergy Immunol.* **2017**, *28*, 170–175. [CrossRef] [PubMed]

104. Wickens, K.; Barthow, C.; Mitchell, E.A.; Kang, J.; van Zyl, N.; Purdie, G.; Stanley, T.; Fitzharris, P.; Murphy, R.; Crane, J. Effects of Lactobacillus rhamnosus HN001 in early life on the cumulative prevalence of allergic disease to 11 years. *Pediatr. Allergy Immunol.* **2018**, *29*, 808–814. [CrossRef] [PubMed]

105. Boyle, R.J.; Tang, M.L.; Chiang, W.C.; Chua, M.C.; Ismail, I.; Nauta, A.; Hourihane, J.O.B.; Smith, P.; Gold, M.; Ziegler, J.; et al. Prebiotic-supplemented partially hydrolysed cow's milk formula for the prevention of eczema in high-risk infants: A randomized controlled trial. *Allergy* **2016**, *71*, 701–710. [CrossRef] [PubMed]

106. Wopereis, H.; Sim, K.; Shaw, A.; Warner, J.O.; Knol, J.; Kroll, J.S. Intestinal microbiota in infants at high risk for allergy: Effects of prebiotics and role in eczema development. *J. Allergy Clin. Immunol.* **2018**, *141*, 1334–1342. [CrossRef] [PubMed]

107. Pajno, G.B.; Fernandez-Rivas, M.; Arasi, S.; Roberts, G.; Akdis, C.A.; Alvaro-Lozano, M.; Beyer, K.; Bindslev-Jensen, C.; Burks, W.; Ebisawa, M.; et al. EAACI Allergen Immunotherapy Guidelines Group. EAACI Guidelines on allergen immunotherapy: IgE-mediated food allergy. *Allergy* **2018**, *73*, 799–815. [CrossRef] [PubMed]

108. Berti, I.; Badina, L.; Cozzi, G.; Giangreco, M.; Bibalo, C.; Ronfani, L.; Barbi, E.; Ventura, A.; Longo, G. Early oral immunotherapy in infants with cow's milk protein allergy. *Pediatr. Allergy Immunol.* **2019**. [CrossRef] [PubMed]

109. Yanagida, N.; Sato, S.; Ebisawa, M. Clinical aspects of oral immunotherapy for the treatment of allergies. *Semin. Immunol.* **2017**, *30*, 45–51. [CrossRef]

110. Taniuchi, S.; Takahashi, M.; Soejima, K.; Hatano, Y.; Minami, H. Immunotherapy for cow's milk allergy. *Hum. Vaccin. Immunother.* **2017**, *13*, 2443–2451. [CrossRef]

111. Longo, G.; Barbi, E.; Berti, I.; Meneghetti, R.; Pittalis, A.; Ronfani, L.; Ventura, A. Specific oral tolerance induction in children with very severe cow's milk-induced reactions. *J. Allergy Clin. Immunol.* **2008**, *121*, 343–347. [CrossRef]

112. Skripak, J.M.; Nash, S.D.; Rowley, H.; Brereton, N.H.; Oh, S.; Hamilton, R.G.; Matsui, E.C.; Burks, A.W.; Wood, R.A. A randomized, double-blind, placebo-controlled study of milk oral immunotherapy for cow's milk allergy. *J. Allergy Clin. Immunol.* **2008**, *122*, 1154–1160. [CrossRef] [PubMed]

113. Pajno, G.B.; Caminiti, L.; Ruggeri, P.; De Luca, R.; Vita, D.; La Rosa, M.; Passalacqua, G. Oral immunotherapy for cow's milk allergy with a weekly up-dosing regimen: A randomized single-blind controlled study. *Ann. Allergy Asthma Immunol.* **2010**, *105*, 376–381. [CrossRef] [PubMed]

114. Martorell, A.; De La Hoz, B.; Ibáñez, M.D.; Boné, J.; Terrados, M.S.; Michavila, A.; Plaza, A.M.; Alonso, E.; Garde, J.; Nevot, S.; et al. Oral desensitization as a useful treatment in 2-year-old children with cow's milk allergy. *Clin. Exp. Allergy* **2011**, *41*, 1297–1304. [CrossRef] [PubMed]

115. Salmivesi, S.; Korppi, M.; Mäkelä, M.J.; Paassilta, M. Milk oral immunotherapy is effective in school-aged children. *Acta Paediatr.* **2013**, *102*, 172–176. [CrossRef] [PubMed]

116. Suárez-Fariñas, M.; Suprun, M.; Chang, H.L.; Gimenez, G.; Grishina, G.; Getts, R.; Nadeau, K.; Wood, R.A.; Sampson, H.A. Predicting development of sustained unresponsiveness to milk oral immunotherapy using epitope-specific antibody binding profiles. *J. Allergy Clin. Immunol.* **2019**, *143*, 1038–1046. [CrossRef]

117. Kauppila, T.K.; Paassilta, M.; Kukkonen, A.K.; Kuitunen, M.; Pelkonen, A.S.; Makela, M.J. Outcome of oral immunotherapy for persistent cow's milk allergy from 11 years of experience in Finland. *Pediatr. Allergy Immunol.* **2019**, *30*, 356–362. [CrossRef] [PubMed]

118. Keet, C.A.; Seopaul, S.; Knorr, S.; Narisety, S.; Skripak, J.; Wood, R.A. Long-term follow-up of oral immunotherapy for cow's milk allergy. *J. Allergy Clin. Immunol.* **2013**, *132*, 737–739. [CrossRef] [PubMed]

119. Manabe, T.; Sato, S.; Yanagida, N.; Hayashi, N.; Nishino, M.; Takahashi, K.; Nagakura, K.I.; Asaumi, T.; Ogura, K.; Ebisawa, M. Long-term outcomes after sustained unresponsiveness in patients who underwent oral immunotherapy for egg, cow's milk, or wheat allergy. *Allergol. Int.* **2019**. [CrossRef] [PubMed]

120. De Schryver, S.; Mazer, B.; Clarke, A.E.; St Pierre, Y.; Lejtenyi, D.; Langlois, A.; Torabi, B.; Zhao, W.w.; Chan, E.S.; Ingrid Baerg, R.N. Adverse Events in Oral Immunotherapy for the Desensitization of Cow's Milk Allergy in Children: A Randomized Controlled Trial. *J. Allergy Clin. Immunol. Pract* **2019**. [CrossRef] [PubMed]

121. Nadeau, K.C.; Schneider, L.C.; Hoyte, L.; Borras, I.; Umetsu, D.T. Rapid oral desensitization in combination with omalizumab therapy in patients with cow's milk allergy. *J. Allergy Clin. Immunol.* **2011**, *127*, 1622–1624. [CrossRef] [PubMed]

122. Takahashi, M.; Taniuchi, S.; Soejima, K.; Hatano, Y.; Yamanouchi, S.; Kaneko, K. Successful desensitization in a boy with severe cow's milk allergy by a combination therapy using omalizumab and rush oral immunotherapy. *Allergy Asthma Clin. Immunol.* **2015**, *11*, 18. [CrossRef] [PubMed]

123. Wood, R.A.; Kim, J.S.; Lindblad, R.; Nadeau, K.; Henning, A.K.; Dawson, P.; Plaut, M.; Sampson, H.A. A randomized, double-blind, placebo-controlled study of omalizumab combined with oral immunotherapy for the treatment of cow's milk allergy. *J. Allergy Clin. Immunol.* **2016**, *137*, 1103–1110. [CrossRef]

124. Martorell-Calatayud, C.; Michavila-Gómez, A.; Martorell-Aragonés, A.; Molini-Menchón, N.; Cerdá-Mir, J.C.; Félix-Toledo, R.; De Las Marinas-Álvarez, M.D. Anti-IgE-assisted desensitization to egg and cow's milk in patients refractory to conventional oral immunotherapy. *Pediatr. Allergy Immunol.* **2016**, *27*, 544–546. [CrossRef] [PubMed]

125. Takahashi, M.; Soejima, K.; Taniuchi, S.; Hatano, Y.; Yamanouchi, S.; Ishikawa, H.; Irahara, M.; Sasaki, Y.; Kido, H.; Kaneko, K. Oral immunotherapy combined with omalizumab for high–risk cow's milk allergy: A randomized controlled trial. *Sci. Rep.* **2017**, *7*, 17453. [CrossRef] [PubMed]

126. Freeland, D.M.H.; Manohar, M.; Andorf, S.; Hobson, B.D.; Zhang, W.; Nadeau, K.C. Oral immunotherapy for food allergy. *Semin. Immunol.* **2017**, *30*, 36–44. [CrossRef] [PubMed]

127. Esmaeilzadeh, H.; Alyasin, S.; Haghighat, M.; Nabavizadeh, H.; Esmaeilzadeh, E.; Mosavat, F. The effect of baked milk on accelerating unheated cow's milk tolerance: A control randomized clinical trial. *Pediatr. Allergy Immunol.* **2018**, *29*, 747–753. [CrossRef] [PubMed]

128. Goldberg, M.R.; Nachshon, L.; Appel, M.Y.; Elizur, A.; Levy, M.B.; Eisenberg, E.; Sampson, H.A.; Katz, Y. Efficacy of baked milk oral immunotherapy in baked milk-reactive allergic patients. *J. Allergy Clin. Immunol.* **2015**, *136*, 1601–1606. [CrossRef]

129. Amat, F.; Kouche, C.; Gaspard, W.; Lemoine, A.; Guiddir, T.; Lambert, N.; Zakariya, M.; Ridray, C.; Nemni, A.; Saint-Pierre, P.; et al. Is a slow-progression baked milk protocol of oral immunotherapy always a safe option for children with cow's milk allergy? A randomized controlled trial. *Clin. Exp. Allergy* **2017**, *47*, 1491–1496. [CrossRef]

130. Inuo, C.; Tanaka, K.; Suzuki, S.; Nakajima, Y.; Yamawaki, K.; Tsuge, I.; Urisu, A.; Kondo, Y. Oral Immunotherapy Using Partially Hydrolyzed Formula for Cow's Milk Protein Allergy: A Randomized, Controlled Trial. *Int. Arch. Allergy Immunol.* **2018**, *177*, 259–268. [CrossRef]

131. de Boissieu, D.; Dupont, C. Sublingual immunotherapy for cow's milk protein allergy: A preliminary report. *Allergy* **2006**, *61*, 1238–1239. [CrossRef]

132. Keet, C.A.; Frischmeyer-Guerrerio, P.A.; Thyagarajan, A.; Schroeder, J.T.; Hamilton, R.G.; Boden, S.; Steele, P.; Driggers, S.; Burks, A.W.; Wood, R.A. The safety and efficacy of sublingual and oral immunotherapy for milk allergy. *J. Allergy Clin. Immunol.* **2012**, *129*, 448–455. [CrossRef] [PubMed]

133. Lanser, B.J.; Leung, D.Y.M. The Current State of Epicutaneous Immunotherapy for Food Allergy: A Comprehensive Review. *Clin. Rev. Allergy Immunol.* **2018**, *55*, 153–161. [CrossRef] [PubMed]

134. Wang, J.; Sampson, H.A. Safety and efficacy of epicutaneous immunotherapy for food allergy. *Pediatr. Allergy Immunol.* **2018**, *29*, 341–349. [CrossRef] [PubMed]

135. Dupont, C.; Kalach, N.; Soulaines, P.; Legoué-Morillon, S.; Piloquet, H.; Benhamou, P.H. Cow's milk epicutaneous immunotherapy in children: A pilot trial of safety, acceptability, and impact on allergic reactivity. *J. Allergy Clin. Immunol.* **2010**, *125*, 1165–1167. [CrossRef] [PubMed]

136. Rutault, K.; Agbotounou, W.; Peillon, A.; Thébault, C.; Vincent, F.; Martin, L. Safety of Viaskin Milk Epicutaneous Immunotherapy (EPIT) in IgE-Mediated Cow's Milk Allergy (CMA) in Children (MILES Study). *J. Allergy Clin. Immunol.* **2016**, *137*, AB132. [CrossRef]

137. Senti, G.; von Moos, S.; Kündig, TM. Epicutaneous Immunotherapy for Aeroallergen and Food Allergy. *Curr. Treat. Options Allergy* **2013**, *1*, 68–78. [CrossRef]

138. Bird, J.A.; Sánchez-Borges, M.; Ansotegui, I.J.; Ebisawa, M.; Ortega Martell, J.A. Skin as an immune organ and clinical applications of skin-based immunotherapy. *World Allergy Organ. J.* **2018**, *11*, 38. [CrossRef]

139. Campana, R.; Moritz, K.; Neubauer, A.; Huber, H.; Henning, R.; Brodie, T.M.; Kaider, A.; Sallusto, F.; Wöhrl, S.; Valenta, R. Epicutaneous allergen application preferentially boosts specific T cell responses in sensitized patients. *Sci. Rep.* **2017**, *7*, 11657. [CrossRef]

140. Zhernov, Y.; Curin, M.; Khaitov, M.; Karaulov, A.; Valenta, R. Recombinant allergens for immunotherapy: State of the art. *Curr. Opin. Allergy Clin. Immunol.* **2019**, in press. [CrossRef]

141. Dahdah, L.; Ceccarelli, S.; Amendola, S.; Campagnano, P.; Cancrini, C.; Mazzina, O.; Fiocchi, A. IgE Immunoadsorption Knocks Down the Risk of Food-Related Anaphylaxis. *Pediatrics* **2015**, *136*, e1617–e1620. [CrossRef]

142. Lupinek, C.; Derfler, K.; Lee, S.; Prikoszovich, T.; Movadat, O.; Wollmann, E.; Cornelius, C.; Weber, M.; Fröschl, R.; Selb, R.; et al. Extracorporeal IgE Immunoadsorption in Allergic Asthma: Safety and Efficacy. *EBioMedicine* **2017**, *17*, 119–133. [CrossRef] [PubMed]

143. Incorvaia, C.; Riario-Sforza, G.G.; Ridolo, E. IgE Depletion in Severe Asthma: What We Have and What Could Be Added in the Near Future. *EBioMedicine* **2017**, *17*, 16–17. [CrossRef] [PubMed]

144. Fiocchi, A.; Artesani, M.C.; Riccardi, C.; Mennini, M.; Pecora, V.; Fierro, V.; Calandrelli, V.; Dahdah, L.; Valluzzi, R.L. Impact of Omalizumab on Food Allergy in Patients Treated for Asthma: A Real-Life Study. *J. Allergy Clin. Immunol. Pract.* **2019**, in press. [CrossRef] [PubMed]

145. Noon, L. Prophylactic inoculation against hayfever. *Lancet* **1911**, *4*, 1572. [CrossRef]

146. Curin, M.; Khaitov, M.; Karaulov, A.; Namazova-Baranova, L.; Campana, R.; Garib, V.; Valenta, R. Next-Generation of Allergen-Specific Immunotherapies: Molecular Approaches. *Curr. Allergy Asthma Rep.* **2018**, *18*, 39. [CrossRef] [PubMed]

147. Valenta, R.; Karaulov, A.; Niederberger, V.; Gattinger, P.; van Hage, M.; Flicker, S.; Linhart, B.; Campana, R.; Focke-Tejkl, M.; Curin, M.; et al. Molecular Aspects of Allergens and Allergy. *Adv. Immunol.* **2018**, *138*, 195–256. [PubMed]

148. Valenta, R.; Campana, R.; Niederberger, V. Recombinant allergy vaccines based on allergen-derived B cell epitopes. *Immunol. Lett.* **2017**, *189*, 19–26. [CrossRef] [PubMed]

149. Focke-Tejkl, M.; Weber, M.; Niespodziana, K.; Neubauer, A.; Huber, H.; Henning, R.; Stegfellner, G.; Maderegger, B.; Hauer, M.; Stolz, F.; et al. Development and characterization of a recombinant, hypoallergenic, peptide-based vaccine for grass pollen allergy. *J. Allergy Clin. Immunol.* **2015**, *135*, 1207–1217. [CrossRef]

150. Weber, M.; Niespodziana, K.; Linhart, B.; Neubauer, A.; Huber, H.; Henning, R.; Valenta, R.; Focke-Tejkl, M. Comparison of the immunogenicity of BM32, a recombinant hypoallergenic B cell epitope-based grass pollen allergy vaccine with allergen extract-based vaccines. *J. Allergy Clin. Immunol.* **2017**, *140*, 1433–1436. [CrossRef]

151. Zieglmayer, P.; Focke-Tejkl, M.; Schmutz, R.; Lemell, P.; Zieglmayer, R.; Weber, M.; Kiss, R.; Blatt, K.; Valent, P.; Stolz, F.; et al. Mechanisms, safety and efficacy of a B cell epitope-based vaccine for immunotherapy of grass pollen allergy. *EBioMedicine* **2016**, *11*, 43–57. [CrossRef]

152. Niederberger, V.; Neubauer, A.; Gevaert, P.; Zidarn, M.; Worm, M.; Aberer, W.; Malling, H.J.; Pfaar, O.; Klimek, L.; Pfützner, W.; et al. Safety and efficacy of immunotherapy with the recombinant B-cell epitope-based grass pollen vaccine BM32. *J. Allergy Clin. Immunol.* **2018**, *142*, 497–509. [CrossRef] [PubMed]

153. Orengo, J.M.; Radin, A.R.; Kamat, V.; Badithe, A.; Ben, L.H.; Bennett, B.L.; Zhong, S.; Birchard, D.; Limnander, A.; Rafique, A.; et al. Treating cat allergy with monoclonal IgG antibodies that bind allergen and prevent IgE engagement. *Nat. Commun.* **2018**, *9*, 1421. [CrossRef] [PubMed]

154. Swoboda, I.; Bugajska-Schretter, A.; Linhart, B.; Verdino, P.; Keller, W.; Schulmeister, U.; Sperr, W.R.; Valent, P.; Peltre, G.; Quirce, S.; et al. A recombinant hypoallergenic parvalbumin mutant for immunotherapy of IgE-mediated fish allergy. *J. Immunol.* **2007**, *178*, 6290–6296. [CrossRef] [PubMed]

155. Zuidmeer-Jongejan, L.; Huber, H.; Swoboda, I.; Rigby, N.; Versteeg, S.A.; Jensen, B.M.; Quaak, S.; Akkerdaas, J.H.; Blom, L.; Asturias, J.; et al. Development of a hypoallergenic recombinant parvalbumin for first-in-man subcutaneous immunotherapy of fish allergy. *Int. Arch. Allergy Immunol.* **2015**, *166*, 41–51. [CrossRef] [PubMed]

156. Clintrials.gov Identifier NCT02017626, NCT02382718. Available online: https://clinicaltrials.gov/ct2/results?cond=Allergy+to+Fish&term=mCyp+c+1&cntry=&state=&city=&dist= (accessed on 28 June 2019).

157. Westman, M.; Lupinek, C.; Bousquet, J.; Andersson, N.; Pahr, S.; Baar, A.; Bergström, A.; Holmström, M.; Stjärne, P.; Lødrup Carlsen, K.C.; et al. Mechanisms for the Development of Allergies Consortium. Early childhood IgE reactivity to pathogenesis-related class 10 proteins predicts allergic rhinitis in adolescence. *J. Allergy Clin. Immunol.* **2015**, *135*, 1199–1206. [CrossRef] [PubMed]

158. Hatzler, L.; Panetta, V.; Lau, S.; Wagner, P.; Bergmann, R.L.; Illi, S.; Bergmann, K.E.; Keil, T.; Hofmaier, S.; Rohrbach, A.; et al. Molecular spreading and predictive value of preclinical IgE response to Phleum pratense in children with hay fever. *J. Allergy Clin. Immunol.* **2012**, *130*, 894–901. [CrossRef] [PubMed]

159. Asarnoj, A.; Hamsten, C.; Wadén, K.; Lupinek, C.; Andersson, N.; Kull, I.; Curin, M.; Anto, J.; Bousquet, J.; Valenta, R.; et al. Sensitization to cat and dog allergen molecules in childhood and prediction of symptoms of cat and dog allergy in adolescence: A BAMSE/MeDALL study. *J. Allergy Clin. Immunol.* **2016**, *137*, 813–821. [CrossRef] [PubMed]

160. Posa, D.; Perna, S.; Resch, Y.; Lupinek, C.; Panetta, V.; Hofmaier, S.; Rohrbach, A.; Hatzler, L.; Grabenhenrich, L.; Tsilochristou, O.; et al. Evolution and predictive value of IgE responses toward a comprehensive panel of house dust mite allergens during the first 2 decades of life. *J. Allergy Clin. Immunol.* **2017**, *139*, 541–549. [CrossRef]

161. Wickman, M.; Lupinek, C.; Andersson, N.; Belgrave, D.; Asarnoj, A.; Benet, M.; Pinart, M.; Wieser, S.; Garcia-Aymerich, J.; Baar, A.; et al. Detection of IgE Reactivity to a Handful of Allergen Molecules in Early Childhood Predicts Respiratory Allergy in Adolescence. *EBioMedicine* **2017**, *26*, 91–99. [CrossRef]

162. Lupinek, C.; Marth, K.; Niederberger, V.; Valenta, R. Analysis of serum IgE reactivity profiles with microarrayed allergens indicates absence of de novo IgE sensitizations in adults. *J. Allergy Clin. Immunol.* **2012**, *130*, 1418–1420. [CrossRef]

163. Westman, M.; Asarnoj, A.; Hamsten, C.; Wickman, M.; van Hage, M. Windows of opportunity for tolerance induction for allergy by studying the evolution of allergic sensitization in birth cohorts. *Semin. Immunol.* **2017**, *30*, 61–66. [CrossRef] [PubMed]

164. Ohsaki, A.; Venturelli, N.; Buccigrosso, T.M.; Osganian, S.K.; Lee, J.; Blumberg, R.S.; Oyoshi, M.K. Maternal IgG immune complexes induce food allergen-specific tolerance in offspring. *J. Exp. Med.* **2018**, *215*, 91–113. [CrossRef] [PubMed]

165. Campana, R.; Marth, K.; Zieglmayer, P.; Weber, M.; Lupinek, C.; Zhernov, Y.; Elisyutina, O.; Khaitov, M.; Rigler, E.; Westritschnig, K.; et al. Vaccination of nonallergic individuals with recombinant hypoallergenic fragments of birch pollen allergen Bet v 1: Safety, effects, and mechanisms. *J. Allergy Clin. Immunol.* **2019**, *143*, 1258–1261. [CrossRef] [PubMed]

166. Glovsky, M.M.; Ghekiere, L.; Rejzek, E. Effect of maternal immunotherapy on immediate skin test reactivity, specific rye I IgG and IgE antibody, and total IgE of the children. *Ann. Allergy* **1991**, *67*, 21–24. [PubMed]

167. Flicker, S.; Linhart, B.; Wild, C.; Wiedermann, U.; Valenta, R. Passive immunization with allergen-specific IgG antibodies for treatment and prevention of allergy. *Immunobiology* **2013**, *218*, 884–891. [CrossRef] [PubMed]

168. Du Toit, G.; Roberts, G.; Sayre, P.H.; Bahnson, H.T.; Radulovic, S.; Santos, A.F.; Brough, H.A.; Phippard, D.; Basting, M.; Feeney, M.; et al. Randomized trial of peanut consumption in infants at risk for peanut allergy. *N. Engl. J. Med.* **2015**, *372*, 803–813. [CrossRef]

169. Perkin, M.R.; Logan, K.; Marrs, T.; Radulovic, S.; Craven, J.; Flohr, C.; Lack, G.; EAT Study Team. Enquiring About Tolerance (EAT) study: Feasibility of an early allergenic food introduction regimen. *J. Allergy Clin. Immunol.* **2016**, *137*, 1477–1486. [CrossRef]

170. Szajewska, H.; Horvath, A. Meta-analysis of the evidence for a partially hydrolyzed 100% whey formula for the prevention of allergic diseases. *Curr. Med. Res. Opin.* **2010**, *26*, 423–437. [CrossRef]

171. Alexander, D.D.; Cabana, M.D. Partially hydrolyzed 100% whey protein infant formula and reduced risk of atopic dermatitis: A meta-analysis. *J. Pediatr. Gastroenterol. Nutr.* **2010**, *50*, 356–358. [CrossRef]

172. Gouw, J.W.; Jo, J.; Meulenbroek, L.A.P.M.; Heijjer, T.S.; Kremer, E.; Sandalova, E.; Knulst, A.C.; Jeurink, P.V.; Garssen, J.; Rijnierse, A.; et al. Identification of peptides with tolerogenic potential in a hydrolysed whey-based infant formula. *Clin. Exp. Allergy* **2018**, *48*, 1345–1353. [CrossRef]

173. Ueno, H.M.; Kato, T.; Ohnishi, H.; Kawamoto, N.; Kato, Z.; Kaneko, H.; Kondo, N.; Nakano, T. T-cell epitope-containing hypoallergenic β-lactoglobulin for oral immunotherapy in milk allergy. *Pediatr. Allergy Immunol.* **2016**, *27*, 818–824. [CrossRef] [PubMed]

174. Ueno, H.M.; Kato, T.; Ohnishi, H.; Kawamoto, N.; Kato, Z.; Kaneko, H.; Kondo, N.; Nakano, T. Hypoallergenic casein hydrolysate for peptide-based oral immunotherapy in cow's milk allergy. *J. Allergy Clin. Immunol.* **2018**, *142*, 330–333. [CrossRef] [PubMed]

175. Rezende, R.M.; Weiner, H.L. History and mechanisms of oral tolerance. *Semin. Immunol.* **2017**, *30*, 3–11. [CrossRef] [PubMed]

176. Campana, R.; Huang, H.J.; Freidl, R.; Linhart, B.; Vrtala, S.; Wekerle, T.; Karaulov, A.; Valenta, R. Recombinant allergen and peptide-based approaches for allergy prevention by oral tolerance. *Semin. Immunol.* **2017**, *30*, 67–80. [CrossRef] [PubMed]

nutrients

MDPI

Communication

Management of Cow's Milk Allergy from an Immunological Perspective: What Are the Options?

Edward F. Knol [1,*], Nicolette W. de Jong [2], Laurien H. Ulfman [3] and Machteld M. Tiemessen [4,5]

[1] Center Translational Immunology, University Medical Center Utrecht, 3584 CT Utrecht, The Netherlands
[2] Department of Internal Medicine, Section Allergology and Clinical Immunology, Erasmus MC,
     3000 CA Rotterdam, The Netherlands; n.w.dejong@erasmusmc.nl
[3] FrieslandCampina, 3818 LE Amersfoort, The Netherlands; Laurien.Ulfman@frieslandcampina.com
[4] Danone Nutricia Research, 3584 CT Utrecht, The Netherlands; Machteld.TIEMESSEN@nutricia.com
[5] Institute of Pharmaceutical Sciences, Utrecht University, 3584 CG Utrecht, The Netherlands
*   Correspondence: e.f.knol@umcutrecht.nl

Received: 22 October 2019; Accepted: 8 November 2019; Published: 11 November 2019

Abstract: The immunological mechanism underlying Immunoglobuline E (IgE)-mediated cow's milk allergy has been subject to investigations for many years. Identification of the key immune cells (mast cells, B cells) and molecules (IgE) in the allergic process has led to the understanding that avoidance of IgE-crosslinking epitopes is effective in the reduction of allergic symptoms but it cannot be envisioned as a treatment. For the treatment and prevention of IgE-mediated cow's milk allergy, it is thought that the induction of a sustained state of immunological tolerance is needed. In this review, we will discuss various approaches aimed at achieving immunological tolerance and their success. Furthermore, we will speculate on the involved immunological mechanism.

Keywords: allergy; cow's milk; formula; therapy; immune cells; Immunoglubuline E

1. Immunological Aspects of Cow's Milk Allergy

Of all known food allergies in infancy, cow's milk allergy (CMA) is of special interest to immunologists as most allergic infants will acquire spontaneous tolerance toward cow's milk before the age of 3 years. A Danish birth cohort showed that children with confirmed cow's milk allergy appeared to be tolerant in 56% and 77% of the children at age 1 and 2 years, respectively [1,2]. While the incidence of cow's milk allergy is estimated to be around 2–3%, less than 0.5% of adults suffer from CMA [3]. Ingestion of cow's milk can lead to acute cutaneous symptoms, such as urticaria, and may also lead to immediate-type pulmonary and/or gastro-intestinal symptoms or, especially at older ages, systemic anaphylaxis [4]. Therefore, there is a need for proper allergy management during the "allergic" years and, in addition, options to accelerate the process of tolerance acquisition would be welcomed by affected families. Besides allergic infants, there is also a population of adults displaying Immunoglobuline E (IgE)-mediated cow's milk allergy [5]. Cow's milk protein (CMP) contains various proteins, of which the whey proteins—β-lactoglobulin and α-lactalbumin, and the caseins (αs1-, αs2-, β-, and κ-casein)—are the most important proteins regarding allergy. Specific IgE antibodies to all the subfractions of both casein and whey proteins can be detected in infants and children with CMA [5]. Besides B cell activity markers, such as IgE antibodies, T cell activity toward the various cow's milk proteins (both whey proteins and caseins) can also be found. Of interest, CM-specific T cells can be isolated from both allergic and non-allergic individuals [6]. Upon T cell activation, the cytokine profile of a T cell will influence the subsequent B cell responses. B cells will class switch their immunoglobulin production under the influence of T-cell-derived cytokines. T-cell-derived Interleukin (IL)-4 production will cause the process of Ig class switching of B cells toward IgE (causing sensitization and subsequent allergy after exposure to the specific allergen), whereas IL-10 promotes the production of Immunoglobuline G4 (IgG4) (possibly involved in the process of immune tolerance). It has been demonstrated that cow's-milk-specific IgG4 is linked to tolerance in children with increased levels of IgE [7,8]. However, there is still a scientific debate whether increased IgG4 might be just an indicator for increased/high exposure. Therefore, the preceding T cell response in cow's-milk-allergic individuals will have a large impact on the immune response toward the proteins in cow's milk, and thereby, the skewing towards an allergic or a tolerogenic response. T cells recognize specific parts of proteins (so called T cell epitopes) only when presented by antigen-presenting cells in the context of an human leucocyte antigen (HLA) molecule. Due to the large variation in genetic profiles between individuals (variation in HLA genotype), many different T cell epitopes may induce an immune response. Within some of the cow's milk proteins, several dominant antigenic regions have been identified. For example, within αs1-casein and β-lactoglobulin, T cell epitopes have been identified [9,10]. Since these T cell epitopes appear to be recognized by T cells of both allergic and non-allergic donors, this suggests that these parts of the protein may potentially be involved in the process of tolerance induction.

Importantly, the size of major histocompatibility complex (MHC) class II presented peptides is only 15–24 amino acids (aa) long, while the size of a protein that can crosslink IgE on mast cells or basophils is much larger, about 5–25 nm [11]. Therefore, as the T cell epitopes have the potential to steer the immune response without causing the detrimental process of IgE crosslinking on mast cells and/or basophils causing the allergic reaction, active immunotherapy using T cell epitopes/peptides may be an attractive option for cow's-milk-allergic patients (both as a preventive as well as a therapeutic approach). One of the challenges is to identify enough diverse numbers of T cell epitopes, which are recognized by the majority of the target population, in order to provide enough stimulation of the immune system and to "re-train" the immune system away from an allergic response toward a tolerogenic response. As mentioned earlier, the HLA genotype diversity of the population will be of importance in the design of the diversity of the peptide mix needed to induce tolerance, as was demonstrated before for birch pollen allergens [12].

For αs1 and β caseins, it has been shown that the IgE binding epitopes are more in the 3D configuration in young children, whereas in adolescents and adults, it is more the linear structure that is recognized [13,14]. Remarkably, this was not the case for the IgG binding epitopes [14]. This might be related to the more immature digestive tract in young children, in which, for instance, the pH of the stomach is somewhat higher than later in life, leaving the more 3D configuration intact [15].

2. Cow's Milk Formula, Including Hydrolysate in Cow's-Milk-Allergic Patients

For infants that cannot be breastfed, other nutritional solutions (in the form of formula) are available. Several different cow's-milk-protein-based formula are currently on the market, varying from a formula based on whole cow's milk protein (CMP), partially hydrolyzed CMP, to extensively hydrolyzed CMP, and even amino-acid-based formulas. Extensively hydrolyzed cow's milk formulas with documented hypoallergenicity are being recommended as a first-choice formula for cow's-milk-allergic infants and young children [16,17]. It must be realized that this is mostly management, but not a cure, of the disease; therefore, in the future, interventions are needed that aim at preventing or curing CMA. For prevention, tolerance inducing partially hydrolyzed formulas have been developed for their potential to reduce the risk of CMA since these formulas contain peptides of a specific length. So far, contradictory results have been reported on the effect of these formulas on prevention. The largest individual study so far did find a lower risk on allergic manifestations, but in a recent meta-analysis, no evidence for a reduction in allergic manifestations (including CMA) was shown [18].

Factors, such as the selection of subjects, set up of study, type of product (hydrolysis degree, type of hydrolyzed cow's milk protein), and studies performed in small groups of patients, could play a role in the observed discrepancies between studies. Furthermore, since the incidence of IgE-mediated CMA is low, and spontaneous tolerance occurrence is relatively high, a preventive study with a primary outcome on CMA (instead of all allergic manifestations) would be important but very challenging to conduct.

Recently, different commercially available infant formulas (intact proteins versus partially hydrolyzed versus extensively hydrolyzed) were investigated on their in vitro immune profile. Not all extensively hydrolyzed formulas reacted in a similar way with respect to IgE reactivity, proliferative responses of immune cells, and cytokine profiles [19]. This was also true for the partial hydrolysates. Direct comparison of in vivo reactivity of the hypoallergenic formulas in an at risk population and/or in CMA patients with immune profiles in these target groups is currently lacking in the literature, but would be needed to investigate whether in vitro profiles can explain the potential differences in effectiveness mechanistically.

By comparing the immune response toward whole protein cow's milk formulas versus partial and extensive hydrolysates, it has been demonstrated that hydrolyzation of proteins reduces allergenicity whilst maintaining immunogenicity (T cell reactivity), depending on the type of hydrolysate [20,21]. However, the degree of hydrolysis not only corresponds to reduced allergenicity, it may also lead to reduced or different T cell reactivity [20,21]. The discrepancy between T cell activation profiles in these studies may be explained by different patient populations included in these studies. The serum and T cells used in the two studies were derived from cow's-milk-allergic children [20] or adults [21], which may have an important impact on IgE binding to conformational versus linear epitopes [13,14]. As the cytokine profile is essential for the subsequent immunological processes, it is important to analyze the capacity of the different T cell epitopes in the different cow's-milk-based formulas with regard to their immunostimulatory capacity.

Another important approach is to consider unprocessed, raw cow's milk instead of the commercially available heated and processed cow's milk. Epidemiological data indicate that consumption of raw milk in the first year of life protects against allergies, including asthma [22]. There are several potent immunomodulating activities in milk that are lost during processing, such as transforming growth factor Bèta (TGF-β), IL-10, vitamin D, and lactoferrin, as well as fatty acids, oligosaccharides, and lipids, some of which are lost during processing [23]. Importantly, these bovine products have potent interspecies effects on human cells. However, although there are several potential advantages to introducing raw milk into the diet of young children, there is a significant risk of bacterial infections that also hampers controlled studies in infants to prove its effect [24]. Therefore, minimal processing techniques of raw milk to preserve the immunomodulating activity but safeguard the microbiological quality are needed. Alternatively, isolating the immune factors that drive the protective effects and provide these isolated components to infants is an interesting way to go.

Additionally, the composition of the intestinal microbiota might influence the immune response toward cow's milk proteins (and the different T cell epitopes). Since it has been shown that the composition of the microbiota affects the development of the mucosal immune system, it is highly likely that differences in microbiota will shape the microenvironment in which the immune response is elicited [25]. Indeed, the microbiome of atopic versus healthy infants was shown to be different, and with specific pre- and/or probiotics, this dysbiosis may be altered toward a beneficial microenvironment in which tolerance toward cow's milk proteins can be induced [26]. Interestingly, combining an extensively hydrolyzed formula with a probiotic strain may further accelerate tolerance development [27].

3. Baked Milk in the Treatment of Cow's Milk Allergy

It has been suggested that the introduction of baked milk into the diet of the child may speed up the resolution of cow's milk allergy [28]. Most children outgrow their cow's milk allergy by the age of 3 years old [29]. These individuals with transient cow's milk allergy produce IgE antibodies that are primarily directed at conformational epitopes. Since high temperatures (baking) reduce allergenicity by destroying these epitopes, the hypothesis is that the transient allergic group of children would tolerate baked milk products. More importantly, the addition of baked-milk products to the diet of these children appears to markedly accelerate the development of tolerance to unheated milk compared to a strict avoidance diet, which is currently the "standard of care" [28]. This approach is being introduced into clinical practice, although the hard evidence to underpin it seems to be lacking.

In a recent systematic review by Lambert et al. to examine the evidence as to whether baked milk introduction into the diet leads to a larger proportion of children outgrowing their milk allergy, only three studies could be included [30]. Although the results are promising, e.g., baked milk was found to be well tolerated in children and no serious adverse reactions were reported, without randomization of the intervention, these studies are at a major risk of confounding by factors that are not equally distributed between the different groups.

Nevertheless, introduction of baked milk into the diet can also increase the quality of life by expanding the diet, boosting nutrition, and promoting participating in social activities [31].

The study of potential biomarkers to predict the tolerability of baked milk, such as allergen-specific IgE or the skin prick test (SPT), are ongoing. A cow's-milk-specific IgE level $\geq$15 KU$_A$/L and SPT $\geq$8 mm in children $\leq$2 years old are highly predictive for a positive oral challenge reaction with baked milk [32]. Using specific IgG4 levels in addition to specific IgE levels may help predict baked milk reactivity. Casein and beta-lactoglobulin-specific IgE/IgG4 ratios appear to be significantly higher in baked-milk-reactive subjects compared to baked-milk-tolerant subjects [33].

Although there are currently no results of randomized controlled studies to determine whether baked milk speeds up the resolution of cow's milk allergy, the opportunity to reduce the child's dietary restrictions can potentially have a major beneficial effect on the food allergic child and their family.

4. Allergen Immunotherapy in the Treatment of Cow's Milk Allergy

A more direct approach to reach tolerance in the cow's milk allergic patients is via allergen immunotherapy (AIT), in which cow's milk is administered for a longer period, starting with micrograms, increasing to milligrams, and finally to grams via either the oral route, including sublingual administration [34], or epicutaneous route [35]. Each of these approaches is intended to induce some level of desensitization with repeated exposure to the allergenic food protein, although the risks and potential benefits of each treatment differ significantly [36]. Permanent tolerance is defined as the ability to ingest food without symptoms despite prolonged periods of avoidance or irregular intake [37]. Both intact cow's milk, as well as partially hydrolyzed cow's milk, has been tested. In a meta-analysis, it was shown that the relative risk for desensitization after allergen immunotherapy (AIT) for cow's milk was 0.12, 95% CI = 0.06–0.25 [34]. At the same time, the safety of AIT is an issue and side effects are found frequently, though merely local [38]. Individual studies have evaluated the immunological parameters changed by AIT. After the treatment with partially hydrolyzed formula of cow's-milk-allergic children and young adults (age range 1–20), a slight reduction in casein-specific IgE was demonstrated after 16 weeks, but no increased concentrations of IgG4, nor changes in casein-induced basophil degranulation [39]. In a younger group (7–12 months) treated with increasing cow's milk concentrations, cow's-milk-specific IgG4 increased 20–40-fold, while IgE decreased about 2-fold [8]. In a cohort of 2-year-old children, AIT strongly reduced the skin reactivity for cow's milk and slightly increased (3-fold) the specific IgE concentrations for caseins and total cow's milk [40]. Comparable levels of IgE decrease and IgG4 increase were noted in a study with older children (5–15 years). However, only casein-specific IgE decreased, but not IgE levels for α-lactalbumin and β-lactoglobulin [41]. Overall, even with the limited number of controlled studies, it seems that the ratio of specific IgE/IgG4 decreased, mostly by the increased amounts of IgG4. The clinical benefit of AIT in cow's-milk-allergic patients is difficult to interpret, because this is only performed at a young age in which the incidence of spontaneous tolerance is significant [1]. Moreover, whereas the adult population needs a solution for cow's milk allergy (CMA), there are no controlled studies performed with AIT in this age group and is therefore is not recommended in adults [38].

One of the challenges in the AIT for cow's milk allergy is that clinical tolerance only remains when AIT is continued [34]. In contrast to AIT for insect venoms or pollen allergens, clinical tolerance to food allergens is lost rapidly after stopping the AIT procedure. The future challenge is therefore to induce sustained clinical tolerance after AIT is stopped.

Nutrients **2019**, *11*, 2734

5. Which Immune Cells Are Involved in Immunotherapy for Cow's Milk Allergy?

To gain insight in the immunological mechanism of tolerance induction, the analysis of various immune cells possibly involved in this process is of high interest. In Figure 1, we have depicted three subsets that we would like to highlight, namely regulatory T-cells (Tregs), regulatory B-cells (Bregs), and innate lymphoid regulatory cells (ILCregs). Tregs are cells with a strong regulatory capacity and are generally referred to as the CD4+CD25+ regulatory T cell subset. In cow's milk allergic adults, it has been demonstrated that the percentage and function of CD4+CD25+ Tregs is intact [42]. Also, in cow's-milk-allergic children, the presence of Tregs has been investigated. A study by Savilahti et al. showed that in the peripheral blood of allergic children, CD4+CD25+ Tregs can be detected and are functionally active [43]. Therefore, specific immunotherapies aiming at the stimulation of these naturally occurring regulatory T cells may contribute to the re-establishment of clinical tolerance in a sustainable way in allergic individuals.

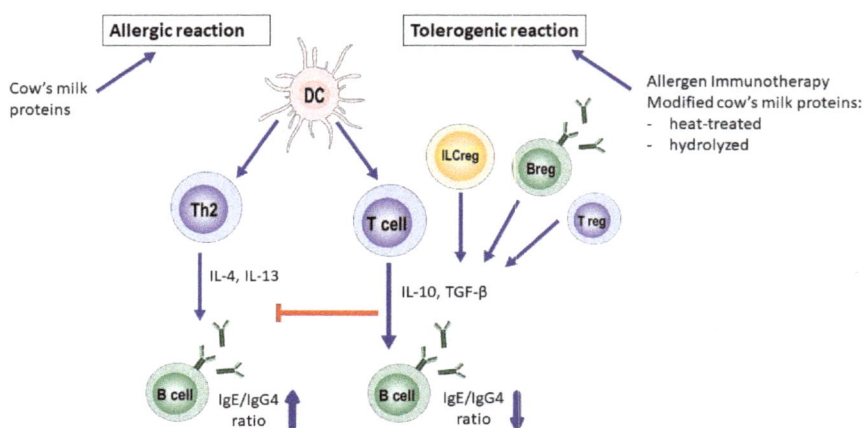

Figure 1. Immune response in the allergic versus tolerant state for cow's milk proteins. Breg: B-regulatory cell; DC: Dendritic cell; ILCreg: Regulatory innate lymphoid cells; IL: interleukin; TGF-β: transforming growth factor Bèta; IgG4: Immunoglobulin G 4; IgE: immunoglobulin E; Treg: T regulatory cell; Th2: T- helper 2 cell.

One of the mechanisms by which CD4+CD25+ (Tregs) may enhance the process of tolerance induction is via the production of the suppressive cytokine IL-10. Results from a study investigating the cow's-milk-specific T cell response in allergic children, which are tolerant to cow's milk, show that their cow's-milk-specific T cell response is dominated by the production of large amounts of IL-10 [44]. This suggests a key role for IL-10-producing T cells in a long-lasting tolerogenic reaction in individuals where the immune system is skewed toward an allergic phenotype. Another potential source of IL-10 is the subset of regulatory innate lymphoid cells. ILCregs have been recently described to be present in the gut and produce IL-10 and TGF-β upon pathogenic stimulation, thereby promoting a tolerogenic environment [45]. Although a direct suppressive effect of ILCregs on ILC2 cells, which produce IL-4 and have been shown to promote food allergy by enhancing mast cell activation and the disruption of Treg function [46] has not been demonstrated, the ILCreg-mediated production of IL-10 may play a role in the process of tolerance induction by affecting other immune cells.

For B cells, it has been suggested that a subpopulation, called regulatory B cells (Bregs), is of importance for the induction and maintenance of tolerance to many different self- and non-self-antigens [47]. Whether Bregs play a role in CMA remains to be established.

6. Conclusions

Important aspects to keep in mind when evaluating the different options for allergy prevention and the treatment of cow's-milk-allergic individuals are the intended target population (infants, children, or adults) and the patient characteristics (IgE and IgG4 serum levels, SPT values). A better immunological understanding of the mechanism underlying a sustained tolerance development in cow's-milk-allergic patients will aid in improving current therapies or developing new therapies.

Author Contributions: Conceptualization, E.F.K. and M.M.T.; methodology, E.F.K. and M.M.T.; writing—original draft preparation, E.F.K., N.W.d.J., L.H.U. and M.M.T.; writing—review and editing, E.F.K., N.W.d.J., L.H.U. and M.M.T.

Funding: This research received no external funding.

Acknowledgments: Part of this work is in cooperation with the research program iAGE/TTW project number 14536, which is partly financed by the Netherlands Organization for Scientific Research (NWO).

Conflicts of Interest: The authors declare no conflict of interest.

References

1. Host, A.; Halken, S.; Jacobsen, H.P.; Christensen, A.E.; Herskind, A.M.; Plesner, K. Clinical course of cow's milk protein allergy/intolerance and atopic diseases in childhood. *Pediatr. Allergy Immunol.* **2002**, *13* (Suppl. 15), 23–28. [CrossRef] [PubMed]

2. Host, A.; Halken, S. Hypoallergenic formulas—When, to whom and how long: After more than 15 years we know the right indication! *Allergy* **2004**, *59* (Suppl. 78), 45–52. [CrossRef] [PubMed]

3. Fiocchi, A.; Schunemann, H.J.; Brozek, J.; Restani, P.; Beyer, K.; Troncone, R.; Martelli, A.; Terracciano, L.; Bahna, S.L.; Rancé, F.; et al. Diagnosis and Rationale for Action Against Cow's Milk Allergy (DRACMA): A summary report. *J. Allergy Clin. Immunol.* **2010**, *126*, 1119–1128. [CrossRef] [PubMed]

4. Sampson, H.A. Food allergy. Part 1: Immunopathogenesis and clinical disorders. *J. Allergy Clin. Immunol.* **1999**, *103*, 717–728. [CrossRef]

5. Lam, H.-Y.; Van Hoffen, E.; Michelsen, A.; Guikers, K.; Van Der Tas, C.H.W.; Knulst, A.C.; Bruijnzeel-Koomen, C.A.F.M.; Bruijnzeel-Koomen, C.A.F.M. Cow's milk allergy in adults is rare but severe: Both casein and whey proteins are involved. *Clin. Exp. Allergy* **2008**, *38*, 995–1002. [CrossRef] [PubMed]

6. Schade, R.P.; Dijk, A.G.V.I.-V.; Van Reijsen, F.C.; Versluis, C.; Kimpen, J.L.; Knol, E.F.; Bruijnzeel-Koomen, C.A.; Van Hoffen, E. Differences in antigen-specific T-cell responses between infants with atopic dermatitis with and without cow's milk allergy: Relevance of TH2 cytokines. *J. Allergy Clin. Immunol.* **2000**, *106*, 1155–1162. [CrossRef] [PubMed]

7. Ruiter, B.; Knol, E.F.; Van Neerven, R.J.J.; Garssen, J.; Bruijnzeel-Koomen, C.A.F.M.; Knulst, A.C.; Van Hoffen, E. Maintenance of tolerance to cow's milk in atopic individuals is characterized by high levels of specific immunoglobulin G4. *Clin. Exp. Allergy* **2007**, *37*, 1103–1110. [CrossRef] [PubMed]

8. Lee, J.H.; Kim, W.S.; Kim, H.; Hahn, Y.S. Increased cow's milk protein-specific IgG4 levels after oral desensitization in 7-to 12-month-old infants. *Ann. Allergy Asthma Immunol.* **2013**, *111*, 523–528. [CrossRef] [PubMed]

9. Ruiter, B.; Trégoat, V.; M'Rabet, L.; Garssen, J.; Knol, E.F.; Hoffen, E.; Bruijnzeel-Koomen, C.A.F.M.; Bruijnzeel-Koomen, C.A.F.M. Characterization of T cell epitopes in alphas1-casein in cow's milk allergic, atopic and non-atopic children. *Clin. Exp. Allergy* **2006**, *36*, 303–310. [CrossRef] [PubMed]

10. Gouw, J.W.; Jo, J.; Meulenbroek, L.A.P.M.; Heijjer, T.S.; Kremer, E.; Sandalova, E.; Knulst, A.C.; Jeurink, P.V.; Garssen, J.; Rijnierse, A.; et al. Identification of peptides with tolerogenic potential in a hydrolysed whey-based infant formula. *Clin. Exp. Allergy* **2018**, *48*, 1345–1353. [CrossRef] [PubMed]

11. Knol, E.F. Requirements for effective IgE cross-linking on mast cells and basophils. *Mol. Nutr. Food Res.* **2006**, *50*, 620–624. [CrossRef] [PubMed]

12. Friedl-Hajek, R.; Spangfort, M.D.; Schou, C.; Breiteneder, H.; Yssel, H.; Joost van Neerven, R.J. Identification of a highly promiscuous and an HLA allele-specific T-cell epitope in the birch major allergen Bet v 1: HLA restriction, epitope mapping and TCR sequence comparisons. *Clin. Exp. Allergy* **1999**, *29*, 478–487. [CrossRef] [PubMed]

13. Vila, L.; Beyer, K.; Jarvinen, K.M.; Chatchatee, P.; Bardina, L.; Sampson, H.A. Role of conformational and linear epitopes in the achievement of tolerance in cow's milk allergy. *Clin. Exp. Allergy* **2001**, *31*, 1599–1606. [CrossRef] [PubMed]

14. Chatchatee, P.; Jarvinen, K.M.; Bardina, L.; Beyer, K.; Sampson, H.A. Identification of IgE- and IgG-binding epitopes on alpha(s1)-casein: Differences in patients with persistent and transient cow's milk allergy. *J. Allergy Clin. Immunol.* **2001**, *107*, 379–383. [CrossRef] [PubMed]

15. Cooke, S.K.; Sampson, H.A. Allergenic properties of ovomucoid in man. *J. Immunol.* **1997**, *159*, 2026–2032. [PubMed]

16. Muraro, A.; Werfel, T.; Hoffmann-Sommergruber, K.; Roberts, G.; Beyer, K.; Bindslev-Jensen, C.; Cardona, V.; Dubois, A.; Dutoit, G.; Eigenmann, P.; et al. EAACI food allergy and anaphylaxis guidelines: Diagnosis and management of food allergy. *Allergy* **2014**, *69*, 1008–1025. [CrossRef] [PubMed]

17. Terheggen-Lagro, S.W.; Khouw, I.M.; Schaafsma, A.; Wauters, E.A. Safety of a new extensively hydrolysed formula in children with cow's milk protein allergy: A double blind crossover study. *BMC Pediatr.* **2002**, *2*, 10. [CrossRef] [PubMed]

18. Osborn, D.A.; Sinn, J.K.; Jones, L.J. Infant formulas containing hydrolysed protein for prevention of allergic disease. *Cochrane Database Syst. Rev.* **2018**, *10*. [CrossRef] [PubMed]

19. Hochwallner, H.; Schulmeister, U.; Swoboda, I.; Focke-Tejkl, M.; Reininger, R.; Civaj, V.; Campana, R.; Thalhamer, J.; Scheiblhofer, S.; Balic NHorak, F. Infant milk formulas differ regarding their allergenic activity and induction of T-cell and cytokine responses. *Allergy* **2017**, *72*, 416–424. [CrossRef] [PubMed]

20. Knipping, K.; Van Esch, B.C.; Dijk, A.G.V.I.-V.; Van Hoffen, E.; Van Baalen, T.; Knippels, L.M.; Van Der Heide, S.; Dubois, A.E.; Garssen, J.; Knol, E.F. Enzymatic treatment of whey proteins in cow's milk results in differential inhibition of IgE-mediated mast cell activation compared to T-cell activation. *Int. Arch. Allergy Immunol.* **2012**, *159*, 263–270. [CrossRef] [PubMed]

21. Meulenbroek, L.A.P.M.; Oliveira, S.; Jager, C.F.D.H.; Klemans, R.J.B.; Lebens, A.F.M.; Van Baalen, T.; Knulst, A.C.; Bruijnzeel-Koomen, C.A.F.M.; Garssen, J.; Knippels, L.M.J.; et al. The degree of whey hydrolysis does not uniformly affect in vitro basophil and T cell responses of cow's milk-allergic patients. *Clin. Exp. Allergy* **2014**, *44*, 529–539. [CrossRef] [PubMed]

22. Riedler, J.; Braun-Fahrlander, C.; Eder, W.; Schreuer, M.; Waser, M.; Maisch, S.; Carr, D.; Schierl, R.; Nowak, D.; Von Mutius, E. Exposure to farming in early life and development of asthma and allergy: A cross-sectional survey. *Lancet* **2001**, *358*, 1129–1133. [CrossRef]

23. Van Neerven, R.J.; Knol, E.F.; Heck, J.M.; Savelkoul, H.F. Which factors in raw cow's milk contribute to protection against allergies? *J. Allergy Clin. Immunol.* **2012**, *130*, 853–858. [CrossRef] [PubMed]

24. Sozanska, B. Raw Cow's Milk and Its Protective Effect on Allergies and Asthma. *Nutrients* **2019**, *11*, 469. [CrossRef] [PubMed]

25. Pabst, O.; Mowat, A.M. Oral tolerance to food protein. *Mucosal Immunol.* **2012**, *5*, 232–239. [CrossRef] [PubMed]

26. Wopereis, H.; Oozeer, R.; Knipping, K.; Belzer, C.; Knol, J. The first thousand days—Intestinal microbiology of early life: Establishing a symbiosis. *Pediatr. Allergy Immunol.* **2014**, *25*, 428–438. [CrossRef] [PubMed]

27. Canani, R.B.; Nocerino, R.; Terrin, G.; Coruzzo, A.; Cosenza, L.; Leone LTroncone, R. Effect of Lactobacillus GG on tolerance acquisition in infants with cow's milk allergy: A randomized trial. *J. Allergy Clin. Immunol.* **2012**, *129*, 580–582. [CrossRef] [PubMed]

28. Kim, J.S.; Nowak-Wegrzyn, A.; Sicherer, S.H.; Noone, S.; Moshier, E.L.; Sampson, H.A. Dietary baked milk accelerates the resolution of cow's milk allergy in children. *J. Allergy Clin. Immunol.* **2011**, *128*, 125–131. [CrossRef] [PubMed]

29. Host, A.; Halken, S. A prospective study of cow milk allergy in Danish infants during the first 3 years of life. Clinical course in relation to clinical and immunological type of hypersensitivity reaction. *Allergy* **1990**, *45*, 587–596. [CrossRef] [PubMed]

30. Lambert, R.; Grimshaw, K.E.C.; Ellis, B.; Jaitly, J.; Roberts, G. Evidence that eating baked egg or milk influences egg or milk allergy resolution: A systematic review. *Clin. Exp. Allergy* **2017**, *47*, 829–837. [CrossRef] [PubMed]

31. Leonard, S.A.; Caubet, J.C.; Kim, J.S.; Groetch, M.; Nowak-Wegrzyn, A. Baked milk and egg-containing diet in the management of milk and egg allergy. *J. Allergy Clin. Immunol Pract.* **2015**, *3*, 13–23. [CrossRef] [PubMed]

32. Nowak-Węgrzyn, A.; Bloom, K.A.; Sicherer, S.H.; Shreffler, W.G.; Noone, S.; Wanich, N.; Sampson, H.A. Tolerance to extensively heated milk in children with cow's milk allergy. *J. Allergy Clin. Immunol.* **2008**, *122*, 342–347. [CrossRef] [PubMed]

33. Caubet, J.C.; Nowak-Wegrzyn, A.; Moshier, E.; Godbold, J.; Wang, J.; Sampson, H.A. Utility of casein-specific IgE levels in predicting reactivity to baked milk. *J. Allergy Clin. Immunol.* **2013**, *131*, 222–224. [CrossRef] [PubMed]

34. Keet, C.A.; Seopaul, S.; Knorr, S.; Narisety, S.; Skripak, J.; Wood, R.A. Long-term follow-up of oral immunotherapy for cow's milk allergy. *J. Allergy Clin. Immunol.* **2013**, *132*, 737–739. [CrossRef] [PubMed]

35. Nurmatov, U.; Dhami, S.; Arasi, S.; Pajno, G.B.; Fernandez-Rivas, M.; Muraro, A.; Roberts, G.; Akdis, C.; Alvaro-Lozano, M.; Beyer, K.; et al. Allergen immunotherapy for IgE-mediated food allergy: A systematic review and meta-analysis. *Allergy* **2017**, *72*, 1133–1147. [CrossRef] [PubMed]

36. Wood, R.A. Food allergen immunotherapy: Current status and prospects for the future. *J. Allergy Clin. Immunol.* **2016**, *137*, 973–982. [CrossRef] [PubMed]

37. Gernez, Y.; Nowak-Wegrzyn, A. Immunotherapy for Food Allergy: Are We There Yet? *J. Allergy Clin. Immunol. Pract.* **2017**, *5*, 250–272. [CrossRef] [PubMed]

38. Pajno, G.B.; Fernandez-Rivas, M.; Arasi, S.; Roberts, G.; Akdis, C.A.; Alvaro-Lozano, M.; Beyer, K.; Bindslev-Jensen, C.; Burks, W.; Ebisawa, M.; et al. EAACI Guidelines on allergen immunotherapy: IgE-mediated food allergy. *Allergy* **2018**, *73*, 799–815. [CrossRef] [PubMed]

39. Inuo, C.; Tanaka, K.; Suzuki, S.; Nakajima, Y.; Yamawaki, K.; Tsuge, I.; Urisu, A.; Kondo, Y. Oral Immunotherapy Using Partially Hydrolyzed Formula for Cow's Milk Protein Allergy: A Randomized, Controlled Trial. *Int. Arch. Allergy Immunol.* **2018**, *177*, 259–268. [CrossRef] [PubMed]

40. Martorell, A.; De La Hoz, B.; Ibáñez, M.D.; Boné, J.; Terrados, M.S.; Michavila, A.; Plaza, A.M.; Alonso, E.; Garde, J.; Nevot, S.; et al. Oral desensitization as a useful treatment in 2-year-old children with cow's milk allergy. *Clin. Exp. Allergy* **2011**, *41*, 1297–1304. [CrossRef] [PubMed]

41. Patriarca, G.; Roncallo, C.; Del Ninno, M.; Pollastrini, E.; Milani, A.; De Pasquale, T.; Schiavino, D.; Nucera, E.; Buonomo, A.; Gasbarrini, G. Oral desensitisation in cow milk allergy: Immunological findings. *Int. J. Immunopathol. Pharmacol.* **2002**, *15*, 53–58. [CrossRef] [PubMed]

42. Tiemessen, M.M.; Van Hoffen, E.; Knulst, A.C.; Van Der Zee, J.A.; Knol, E.F.; Taams, L.S. CD4 CD25 regulatory T cells are not functionally impaired in adult patients with IgE-mediated cow's milk allergy. *J. Allergy Clin. Immunol.* **2002**, *110*, 934–936. [CrossRef] [PubMed]

43. Savilahti, E.M.; Karinen, S.; Salo, H.M.; Klemetti, P.; Saarinen, K.M.; Klemola, T.; Kuitunen, M.; Hautaniemi, S.; Savilahti, E.; Vaarala, O. Combined T regulatory cell and Th2 expression profile identifies children with cow's milk allergy. *Clin Immunol.* **2010**, *136*, 16–20. [CrossRef] [PubMed]

44. Tiemessen, M.M.; Van Ieperen-Van Dijk, A.G.; Bruijnzeel-Koomen, C.A.; Garssen, J.; Knol, E.F.; Van Hoffen, E. Cow's milk-specific T-cell reactivity of children with and without persistent cow's milk allergy: Key role for IL-10. *J. Allergy Clin. Immunol.* **2004**, *113*, 932–939. [CrossRef] [PubMed]

45. Wang, S.; Xia, P.; Chen, Y.; Qu, Y.; Xiong, Z.; Ye, B.; Du, Y.; Tian, Y.; Yin, Z.; Xu, Z.; et al. Regulatory Innate Lymphoid Cells Control Innate Intestinal Inflammation. *Cell* **2017**, *171*, 201–216. [CrossRef] [PubMed]

46. Noval Rivas, M.; Burton, O.T.; Oettgen, H.C.; Chatila, T. IL-4 production by group 2 innate lymphoid cells promotes food allergy by blocking regulatory T-cell function. *J. Allergy Clin. Immunol.* **2016**, *138*, 801–811. [CrossRef] [PubMed]

47. Van de Veen, W.; Stanic, B.; Wirz, O.F.; Jansen, K.; Globinska, A.; Akdis, M. Role of regulatory B cells in immune tolerance to allergens and beyond. *J. Allergy Clin. Immunol.* **2016**, *138*, 654–665. [CrossRef] [PubMed]

nutrients MDPI

Concept Paper

Prevention of Allergic Sensitization and Treatment of Cow's Milk Protein Allergy in Early Life: The Middle-East Step-Down Consensus

Yvan Vandenplas [1,*], Bakr Al-Hussaini [2], Khaled Al-Mannaei [3], Areej Al-Sunaid [4], Wafaa Helmi Ayesh [5], Manal El-Degeir [6], Nevine El-Kabbany [7], Joseph Haddad [8], Aziza Hashmi [9], Furat Kreishan [10] and Eslam Tawfik [11]

[1]   KidZ Health Castle, UZ Brussel, Vrijne Unversiteit Brussel, 1090 Brussels, Belgium
[2]   Department of Paediatrics, King Abdulaziz University Hospital, Jeddah 22252, Saudi Arabia
[3]   Department of Paediatrics, Al Salam International Hospital, Dasma 35151, Kuwait
[4]   Department of Paediatric Gastroenterology, King Abdullah Specialized Children's Hospital, Ministry of National Guard Health Affairs, Riyadh 11426, Saudi Arabia
[5]   Clinical Nutrition Department, Dubai Health Authority, P.O. Box, Dubai 4545, UAE
[6]   Department of Paediatrics, National Guard Hospital, Dammam 31412, Saudi Arabia
[7]   Department of Paediatrics, Mediclinic Welcare Hospital, P.O. Box, Dubai 31500, UAE
[8]   Department of Paediatrics, Saint George Hospital University Medical Center, Balamand University, P.O. Box, Beirut 166378, Lebanon
[9]   Department of Clinical Nutrition Services, King Abdulaziz Medical City-Jeddah, Ministry of National Guard Health Affairs, Jeddah 21423, Saudi Arabia
[10] Department of Paediatrics, Alhakeem Furat Clinic, Amman 11942, Jordan
[11] Department of Paediatrics, Sheikh Khalifa Medical City, P.O. Box, Abu Dhabi 51900, UAE
*   Correspondence: yvan.vandenplas@uzbrussel.be; Tel.: +3224775794

Received: 7 May 2019; Accepted: 20 June 2019; Published: 26 June 2019

Abstract: Allergy risk has become a significant public health issue with increasing prevalence. Exclusive breastfeeding is recommended for the first six months of life, but this recommendation is poorly adhered to in many parts of the world, including the Middle-East region, putting infants at risk of developing allergic sensitization and disorders. When breastfeeding is not possible or not adequate, a partially hydrolyzed whey formula (pHF-W) has shown proven benefits of preventing allergy, mainly atopic eczema, in children with a genetic risk. Therefore, besides stimulating breastfeeding, early identification of infants at risk for developing atopic disease and replacing commonly used formula based on intact cow milk protein (CMP) with a clinically proven pHF-W formula is of paramount importance for allergy prevention. If the child is affected by cow's milk protein allergy (CMPA), expert guidelines recommend extensively hydrolyzed formula (eHF), or an amino acid formula (AAF) in case of severe symptoms. The Middle-East region has a unique practice of utilizing pHF-W as a step-down between eHF or AAF and intact CMP, which could be of benefit. The region is very heterogeneous with different levels of clinical practice, and as allergic disorders may be seen by healthcare professionals of different specialties with different levels of expertise, there is a great variability in preventive and treatment approaches within the region itself. During a consensus meeting, a new approach was discussed and unanimously approved by all participants, introducing the use of pHF-W in the therapeutic management of CMPA. This novel approach could be of worldwide benefit.

Keywords: partial hydrolysate; cow's milk protein allergy; hydrolysate; infant feeding; Middle-East; step-down; infant allergy

---

1. Introduction

Cow's milk protein allergy (CMPA) is the most common form of food allergy in early childhood, and its prevalence has been on a steady rise over the years [1]. Intact cow milk protein (CMP) is usually the first food exposure given to an infant, and an adverse reaction to CMP is often the first symptom of an atopic condition in children [2]. CMPA in infancy is closely associated with other atopic manifestations, including 3–6 times higher risk of atopic eczema, allergic rhinitis, and asthma at 10 years of age [3].

2. Allergy Risk in Early Life

A positive family history, including history of allergic disorders in parents and/or siblings, is considered to be a strong determinant of allergy risk in an infant. The risk is shown to be even higher in the case of a history of atopic eczema or asthma in the family [4]. In addition, environmental factors in the pre-, peri-, and postnatal periods also seem to influence the risk of allergies in early life (Table 1) [4]. However, a negative family history at birth does not rule out the future risk of allergy; the child is demonstrated to have similar levels of allergy risk if an immediate family member becomes allergic after the birth of the child [5].

Table 1. Risk factors for allergy [4,6].

• Family history
• Environmental factors
• Formula feeding (with intact protein)
• Shorter duration of breastfeeding
• Older maternal age
• Higher parity
• Prematurity
• Caesarean delivery

3. Clinical Presentation and Diagnosis of CMPA

The first step in diagnosing CMPA involves a thorough medical history and physical examination, which can be completed with diagnostic tests to be interpreted in the context of a medical history. CMPA can induce a wide range of symptoms that could be "immediate" (from minutes up to 2 hours) and "delayed" (up to 48 hours or even 1 week) after the exposure, or a combination of both [7]. Immediate reactions are more likely to be IgE-mediated, whereas delayed reactions may also involve non-IgE-mediated immune mechanisms. Symptoms and signs of CMPA commonly involve the skin, digestive, and respiratory systems [7]. There may be an overlap between IgE-positive and IgE-negative symptomatology, especially in cases of symptoms involving the gastrointestinal system, such as allergic proctitis or proctocolitis [7]. However, certain symptoms such as angioedema and atopic eczema are relatively specific to positive CMP-specific IgE [7]. Some clinically useful diagnostic tests for CMPA are as follows:

3.1. IgE-Mediated

If IgE-mediated allergy is suspected based on a focused clinical history, a skin prick test or blood tests for specific IgE antibodies to the suspected foods and likely co-allergens are indicated for diagnosis [6–8]. The two tests show variable sensitivity and specificity [6]. It is important to acknowledge that a positive skin prick test or a positive serum specific IgE blood test shows sensitization (i.e., presence of IgE antibodies) to a food allergen, but, on its own, does not confirm an allergy [8]. Oral challenge with cow milk protein is still considered the best confirmatory method [9].

3.2. Non-IgE-Mediated

If non-IgE-mediated allergy is suspected based on the clinical history, a trial elimination of cow milk protein (normally for between 2 and 6 weeks) and reintroduction after the trial period is indicated for diagnosis [6]. The cow's milk-related symptom score (CoMiSS) is a simple, fast, and easy-to-use awareness tool for cow's milk-related symptoms including general, dermatological, gastrointestinal, and respiratory symptoms. However, it does not diagnose CMPA and does not replace the food challenge [10].

4. Allergy Prevention and CMPA Treatment in the Middle-East

4.1. Allergy Prevention

Exclusive breastfeeding up to 6 months of age is the preferred feeding for all infants. The World Health Organization recommends exclusive breastfeeding up to 6 months of age, with continued breastfeeding along with appropriate complementary foods up to two years of age or beyond [11]. However, the breastfeeding recommendations are poorly adhered to in the Middle-East region [12–15]. Cow-milk-based formulas are offered to infants when breastfeeding is not possible or not sufficient, placing the vulnerable children at risk of developing CMPA and at increased risk of atopic eczema. The region is also known for consuming other types of milk that have proven cross-reactivity to cow milk, including that from goat, sheep, and buffalo [16–18]. Camel's milk has shown low cross-reactivity with cow milk and may be a safer alternative than other types of milk such as goat milk [19]. However, although rare, cutaneous and systemic allergic reactions to camel's milk have been reported in the literature [20,21].

4.2. CMPA Treatment

The recommended management of CMPA involves strict avoidance of intact CMP, by replacing it with extensively hydrolyzed formula (eHF) or amino acid formula (AAF) in case of severe symptoms such as anaphylaxis. The diagnosis is confirmed with a positive challenge test. Later, CMP is reintroduced when tolerated after a successful challenge [7,22]. However, there is a unique approach adopted in certain institutes within the Middle-East, which involves using partially hydrolyzed whey formula (pHF-W) as a bridge between eHF or AAF and the intact CMP. There is some limited data on the benefits of oral immunotherapy in CMPA using pHF vs eHF in improving the tolerance to intact CMP [23].

The Middle-East Step-Down Consensus meeting was organized to evaluate the potential of this unique approach and provide practical recommendations to clinicians on the prevention of allergy and the management of CMPA.

5. Methods

For the development of a regional consensus, 10 leading experts from the Kingdom of Saudi Arabia, UAE, Lebanon, Jordan, and Kuwait convened in a meeting. A structured quantitative method was employed to facilitate the discussion and reach a consensus [24]. Statements were prepared before the consensus meeting, based on local clinical practice and discussions with experts from the region. Before the voting, each of the statements was extensively discussed within the group and amended. All group members voted anonymously, and a nine-point scale was used to quantify the consensus (1 for strongly disagree to 9 for fully agree). A vote of 6 and above meant "agreement", and a vote of 9 was considered an expression of stronger agreement than 6. Consensus was considered to be achieved if over 75% of the votes were of the scale of "6, 7, 8, or 9".

6. Consensus Recommendations

6.1. Prevention

Even though breast milk contains intact human proteins, they are most likely partially pre-digested by proteases within the human mammary gland; CMP is present as peptides. Hence the breastfed infant receives partially pre-digested proteins [25]. When breastfeeding is not possible or sufficient, certain pHF-W have shown benefits in prevention of allergy, especially atopic eczema in at-risk infants [26–29]. Animal models have shown that pHF-W is also able to induce oral tolerance, whereas extensively hydrolyzed proteins are less likely to do so [30,31]. In addition, pHF-W also offers better gastrointestinal tolerance and better digestibility compared to cow's milk and whey- or casein-predominant standard formula with intact proteins [32,33]. However, not all pHFs have been able to demonstrate the same clinical benefits in allergy prevention [34]. Healthcare professionals should critically evaluate clinical evidence for hydrolyzed protein used in each formulation before recommending it. Goat, sheep, and buffalo milk have no indication in the prevention of atopic disease. Moreover, not all healthcare facilities have the resources to obtain an allergy risk assessment immediately at birth, and the infant should be considered to be at risk of allergy until the risk assessment has been done. The participants agreed that a clinically proven pHF-W formula can play an important role in allergy prevention (Table 2, Figure 1).

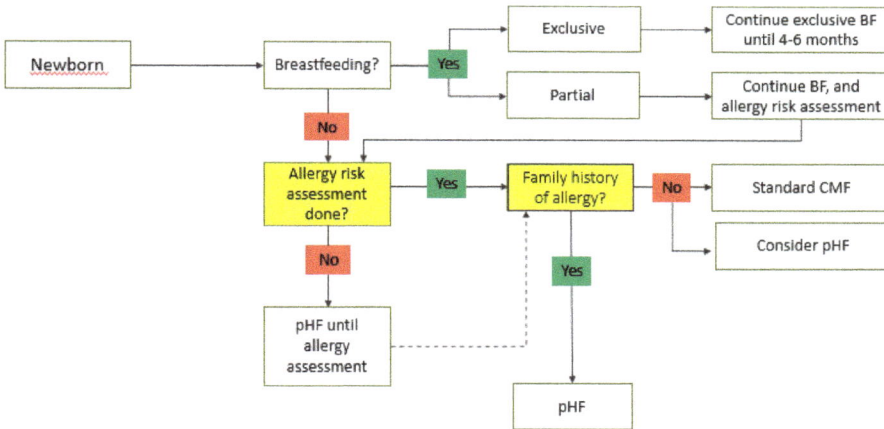

Figure 1. Middle-East Consensus algorithm for the prevention of allergy. BF: breastfeeding; CMF: cow milk protein; pHF: partially hydrolyzed formula.

Table 2. Middle-East Consensus statements on the prevention of allergy.

Sr. No.	Statement	Agreement
1	Exclusive breastfeeding up to 6 months is the best feeding for every infant to achieve optimal growth, development, and health (WHO statement).	100% (rating 9)
2	When breastfeeding is not possible or when breast milk is not available, partially hydrolyzed whey formula (pHF-W) with documented safety and efficacy should be recommended for infants at risk of allergy.	100% (rating 9)
3	Not all pHFs are the same, as different formulations have different peptide compositions and production methods and have demonstrated different outcomes.	100% (rating 9)
4	When breastfeeding is not possible or when breast milk is not available, pHF-W with documented safety and efficacy could be considered for all infants.	100% (rating 9)

6.2. Treatment

In non-breastfed infants, cow-milk-based formula and supplementary foods containing CMP or other unmodified animal milk proteins such as goat milk and sheep milk should be strictly avoided [7]. An elimination diet in formula-fed infants usually involves an eHF with proven efficacy in infants with CMPA, which should be tolerated by at least 90% of children with a proven CMPA [7,35].

Soy protein-based formula may be an option in infants older than 6 months if an alternative to eHF is needed, provided that the tolerance to soy protein has been established. Soy contains isoflavones and phytate, which may affect nutrient absorption that makes soy not suitable under the age of 6 months [7]. The American Academy of Pediatrics acknowledges that 10 to 14% of infants with CMPA will also become allergic to soy [36]. Rice drinks are not recommended because of the high arsenic content and since they are nutritionally not adapted to the need of infants [37,38]. However, hydrolyzed rice-based infant formulae, which are nutritionally adapted and have an arsenic content similar to that of cow milk-based infant formula, are on the market in some countries [39].

When a child is put on eHF, and if there is no improvement within 2 to 4 weeks, an allergic reaction to the remaining peptides in the eHF can be considered, particularly in infants with sensitization against multiple foods. In these cases, an AAF should be tried before CMPA is ruled out as cause of the symptoms [7]. In infants with extremely severe or life-threatening symptoms, an AAF is considered as the first choice [7].

The Step-Down Approach for CMPA Treatment

A step-down approach can be considered while managing children with CMPA, using pHF-W as a bridge between eHF or AAF and the intact CMP (Table 3, Figures 2–4). However, strict protocols need to be adhered to, including a carefully conducted pHF-W challenge, before initiating this approach (Table 4).

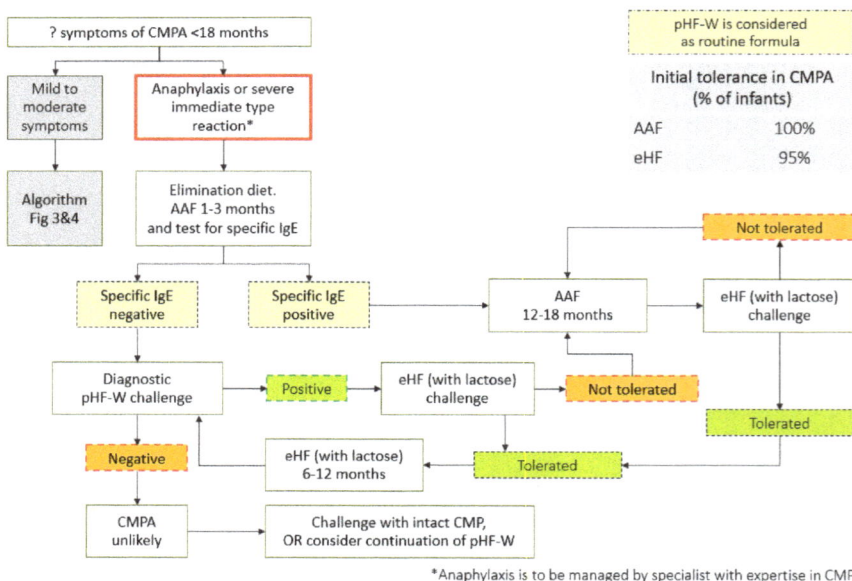

Figure 2. CMPA step-down treatment algorithm (anaphylaxis). AAF: amino acid based formula; eHF: extensively hydrolyzed formula; pHF-W: partially hydrolyzed formula-Whey; CMPA: cow's milk protein allergy; CMP: cow milk protein. *Anaphylaxis is to be managed by specialist with expertise in CMPA.

Table 3. Consensus statements on cow's milk protein allergy (CMPA) treatment: The Middle-East step-down approach.

Sr. No.	Statement	Agreement
1	Management of cow milk protein allergy involves avoidance of cow milk protein, through extensively hydrolyzed formula (eHF) in most of the infants, or if it is not tolerated, amino acid formula (AAF).	100% (rating 9)
2	In case of anaphylaxis, start with AA-based formula	100% (rating 9)
3	pHF-W can be used in the transition from eHF or AAF to intact CMP, if the initial pHF-W challenge is tolerated by the child.	100% (rating 9)
4	pHF formulas should not be interchanged, as the formulas differ in their clinical outcomes.	100% (rating 9)

Table 4. Oral challenge protocol [7].

- Place a drop of milk on the inside of lower lip and observe for reaction; if no reaction, the dose can be increased every 30 minutes until 100 mL
- If severe reactions expected: Stepwise dosing of 0.1, 0.3, 1.0, 3.0, 10.0, 30.0, and 100 mL given at 30-minute intervals
- If delayed reactions expected: Stepwise dosing of 1, 3.0, 10.0, 30.0, and 100 mL given at 30-minute intervals
- Patients should be observed for at least 2 hours following the maximum dose
- If no reaction, then the milk should be continued at home every day with at least 200 mL/day for at least 2 weeks
- Parents should be contacted by telephone to document any potential late reactions

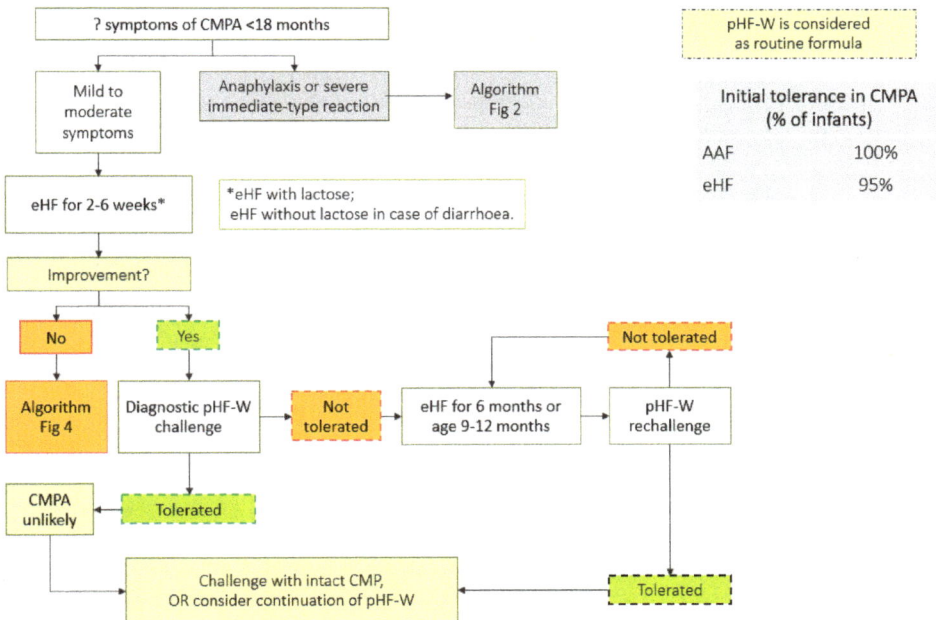

Figure 3. CMPA step-down treatment algorithm (mild-to-moderate symptoms). AAF: amino acid based formula; eHF: extensively hydrolyzed formula; pHF-W: partially hydrolyzed formula-Whey; CMPA: cow's milk protein allergy; CMP: cow milk protein. * use eHF without lactose in case of diarrhoea.

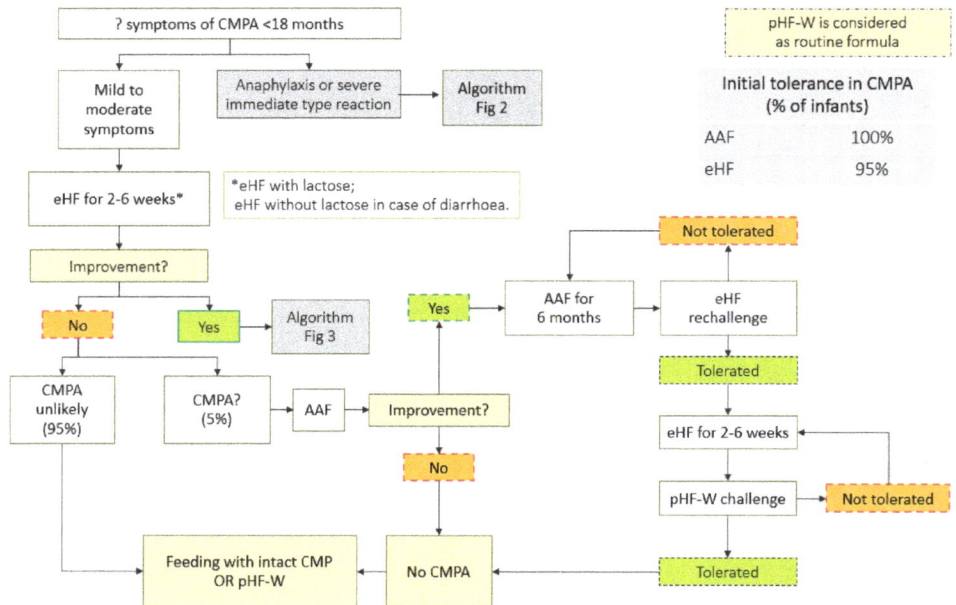

Figure 4. CMPA step-down treatment algorithm (mild-to-moderate symptoms). AAF: amino acid based formula; eHF: extensively hydrolyzed formula; pHF-W: partially hydrolyzed formula-Whey; CMPA: cow's milk protein allergy; CMP: cow milk protein. * use eHF without lactose in case of diarrhoea.

Open oral challenge is usually the first step especially in low-risk groups. A standardized oral challenge test is performed under medical supervision. A double-blind, placebo-controlled food challenge is the reference standard and the most specific test; however, the test is time-consuming and expensive. Therefore, an open oral challenge is usually the first step, particularly if the history indicates a low likelihood of a reaction. The oral challenge should be performed with an infant formula based on cow milk in the first year of life [7].

7. Discussion

This step-down protocol for CMPA management includes the use of pHF-W, considered as an alternative standard infant formula. Although evidence from the literature is not available, it is well known that many healthcare professionals advice the use of pHF-W when CMPA is suspected. Many HCPs recommend "hydrolysates" in the management of infants suspected to suffer CMPA, but mix partial and extensive hydrolysates. A European survey discovered major deficits in the management of CMPA, including limited knowledge of diagnostic tests, eliminations, and selection of formula for the management of CMPA in non-breastfed infants [40].

All existing guidelines recommend eHF in the management of CMPA, but none mention pHF-W in the algorithm. In this algorithm, we propose pHF as an alternative to standard infant formula with intact protein, as we recommend pHF-W as the formula to be used in the challenge test (after strong improvement or disappearance of the symptoms with 2–4 weeks of eHF). We hypothesize that this algorithm will result in a decrease in the misuse of pHF in the treatment of CMPA. However, non-IgE mediated allergy might be difficult to distinguish from functional gastro-intestinal disorders. Nutritional treatment with pHF-W is a recommended approach in the management of functional gastro-intestinal disorders [41]. Complete elimination of lactose from the infant's diet is disadvantageous for the development of a healthy gut microbiota and does result in a decreased calcium absorption [42–44].

An allergy-focused clinical history is a must before carrying out allergy testing. The skin and blood tests as well as oral challenge should be undertaken by healthcare professionals with appropriate competencies to select, perform, and interpret them. The oral challenge should be undertaken under medical supervision, and in an inpatient setting in case a severe reaction could occur [6]. The role of soy-based formula in treatment of CMPA is still debated, but is not recommended below 6 months of age [7]. Goat, sheep, and buffalo milk are not suitable alternatives for CMPA prevention or treatment because they are not nutritionally adapted (if not marketed as infant formula) and cross-react with cow milk protein [45].

8. Conclusions

The Middle-East Step-Down approach of managing CMPA using pHF-W as a bridge between eHF or AAF and intact CMP has great potential to improve the lives of affected children and families. This uncovers another clinically relevant facet of a specific pHF-W that was shown to prevent atopic eczema in children at risk of allergy. However, clinicians should keep in mind that not all pHFs are the same and choose the pHF-W formula according to published evidence.

Author Contributions: This consensus was led by Y.V., who contributed significantly to the review and finalization of the algorithms and manuscript. All other authors contributed equally to the development and finalization of the manuscript.

Funding: The consensus statements, algorithms, and supporting evidence presented in this paper were discussed and formulated at a consensus meeting in Dubai, UAE. All authors received support to travel and attend the meeting from the Nestle Nutrition Institute. The views expressed in this paper are purely those of the authors without any influence from Nestle Nutrition Institute.

Acknowledgments: Sanjith Soman of McCann Health Dubai compiled the comments of the authors and supported the writing of this paper.

Conflicts of Interest: Yvan Vandenplas has participated as a clinical investigator, and/or advisory board member, and/or consultant, and/or speaker for Abbott Nutrition, Astel Medica, Danone, Nestle Health Science, Nestle Nutrition Institute, Nutricia, Mead Johnson Nutrition, United Pharmaceuticals, and Wyeth. None of the other authors have a conflict of interest.

References

1. Savage, J.; Johns, C.B. Food allergy: Epidemiology and natural history. *Immunol. Allergy Clin. North Am.* **2015**, *35*, 45–59. [CrossRef] [PubMed]
2. Oranje, A.P.; Wolkerstorfer, A.; de Waard-van der Spek, F.B. Natural course of cow's milk allergy in childhood atopic eczema/dermatitis syndrome. *Ann. Allergy Asthma Immunol.* **2002**, *89*, 52–55. [CrossRef]
3. Tikkanen, S.; Kokkonen, J.; Juntti, H.; Niinimäki, A. Status of children with cow's milk allergy in infancy by 10 years of age. *Acta Paediatr.* **2000**, *89*, 1174–1180. [CrossRef] [PubMed]
4. Sardecka, I.; Łoś-Rycharska, E.; Ludwig, H.; Gawryjołek, J.; Krogulska, A. Early risk factors for cow's milk allergy in children in the first year of life. *Allergy Asthma Proc.* **2018**, *39*, e44–e54. [CrossRef] [PubMed]
5. Fuertes, E.; Standl, M.; von Berg, A.; Lehmann, I.; Hoffmann, B.; Bauer, C.-P.; Koletzko, S.; Berdel, D.; Heinrich, J. Parental allergic disease before and after child birth poses similar risk for childhood allergies. *Allergy* **2015**, *70*, 873–876. [CrossRef] [PubMed]
6. National Institute for Health and Clinical Excellence. *Food Allergy in Children and Young People*; National Institute for Health and Clinical Excellence: London, UK, 2011.
7. Koletzko, S.; Niggemann, B.; Arato, A.; Dias, J.A.; Heuschkel, R.; Husby, S.; Mearin, M.L.; Papadopoulou, A.; Ruemmele, F.M.; Staiano, A.; et al. Diagnostic approach and management of cow's-milk protein allergy in infants and children: ESPGHAN GI Committee practical guidelines. *J. Pediatr. Gastroenterol. Nutr.* **2012**, *55*, 221–229. [CrossRef] [PubMed]
8. Venter, C.; Brown, T.; Meyer, R.; Walsh, J.; Shah, N.; Nowak-Węgrzyn, A.; Chen, T.-X.; Fleischer, D.M.; Heine, R.G.; Levin, M.; et al. Better recognition, diagnosis and management of non-IgE-mediated cow's milk allergy in infancy: iMAP—An international interpretation of the MAP (Milk Allergy in Primary Care) guideline. *Clin. Transl. Allergy* **2017**, *7*, 26. [CrossRef] [PubMed]

9. Costa, A.J.F.; Sarinho, E.S.C.; Motta, M.E.F.A.; Gomes, P.N.; de Oliveira de Melo, S.M.; da Silva, G.A.P. Allergy to cow's milk proteins: What contribution does hypersensitivity in skin tests have to this diagnosis? *Pediatr. Allergy Immunol.* **2011**, *22*, e133–e138. [CrossRef]

10. Vandenplas, Y.; Dupont, C.; Eigenmann, P.; Host, A.; Kuitunen, M.; Ribes-Koninckx, C.; Shah, N.; Shamir, R.; Staiano, A.; Szajewska, H.; et al. A workshop report on the development of the Cow's Milk-related Symptom Score awareness tool for young children. *Acta Paediatr.* **2015**, *104*, 334–339. [CrossRef]

11. Breastfeeding. Available online: https://www.who.int/topics/breastfeeding/en/ (accessed on 13 March 2019).

12. Alfaleh, K.; Alluwaimi, E.; Aljefri, S.; Alosaimi, A.; Behaisi, M. Infant formula in saudi arabia: A cross sectional survey. *Kuwait Med. J.* **2014**, *46*, 328–332.

13. Al-Nuaimi, N.; Katende, G.; Arulappan, J. Breastfeeding Trends and Determinants: Implications and recommendations for Gulf Cooperation Council countries. *Sultan Qaboos Univ. Med. J.* **2017**, *17*, e155–e161. [CrossRef] [PubMed]

14. Alzaheb, R.A. A review of the factors associated with the timely initiation of breastfeeding and exclusive breastfeeding in the Middle East. *Clin. Med. Insights Pediatr.* **2017**, *11*, 1–15. [CrossRef] [PubMed]

15. Gardner, H.; Green, K.; Gardner, A. Infant feeding practices of emirati women in the rapidly developing city of Abu Dhabi, United Arab Emirates. *Int. J. Environ. Res. Public Health* **2015**, *12*, 10923–10940. [CrossRef]

16. Sheehan, W.J.; Gardynski, A.; Phipatanakul, W. Skin testing with water buffalo's milk in children with cow's milk allergy. *Pediatr. Asthma. Allergy Immunol.* **2009**, *22*, 121–125. [CrossRef]

17. Restani, P.; Beretta, B.; Fiocchi, A.; Ballabio, C.; Galli, C.L. Cross-reactivity between mammalian proteins. *Ann. Allergy Asthma Immunol.* **2002**, *89*, 11–15. [CrossRef]

18. Bellioni-Businco, B.; Paganelli, R.; Lucenti, P.; Giampietro, P.G.; Perborn, H.; Businco, L. Allergenicity of goat's milk in children with cow's milk allergy. *J. Allergy Clin. Immunol.* **1999**, *103*, 1191–1194. [CrossRef]

19. Ehlayel, M.; Bener, A.; Abu Hazeima, K.; Al-Mesaifri, F. Camel milk is a safer choice than goat milk for feeding children with cow milk allergy. *ISRN Allergy* **2011**, *2011*, 1–5. [CrossRef] [PubMed]

20. Ehlayel, M.; Bener, A. Camel's milk allergy. *Allergy Asthma Proc.* **2018**, *39*, 384–388. [CrossRef]

21. Al-Hammadi, S.; El-Hassan, T.; Al-Reyami, L. Anaphylaxis to camel milk in an atopic child. *Allergy* **2010**, *65*, 1623–1625. [CrossRef]

22. Vandenplas, Y.; Abuabat, A.; Al-Hammadi, S.; Aly, G.S.; Miqdady, M.S.; Shaaban, S.Y.; Torbey, P.-H. Middle east consensus statement on the prevention, diagnosis, and management of cow's milk protein allergy. *Pediatr. Gastroenterol. Hepatol. Nutr.* **2014**, *17*, 61–73. [CrossRef]

23. Inuo, C.; Tanaka, K.; Suzuki, S.; Nakajima, Y.; Yamawaki, K.; Tsuge, I.; Urisu, A.; Kondo, Y. Oral immunotherapy using partially hydrolyzed formula for cow's milk protein allergy: A randomized, controlled trial. *Int. Arch. Allergy Immunol.* **2018**, *177*, 259–268. [CrossRef] [PubMed]

24. CDC Evaluation Research Team. Gaining Consensus among Stakeholders Through the Nominal Group Technique. Available online: https://www.cdc.gov/healthyyouth/evaluation/pdf/brief7.pdf (accessed on 13 March 2019).

25. Nielsen, S.D.; Beverly, R.L.; Dallas, D.C. Milk proteins are predigested within the human mammary gland. *J. Mammary Gland Biol. Neoplasia* **2017**, *22*, 251–261. [CrossRef] [PubMed]

26. von Berg, A.; Filipiak-Pittroff, B.; Krämer, U.; Hoffmann, B.; Link, E.; Beckmann, C.; Hoffmann, U.; Reinhardt, D.; Grübl, A.; Heinrich, J.; et al. Allergies in high-risk schoolchildren after early intervention with cow's milk protein hydrolysates: 10-year results from the German Infant Nutritional Intervention (GINI) study. *J. Allergy Clin. Immunol.* **2013**, *131*, 1565–1573. [CrossRef] [PubMed]

27. von Berg, A.; Koletzko, S.; Grübl, A.; Filipiak-Pittroff, B.; Wichmann, H.-E.; Bauer, C.P.; Reinhardt, D.; Berdel, D.; German Infant Nutritional Intervention Study Group. The effect of hydrolyzed cow's milk formula for allergy prevention in the first year of life: the German Infant Nutritional Intervention Study, a randomized double-blind trial. *J. Allergy Clin. Immunol.* **2003**, *111*, 533–540. [CrossRef] [PubMed]

28. von Berg, A.; Filipiak-Pittroff, B.; Schulz, H.; Hoffmann, U.; Link, E.; Sußmann, M.; Schnappinger, M.; Brüske, I.; Standl, M.; Krämer, U.; et al. Allergic manifestation 15 years after early intervention with hydrolyzed formulas–the GINI Study. *Allergy* **2016**, *71*, 210–219. [CrossRef] [PubMed]

29. Szajewska, H.; Horvath, A. A partially hydrolyzed 100% whey formula and the risk of eczema and any allergy: an updated meta-analysis. *World Allergy Organ. J.* **2017**, *10*, 27. [CrossRef] [PubMed]

30. Pecquet, S.; Bovetto, L.; Maynard, F.; Fritsché, R. Peptides obtained by tryptic hydrolysis of bovine β-lactoglobulin induce specific oral tolerance in mice. *J. Allergy Clin. Immunol.* **2000**, *105*, 514–521. [CrossRef] [PubMed]

31. Fritsché, R.; Pahud, J.J.; Pecquet, S.; Pfeifer, A. Induction of systemic immunologic tolerance to beta-lactoglobulin by oral administration of a whey protein hydrolysate. *J. Allergy Clin. Immunol.* **1997**, *100*, 266–273. [CrossRef]

32. Billeaud, C.; Guillet, J.; Sandler, B. Gastric emptying in infants with or without gastro-oesophageal reflux according to the type of milk. *Eur. J. Clin. Nutr.* **1990**, *44*, 577–583.

33. Exl, B.M.; Deland, U.; Secretin, M.C.; Preysch, U.; Wall, M.; Shmerling, D.H. Improved general health status in an unselected infant population following an allergen-reduced dietary intervention programme: the ZUFF-STUDY-PROGRAMME. Part II: Infant growth and health status to age 6 months. ZUg-FrauenFeld. *Eur. J. Nutr.* **2000**, *39*, 145–156. [CrossRef]

34. Boyle, R.J.; Ierodiakonou, D.; Khan, T.; Chivinge, J.; Robinson, Z.; Geoghegan, N.; Jarrold, K.; Afxentiou, T.; Reeves, T.; Cunha, S.; et al. Hydrolysed formula and risk of allergic or autoimmune disease: Systematic review and meta-analysis. *BMJ* **2016**, *352*, i974. [CrossRef] [PubMed]

35. Giampietro, P.G.; Kjellman, N.I.; Oldaeus, G.; Wouters-Wesseling, W.; Businco, L. Hypoallergenicity of an extensively hydrolyzed whey formula. *Pediatr. Allergy Immunol.* **2001**, *12*, 83–86. [CrossRef] [PubMed]

36. Bhatia, J.; Greer, F.; American Academy of Pediatrics Committee on Nutrition. Use of soy protein-based formulas in infant feeding. *Pediatrics* **2008**, *121*, 1062–1068. [CrossRef] [PubMed]

37. Dennis, S.; Fitzpatrick, S.; Egan, K.; Flannery, B.; Kanwal, R.; Smegal, D.; Spungen, J.; Tao, S. *Arsenic in Rice and Rice Products Risk Assessment Report*; Center for Food Safety and Applied Nutrition, Food and Drug Administration, U.S. Department of Health and Human Services: Washington, DC, USA, 2016.

38. Hojsak, I.; Braegger, C.; Bronsky, J.; Campoy, C.; Colomb, V.; Decsi, T.; Domellöf, M.; Fewtrell, M.; Mis, N.F.; Mihatsch, W.; et al. Arsenic in rice: A cause for concern. *J. Pediatr. Gastroenterol. Nutr.* **2015**, *60*, 142–145. [CrossRef] [PubMed]

39. Meyer, R.; Carey, M.P.; Turner, P.J.; Meharg, A.A. Low inorganic arsenic in hydrolyzed rice formula used for cow's milk protein allergy. *Pediatr. Allergy Immunol.* **2018**, *29*, 561–563. [CrossRef]

40. Werkstetter, K.; Chmielewska, A.; Dolinšek, J.; Burk, F.E.; Korponay-Szabó, I.; Kurppa, K.; Mišak, Z.; Papadopoulou, A.; Popp, A.; Ribes-Konickx, C.; et al. Diagnosis and management of cow's milk protein allergy—How big is the gap between ideal and reality? A quality-of-care survey in Europe. *J. Pediatr. Gastroenterol. Nutr.* **2018**, *66*, 399–400.

41. Salvatore, S.; Abkari, A.; Cai, W.; Catto-Smith, A.; Cruchet, S.; Gottrand, F.; Hegar, B.; Lifschitz, C.; Ludwig, T.; Shah, N.; et al. Review shows that parental reassurance and nutritional advice help to optimise the management of functional gastrointestinal disorders in infants. *Acta Paediatr.* **2018**, *107*, 1512–1520. [CrossRef]

42. Francavilla, R.; Calasso, M.; Calace, L.; Siragusa, S.; Ndagijimana, M.; Vernocchi, P.; Brunetti, L.; Mancino, G.; Tedeschi, G.; Guerzoni, E.; et al. Effect of lactose on gut microbiota and metabolome of infants with cow's milk allergy. *Pediatr. Allergy Immunol.* **2012**, *23*, 420–427. [CrossRef]

43. Vandenplas, Y. Lactose intolerance. *Asia Pac. J. Clin. Nutr.* **2015**, *24* (Suppl. 1), S9–S13.

44. Heine, R.G.; AlRefaee, F.; Bachina, P.; De Leon, J.C.; Geng, L.; Gong, S.; Madrazo, J.A.; Ngamphaiboon, J.; Ong, C.; Rogacion, J.M. Lactose intolerance and gastrointestinal cow's milk allergy in infants and children—Common misconceptions revisited. *World Allergy Organ. J.* **2017**, *10*, 41. [CrossRef]

45. Järvinen, K.M.; Chatchatee, P. Mammalian milk allergy: Clinical suspicion, cross-reactivities and diagnosis. *Curr. Opin. Allergy Clin. Immunol.* **2009**, *9*, 251–258. [CrossRef] [PubMed]

nutrients

MDPI

Review

Raw Cow's Milk and Its Protective Effect on Allergies and Asthma

Barbara Sozańska

1st Department of Pediatrics, Allergology and Cardiology, Wroclaw Medical University, Wroclaw 50-367, Poland; bsoz@go2.pl; Tel.: +48605370686

Received: 3 February 2019; Accepted: 18 February 2019; Published: 22 February 2019

Abstract: Living on a farm and having contact with rural exposures have been proposed as one of the most promising ways to be protected against allergy and asthma development. There is a significant body of epidemiological evidence that consumption of raw milk in childhood and adulthood in farm but also nonfarm populations can be one of the most effective protective factors. The observation is even more intriguing when considering the fact that milk is one of the most common food allergens in childhood. The exact mechanisms underlying this association are still not well understood, but the role of raw milk ingredients such as proteins, fat and fatty acids, and bacterial components has been recently studied and its influence on the immune function has been documented. In this review, we present the current understanding of the protective effect of raw milk on allergies and asthma.

Keywords: raw milk; allergy; asthma; protection

1. Introduction

Cow's milk and its preserves are basic ingredients in our diet. For centuries it was introduced as the first infant food as an alternative to breast milk and treated as necessary for growth and development. Milk is now processed on an industrial scale to avoid the risk of pathogenic bacteria infection from unpasteurized milk. The consumption of raw milk is much less common than in the past, but a considerable number of consumers, especially living on farms, still drink milk directly from cows [1].

In the last few decades, a significant increase in the prevalence of allergic diseases and asthma has been observed. For over a decade, an accumulating body of evidence has indicated that living on a farm can reduce the risk of allergen sensitization and allergic diseases in children. Among the first observations were those in alpine villages where children of full-time farming parents had lower rates of atopy and hay fever than their peers living in the same rural community [2]. Further studies from the region, and in several other parts of the world, confirmed a protective effect of an early farming environment in children [3–5]. Findings have not been entirely consistent [6,7], suggesting that protective effects may depend on a particular type of farming environment. Although exposures in the first years of life are believed to exert the most important influence, and farming protective effects are best documented in children, it has been proposed that they may persist into adult life, and a lower prevalence of atopic sensitization among adult farmers has been observed [8–10]. Considerable effort has been put into deciphering which particular aspect(s) of a farm environment exert the protective effect. Success has so far been very limited, probably because it has proved impossible to separate individual exposures. Most proposed explanations have been based on variations in the "hygiene hypothesis" and a possible immunomodulatory effect of a farm environment. Specific types of exposure have been proposed, such as contact with farm animals, exposure to barns and animal sheds, and the diversity of microorganisms in farm environments [5,11,12]. One of the best

Nutrients **2019**, *11*, 469

documented protective effects on allergies and asthma is the consumption of unpasteurized milk products both in farming and non-farming environments [13]. The observation is even more intriguing, considering the fact that milk is one of the most common food allergens in childhood.

2. The Protective Effect of Raw Milk on Allergies in Farming and Non-Farming Populations—Epidemiological Studies

There are many cross-sectional surveys that confirm the protective effect of raw milk on allergies and asthma. First epidemiological evidence came from alpine regions. The ALEX study reported that children at school age were less likely to suffer from asthma or atopic sensitization when they consumed unpasteurized milk from the farm in their first year of life. This association was independent of living on a farm, although additional exposure to farm stables strengthened the effect [3]. These observations were confirmed in the next alpine study by Waser et al., where a significant inverse association with doctor diagnosed asthma, diagnosed rhinoconjunctivitis, and atopy was observed with respect to the consumption of farm milk in children from farming and non-farming families. The effect was independent of concomitant exposures to microbial compounds present in animal sheds and farm homes [14]. The protective effect of current farm milk consumption on atopy and eczema, but not asthma, in English children living in a rural environment was also observed [15]. In Crete, the consumption of unpasteurized milk has proven to protect against atopy, independently of having contact with farm animals and being a rural child [16]. In Poland, the consumption of raw milk in the first year of life reduced atopy risk even in adulthood, even more so in children, independent of where they resided. Current milk consumption also provided protection, albeit weaker. Protective effects on asthma were observed both in village and town inhabitants and at all ages, but for doctor-diagnosed hay fever the effect was only present for nonfarmers and those living in a town [17]. In a study of older adults in a US farming cohort, the consumption of raw milk was not associated with atopy and asthma, but raw milk as the primary milk consumed in childhood reduced odds of atopy [10].

In GABRIELA, the first study designed to find the biological components of cow's milk that might explain the protective effect of farm milk on asthma and atopy in children, the reported consumption of raw milk in early life and currently was inversely related to atopy, hay fever, and asthma in school-aged children [18]. Heated farm milk was not associated with asthma outcomes. Total fat and protein content, total bacterial counts, and lactose levels had no association with allergies, atopy, or asthma. Inverse associations with asthma were found for α-lactalbumin, β-lactoglobulin, and bovine serum albumin (BSA) whey protein, but not with atopy. Although higher counts of microbes were detected in raw milk compared with pasteurized milk or heated farm milk, surprisingly there was no association between total bacterial counts and health outcomes. The cross-sectional design of the survey could not determine the long-term bacterial exposure due to raw milk consumption or how it could influence the gut microflora and immune system.

There are few publications dealing with the protective effect of raw milk on allergies and asthma. Most concern a birth cohort study, Protection against Allergy Study in Rural Environments (PASTURE), which included over 900 children from rural areas in five European countries [19]. Regular consumption of unpasteurized milk was inversely related to asthma onset at 6 years of age. This protective effect was stronger with recent exposure and in higher fat content milk. Higher ω-3 polyunsaturated fatty acids (PUFA) levels in milk was protective [20]. In the same cohort, maternal consumption of butter and unskimmed cow's milk during pregnancy affected the fetal immune system for the production of interferon (INF)-gamma [21]. Interestingly, early consumption of raw farm milk not only protected from allergies and asthma but also exerted strong protection against rhinitis, otitis, and other respiratory track infections. This effect was strongest in raw milk. Boiled milk exhibited an attenuated effect [22]. In a recent study in the PASTURE cohort, cheese consumption and its diversity at 18 months of age had a protective effect on atopic eczema and food allergy but not on atopy or asthma at age 6. The authors proposed two possible explanations: (1) a positive effect of the microbial diversity of cheeses as a fermented food and its influence on the rich diversity of gut microbiota or (2) a potential anti-inflammatory effect by the inhibition of proinflammatory cytokines and intestinal microbiota metabolite production after dairy product consumption [23].

3. Raw Milk Versus Commercial Milk—What Are the Differences?

The epidemiological evidence showed that the protective effect on allergies and asthma is exerted by raw milk, but not (or weaker) by processed or commercial milk. The commercial milk differs in many aspects from raw farm milk, as the processing of milk changes its composition.

Raw milk is subjected to two basic processes: UHT homogenization and sterilization, which affect the content and functionality of fats, bacteria, and proteins in milk. Homogenization is the breaking of fat pellets into very small ones to prevent the formation of cream on the milk surface. This process changes the physical structure of fats but also casein and whey proteins. Splitting large spheres of fat into many small ones increases the total surface on which casein proteins are more easily adsorbed. In an experimental study on sensitized mice, only homogenized milk induced an allergic reaction in the intestinal wall of the animals, indicating that such milk processing may predispose one to an allergic reaction [24]. The quality of fat in milk depends on the ingredients cows are fed—if it is only grass, without any industrial mixtures, the content of polyunsaturated omega-3 fatty acids is higher. A prospective study by KOALA showed that the higher concentration of these acids in the milk of breastfeeding mothers reduced the risk of eczema at 2 years of age and 1 year of allergy in children [25].

Heating is another milk processing procedure. It might be the pasteurization process (heating up to about 75° for about 30 s) or the industrial, UHT sterilization (heating up to 130–160° and then quickly cooling), which reduces the bacterial content and enzymes that allow such milk to be stored without a refrigerator for several months.

Epidemiological evidence confirms that heating raw milk influences its allergy protective properties [18]. Although the main aim of heating is to reduce bacterial content and slow microbial growth, it may also affect the heat-sensitive milk components. Whey proteins are the most sensitive to this process. They lose their biological functionality by denaturation, aggregation, and glycation after heating [26]. Pasteurization at 72 °C denatures only part of the bovine IgG, but sterilization and homogenization denatures all immunoglobulins. Short heating up to 72 °C can change the lactoferrin structure and decrease its level in milk. Bovine transforming growth factor (TGF)-β1 concentrations were decreased in commercial pasteurized milk compared to raw milk [27]. The level of TGF-β2 did not differ between pasteurized and raw milk, but with increasing temperature a reduction of TGF-β2 concentrations in both was observed [28]. In the murine model of house dust mites (HDM) induced asthma, the effects of raw and heated raw cow's milk on asthma prevention were compared. Airway hyperresponsiveness was prevented by raw milk but not heated milk. Raw milk reduced the total number of inflammatory cells such as eosinophils, lymphocytes, neutrophils, and macrophages in bronchoalveolar lavage fluid (BALF). Th2 and Th17 cells and their cytokine production of Interleukin (IL)-4 and IL-13 were reduced in lung cell suspensions both by raw and heated milk. Raw milk reduced IL-5 and IL-13 production after ex vivo restimulation of lung T cells with HDM [29]. These associations may confirm a causal relationship between raw cow's milk consumption and the prevention of allergic asthma.

4. Raw Milk Proteins

The main fractions of milk proteins are casein (82% of total protein content), serum, and whey proteins (18% of total protein content). Casein is thermostable and is not destroyed during industrial processing. Whey protein is a group of a dozen proteins, including the major β-lactoglobulin milk allergen, α-lactalbumin, serum albumin, as well as immunoglobulins, lactoferrin, enzymes, and cytokines such as TGF-β and IL-10. Heating can change the physicochemical characteristics of these proteins and affect their biological impact. They are bioactive compounds that can influence the immunological system and prevent allergic reaction. The role of bioactive whey proteins from raw milk in allergic diseases is presented in detail in a recently published review [30]. The first study, which documented the effect of whey protein's protective properties, was a Gabriela survey conducted on rural farm children [18]. The inverse associations with asthma were found for α-lactalbumin, β-lactoglobulin, BSA whey protein content, but not with atopy. Total protein content had no association with allergies, atopy, or asthma.

The caseins, β-lactoglobulin, and α-lactalbumin cannot be linked directly to immune functioning [31]. Other cow's milk protein components such as lactoferrin, immunoglobulins, lysozyme, and cytokines (TGF-β, IL-10) may have a direct immunity-related effect.

Lactoferrin is an immunostimulator and immunoregulator of antigen presenting cells in culture [32]. It has inhibited the cytokine production of Th1 but not Th2 cell lines [33]. It has an antimicrobial effect (via direct interaction with the bacteria cell wall or by the binding iron needed by some of bacteria for growth), by which it can modulate microbial composition [31,34]. Bacteria with low iron requirements such as *Bifidobacteria* and *Lactobacilli* are promoted in the gut [35]. Their presence in an infant's gut has been correlated with protection against allergies [36]. Moreover, lactoferrin reduced allergen-induced airway inflammation in a murine asthma model [37]. Its presence stimulated the production of TGF-β and IL-10 in the gut [38].

TGF-β is a multifunctional cytokine, of which higher levels (mainly TGF-β2 and TGF-β1 isoforms) were found in raw milk and in human breast milk of mothers from farm environments [27]. It plays a key role in the development and maturation of the mucosal immune system [39]. TGF-β1 increases the expression of intestinal tight junctions, which enhances barrier function of the gut [40]. This may potentially protect against food allergic sensitization and reduce allergy-related symptoms in infancy [39]. It has been shown in an animal model that this cytokine present in breast milk induced oral tolerance to allergens and protected from allergic asthma [41]. The lack of TGF-β in the milk formula promoted the production of the proinflammatory cytokine profile and increased numbers of eosinophils, activated mast cells, and dendritic cells in gut in another animal model. Supplementation of TGF-β induced oral tolerance to β-lactoglobulin from cow's milk and increased IL-10 production [42]. IL-10 is a regulatory cytokine present in bovine milk. It inhibited immunoglobulin E (IgE) induced mast cell activation, Th2 cell activation, and eosinophil function [43]. The inverse correlation between IL-10 and allergic diseases and asthma were observed [44]. TGF-β and IL-10 induced conversion of naive peripheral T cells into FoxP3 regulatory T cells [45].

As was proposed by Neerven et al. [31], the consumption of immunomodulatory cytokines (such as TGF-β and IL-10) in unprocessed bovine milk may create the environment, which promotes regulatory T cell production, needed for developing and maintaining oral tolerance in the gut with IgA and IgG4, but not IgE production. IgA present in intestinal secretion can prevent binding food allergens to IgE. Low levels of human milk IgA correlated with allergy development [46]. To date, the role of immunoglobulins present in bovine milk (predominantly IgG) in allergy prevention is not well understood. We can speculate about the role of IgG in breast milk in forming immune complexes with allergens, but it has never been studied [47]. However, bovine milk may contain IgG antibodies specific for human allergens [31]. Theoretically they might suppress allergic responses by blocking Ig-E mediated activation of mast cells and basophils [30]. Recently, it has been shown that bovine IgG and raw milk can induce an innate immune memory in human monocytes modulating the responsiveness of the innate immune system to pathogen-related stimuli [48].

5. Fat and Fatty Acids

Milk fat content depends on the type of feeding and on the age and breed of cows, and it is a precious source of saturated and mono- and polyunsaturated fatty acids. Fat separation for adjusting fat milk levels and homogenization for fat creaming prevention are the main fat-changing processes in commercial milk production. The protective effect of high fat containing products such as full cream milk and butter on asthma was presented in the PASTURE cohort [14]. A higher content of anti-inflammatory ω-3 fatty acids in raw milk was found to be protective. The fat content was associated with asthma severity: high fat milk exerted a stronger protective effect on milder forms of asthma. Authors suggested that ω-3 PUFA may exert its effect by shifting the metabolic balance of eicosanoid synthesis from proinflammatory to anti-inflammatory mediators [20]. Short-chain fatty acids (SCFA) are metabolites present in relatively high concentration in bovine milk (not present in human milk fat), but may also be produced by microbes in the gut following the fermentation of fibers [31]. They may exert an anti-inflammatory effect by the inhibition of histone deacetylation, which influences the expansion of regulatory T cells, and may increase the production of IL-10 [49,50]. In a recent study in the PASTURE cohort, yogurt introduction in the first year of life increased the level of butyrate SCFA in fecal samples at one year of age, and the children with a high level of butyrate or propionate were protected from asthma and food allergy later in life and against atopy at age 6 [51]. Oral administration of butyrate, propionate, and acetate in mice experimental models reduced airway hyperresponsiveness during metacholine challenge and the number of inflammatory cells such as eosinophils in bronchoalveolar lavages [51]. Similarly, in a study among adults, the fecal butyrate was increased after yogurt consumption [52]. In a recent prospective observational study in an urban population, the inverse association between yogurt consumption in the first year of life and atopic eczema and food sensitization at five years of age was shown [53]. In another prospective study in New Zealand, a similar effect was observed at 12 months of age [54]. Authors speculate that the protective effect might be connected with probiotic bacteria in yogurt.

6. Microbial Content in Raw Milk

The presence and composition of bacteria is the clearest difference between unpasteurized and processed milk. Data from British laboratories presented by Perkin documented a greater total number of bacteria (including *Escherichia coli* and coagulase-positive *staphylococci)* in unpasteurized milk. Non-pathogenic *Listeria* species were found in 37% of raw milk samples and only in 0.4% of pasteurized samples [55]. The prevalence of pathogens in milk depends on numerous factors, but raw milk can be contaminated with pathogens even coming from healthy animals, as dairy farms may be the reservoir of various foodborne pathogens [1].

Surprisingly, the levels of lipopolysaccharides (LPS) endotoxins measured in raw and commercial milk did not differ in the PASTURE study [56]. On the contrary, in another study, the level of endotoxins was much higher in the samples of whole raw milk than in the processed shop milk and cold storage or heating increased the endotoxin concentrations in farm but not in the processed milk [57]. In the GABRIELA study, viable bacterial cell counts were higher in raw milk than in shop milk, but these differences were not associated with asthma and atopy [18]. In a recent study by Brick et al., the presence of the CD14 molecule in raw milk, a receptor of bacterial endotoxin, was confirmed. [26]. Such bacterial endotoxins present in farm dust modified the mechanisms of primarily non-specific immune response to allergens. House dust mites induce the activation of airway epithelial cells mediated by toll-like receptor 4. Airway exposure to endotoxins inhibited activation of NF-kB by the increase in the synthesis of its attenuator, enzyme A20. These associations, observed in an experimental model in mice, have been confirmed in further experiments on human bronchial epithelial cultures and in a case–control study of asthmatics [58]. In other studies, cluster of differentiation 14 (CD14) receptor polymorphisms at exposure to bacterial endotoxins in the first year of life in children from rural households reduced the risk of developing bronchial asthma [59] and atopic eczema [60].

Bieli et al. investigated whether the CD14 receptor polymorphism for bacterial endotoxins modifies the appearance of a protective effect on the consumption of unpasteurized milk in asthma and hay fever in children. Polymorphisms of the promoter gene for CD14 (CD14/-1721) had an effect on the occurrence of a protective effect. The inverse relationship between drinking unpasteurized milk and the occurrence of asthma, allergic rhinoconjunctivitis, and wheezing was most strongly expressed in the homozygote AA (adenine-adenine), less expressed in heterozygous AG (adenine-guanine), and did not occur in children with the genotype GG (guanine-guanine). This effect was independent of the children's place of residence (rural farm or not). The authors speculate that the effect on the immune system through CD14 may proceed either due tothe microbiological components of unpasteurized milk or intestinal microflora altered under the influence of probiotics delivered with milk. Alternatively, as it is known that CD14 is also a phospholipid receptor, fatty components of milk, including omega-3 fatty acids, can also play a role [61].

In the experimental murine model of gastrointestinal allergy, feeding with the untreated raw milk containing bacteria induced a greater allergic response (measured by specific IgE antibody level and MMCP-1) than with sterilized milk and heated milk. A higher in vitro production of IL-10 by splenocytes, regulatory cytokine playing an important role in suppressing allergy responses, was also found [28].

The protective properties of raw milk against allergies and asthma may be connected not only with the direct content of bacteria but also with the influence of different raw milk ingredients on gut microbiome. Microbial gut composition may be associated with increased risk for atopy, eczema, or wheeze, and it may differ depending on allergy status [62,63]. Early gut microbiome composition may also be associated with allergy resolution later in life, as was shown in a study of milk-allergic children [64]. Some factors present in raw milk may potentially modulate microbiota composition. Proteins such as lactoferrin and lysozyme have antimicrobial activity. Saccharides can promote the growth of bifidobacteria. This may influence the production of SCFAs, such as acetate, butyrate, and propionate, and enhance the epithelial barrier function of the gut [31]. Moreover, it was shown that short chain fatty acids affected the bone marrow dendritic cell maturation and inhibited the Th2-dependent response, which reduced allergic airway inflammation after allergen exposure. Decreased fatty acid metabolism by gut microbiota was associated with milk allergy resolution with time [64].

7. Milk Exosomes

Bovine and human milk contain substantial amounts of exosomal miRNAs (miRs) that may be transferred to the infant to promote immune regulatory functions. Raw cow's milk contains high amounts of bioactive miRs. Pasteurization process decreases its level. Boiling milk results in complete miRs degradation. Milk exosomes are of critical importance for the maturation of the immune system during the postnatal period and early infancy. Milk-derived exosomal microRNAs may be potential stimuli for thymic Treg maturation and raw milk-mediated atopy prevention [65]. Farm milk exposure has been associated with increased numbers of CD4+CD25+FoxP3+ regulatory T cells (Tregs), lower atopic sensitization, and asthma in 4.5-year-old children [66]. Milk miRs may promote a selection process turning self-reactive thymocytes into stable Treg cells, and functionally active FoxP3 Treg cells suppress the development of Th2 cell-dependent immune responses. It has been suggested that milk's exosomal miR system may represent "the missing candidate" inducing atopy-preventive effects of raw cow's milk consumption, and the future prevention of atopic diseases might be possible by an addition of appropriate miR-155-enriched exosomes to artificial infant formula [67].

8. Genetics, Epigenetics, and Raw Cow's Milk Exposure

Allergic diseases and asthma have a strong genetic background. It has been shown that environmental factors may interact with the genome both by modifying the environmental effect by the underlying genome or, vice versa, the genetic effect by environmental exposure. As was presented above, polymorphisms in the CD14 receptor influenced a protective effect of unpasteurized milk consumption on allergy and asthma [61]. In a genome-wide association study in the GABRIEL population, the gene–environment interaction for atopy, asthma, and farming exposures were tested. The common genetic polymorphisms previously described in asthma did not modify the protective effect of farming exposures on asthma, although the interaction was detected in rare SNPs [68]. The exposure to raw farm milk in pregnancy and the first year of life was associated with changes in the gene expression of innate immunity receptors [69]. Recent epigenetic studies of genomic

Nutrients **2019**, *11*, 469

adaptation to the environment that was not caused by changes in the nucleotide sequence of the genetic code itself revealed additional methods of interaction. Large-scale genome-wide meta-analysis of DNA methylation and childhood asthma identified novel epigenetic variations, which might be potential biomarkers of later asthma risk [70,71]. In a prospective PASTURE study, DNA methylation patterns changed significantly in the first year of life in asthma-related genes, and exposure to farm environments seemed to influence methylation patterns in this population [72]. Further studies on the effects of raw milk exposure as well as genome and epigenome interactions on asthma and allergy phenotypes are needed to disentangle these complicated associations.

9. Conclusions

There is a debate about the role of raw cow's milk role in human health. Sceptics say that raw milk carries a significant risk of bacterial pathogens infection and there is no clear evidence that raw milk has any nutritional benefits compared to pasteurized milk. Enthusiasts see in milk the hope for effective prevention of allergic diseases and even respiratory tract infections [73]. There is no doubt that the components of raw milk can influence the immune function, but the final proof based on controlled studies in infants is not possible due toethical reasons. Undoubtedly, even if the final understanding of the role of raw cow's milk seems to be a distant prospect, it is one of the most intriguing and promising paths to be studied in allergy prevention.

Funding: This research received no external funding.

Conflicts of Interest: The author declares no conflict of interest.

References

1. Lucey, J.A. Raw milk consumption. Risks and benefits. *Nutr. Today* **2015**, *50*, 189–193. [CrossRef] [PubMed]
2. Braun-Fahrländer, C.; Gassner, M.; Grize, L.; Neu, U.; Sennhauser, F.H.; Varonier, H.S.; Vuille, J.C.; Wüthrich, B. Prevalence of hay fever and allergic sensitization in farmer's children and their peers living in the same rural community. SCARPOL team. Swiss Study on Childhood Allergy and Respiratory Symptoms with Respect to Air Pollution. *Clin. Exp. Allergy* **1999**, *29*, 28–34. [CrossRef] [PubMed]
3. Riedler, J.; Braun-Fahrländer, C.; Eder, W.; Schreuer, M.; Waser, M.; Maisch, S. Exposure to farming in early life and development of asthma and allergy: A cross-sectional survey. *Lancet* **2001**, *358*, 1129–1133. [CrossRef]
4. Remes, S.T.; Koskela, H.O.; Iivanainen, K.; Pekkanen, J. Allergen-specific sensitization in asthma and allergic diseases in children: The study on farmers' and non-farmers' children. *Clin. Exp. Allergy* **2005**, *35*, 160–166. [CrossRef] [PubMed]
5. Illi, S.; Depner, M.; Genuneit, J.; Horak, E.; Loss, G.; Strunz-Lehner, C.; Buchele, G.; Boznanski, A.; Danielewicz, H.; Cullinan, P.; et al. Protection from asthma and allergy in farm environments –The GABRIEL Advanced Studies. *J. Allergy Clin. Immunol.* **2012**, *129*, 1470–1477. [CrossRef] [PubMed]
6. Wickens, K.; Lane, J.M.; Fitzharris, P.; Siebers, R.; Riley, G.; Douwes, J.; Smith, T.; Crane, J. Farm residence and exposures and the risk of allergic diseases in New Zealand children. *Allergy* **2002**, *57*, 1171–1179. [CrossRef] [PubMed]
7. Downs, S.H.; Marks, G.B.; Mitakakis, Z.; Leuppi, J.D.; Car, N.G.; Peat, J.K. Having lived on a farm and protection against allergic diseases in Australia. *Clin. Exp. Allergy* **2001**, *31*, 570–575. [CrossRef] [PubMed]
8. Lampi, J.; Canoy, D.; Jarvis, D.; Hartikainen, A.L.; Keski-Nisula, L.; Jarvelin, M.R.; Pekkanen, J. Farming environment and prevalence of atopy at age 31: Prospective birth cohort study in Finland. *Clin. Exp. Allergy* **2011**, *41*, 987–993. [CrossRef] [PubMed]
9. Filipiak, B.; Heinrich, J.; Schäfer, T.; Ring, J.; Wichmann, H.E. Farming, rural lifestyle and atopy in adults from southern Germany–results from the MONICA/KORA study Augsburg. *Clin. Exp. Allergy* **2001**, *31*, 1829–1838. [CrossRef] [PubMed]
10. House, J.S.; Wyss, A.B.; Hoppin, J.A.; Richards, M.; Long, S.; Umbach, D.M.; Henneberger, P.K.; Beane Freeman, L.E.; Sandler, D.P.; Long, O'Connell, E. Early-life farm exposures and adult asthma and atopy in the Agricultural Lung Health Study. *J. Allergy Clin. Immunol.* **2017**, *140*, 249–256. [CrossRef] [PubMed]

11. Ege, M.J.; Frei, R.; Bieli, C.; Schram-Bijerk, D.; Waser, M.; Benz, M.; Weiss, G.; Nyberg, F.; van Hage, M.; Pershagen, G.; et al. Not all farming environments protect against the development of asthma and wheeze in children. *J. Allergy Clin. Immunol.* **2007**, *119*, 1140–1147. [CrossRef] [PubMed]

12. Ege, M.J.; Mayer, M.; Normand, A.C.; Genuneit, J.; Cookson, W.O.; Braun-Fahrlander, C.; Heederik, D.; Piarroux, R.; von Mutius, E. GABRIELA Transregio 22 Study Group. Exposure to environmental microorganisms and childhood asthma. *N. Engl. J. Med.* **2011**, *364*, 701–709. [CrossRef] [PubMed]

13. Braun-Fahrländer, C.; von Mutius, E. Can farm milk consumption prevent allergic diseases? *Clin. Exp. Allergy* **2011**, *41*, 29–35. [CrossRef] [PubMed]

14. Waser, M.; Michels, K.B.; Bieli, C.; Floistrup, H.; Pershagen, G.; von Mutius, E.; Ege, M.; Riedler, J.; Schram-Bijerk, D.; Brunekreef, B.; et al. Inverse association of farm milk consumption with asthma and allergy in rural and suburban populations across Europe. *Clin. Exp. Allergy* **2007**, *37*, 661–670. [CrossRef] [PubMed]

15. Perkin, M.; Strachan, D. Which aspects of the farming lifestyle explain the inverse association with childhood allergy? *J. Allergy Clin. Immunol.* **2006**, *117*, 1374–1381. [CrossRef] [PubMed]

16. Barnes, M.; Cullinan, P.; Athanasaki, P.; MacNeill, S.; Hole, A.M.; Harris, J.; Kalogeraki, S.; Chatzinikolaou, M.; Drakonakis, N.; Bibaki-Liakou, V.; et al. Crete: Does farming explain urban and rural differences in atopy? *Clin. Exp. Allergy* **2001**, *31*, 1822–1828. [CrossRef] [PubMed]

17. Sozańska, B.; Pearce, N.; Dudek, K.; Cullinan, P. Consumption of unpasteurized milk and its effects on atopy and asthma in children and adult inhabitants in rural Poland. *Allergy* **2013**, *68*, 644–650. [CrossRef] [PubMed]

18. Loss, G.; Apprich, S.; Waser, M.; Kneifel, W.; von Mutius, E.; Genuneit, J.; Buchele, G.; Weber, J.; Sozańska, B.; Danielewicz, H.; et al. The protective effect of farm milk consumption on childhood asthma and atopy: The GABRIELA study. *J. Allergy Clin. Immunol.* **2011**, *128*, 766–773. [CrossRef] [PubMed]

19. Von Mutius, E.; Schmid, S. The PASTURE project: EU support for the improvement of knowledge about risk factors and preventive factors for atopy in Europe. *Allergy* **2006**, *61*, 407–413. [CrossRef] [PubMed]

20. Brick, T.; Schober, Y.; Böcking, C.; Pekkanen, J.; Genuneit, J.; Loss, G.; Dalphin, J.C.; Riedler, J.; Lauener, R.; Nockher, W.A.; et al. ω-3 fatty acids contribute to the asthma-protective effect of unprocessed cow's milk. *J. Allergy Clin. Immunol.* **2016**, *137*, 1699–1706. [CrossRef] [PubMed]

21. Pfefferle, P.I.; Buchele, G.; Blumer, N.; Roponen, M.; Ege, M.; Krauss-Etschmann, S.; Genuneit, J.; Hyvärinen, A.; Hirvonen, M.R.; Lauener, R.; et al. Cord blood cytokines are modulated by maternal farming activities and consumption of farm dairy products during pregnancy: The PASTURE Study. *J. Allergy Clin. Immunol.* **2010**, *125*, 108–115. [CrossRef] [PubMed]

22. Loss, G.; Depner, M.; Ulfman, L.H.; van Neerven, R.J.; Hose, A.J.; Genuneit, J.; Karvonen, A.M.; Hyvärinen, A.; Kaulek, V.; Roduit, C.; et al. Consumption of unprocessed cow's milk protects infants from common respiratory infections. *J. Allergy Clin. Immunol.* **2015**, *135*, 56–62. [CrossRef] [PubMed]

23. Nicklaus, S.; Divaret-Chauveau, A.; Chardon, M.L.; Roduit, C.; Kaulek, V.; Ksiazek, E.; Dalphin, M.L.; Karvonen, A.M.; Kirjavainen, P.; Pekkanen, J.; et al. The protective effect of cheese consumption at 18 months on allergic diseases in the first 6 years. *Allergy* **2018**. [CrossRef] [PubMed]

24. Poulsen, O.M.; Nielsen, B.R.; Basee, A.; Hau, J. Comparison of intestinal anaphylactic reactions in sensitized mice challenged with intreated bovine milk and homogenized bovine milk. *Allergy* **1990**, *45*, 321–326. [CrossRef] [PubMed]

25. Thijs, C.; Muller, A.; Rist, L.; Kummeling, I.; Snijders, B.E.; Huber, M.; van Ree, R.; Simões-Wüst, A.P.; Dagnelie, P.C.; van den Brandt, P.A. Fatty acids in breast milk and development of atopic eczema and allergic sensitization in infancy. *Allergy* **2011**, *66*, 58–67. [CrossRef] [PubMed]

26. Brick, T.; Ege, M.; Boeren, S.; Böck, A.; von Mutius, E.; Vervoort, J.; Hettinga, K. Effect of Processing Intensity on Immunologically Active Bovine Milk Serum Proteins. *Nutrients* **2017**, *31*, 963. [CrossRef] [PubMed]

27. Peroni, D.G.; Piacentini, G.L.; Bodini, A.; Pigozzi, R.; Boner, A.L. Transforming growth factor-beta is elevated in unpasteurized cow's milk. *Pediatr. Allergy Immunol.* **2009**, *20*, 42–44. [CrossRef] [PubMed]

28. Hodgkinson, A.J.; McDonald, N.A.; Hine, B. Effect of raw milk on allergic responses in a murine model of gastrointestinal allergy. *Br. J. Nutr.* **2014**, *14*, 390–397. [CrossRef] [PubMed]

29. Abbring, S.; Verheijden, K.A.T.; Diks, M.A.P.; Leusink-Muis, A.; Hols, G.; Baars, T.; Garssen, J.; van Esch, B.C.A.M. Raw Cow's Milk Prevents the Development of Airway Inflammation in a Murine House Dust Mite-Induced Asthma Model. *Front. Immunol* **2017**, *8*, 1045. [CrossRef] [PubMed]

30. Abbring, S.; Hols, G.; Garssen, J.; van Esch, B.C.A.M. Raw cow's milk consumption and allergic diseases–the potential role of bioactive whey proteins. *Eur. J. Pharmacol.* **2019**, *843*, 55–65. [CrossRef] [PubMed]

31. Van Neerven, R.J.J.; Knol, E.F.; Heck, J.M.L.; Savelkoul, H.F.J. Which factors in raw cow's milk contribute to protection against allergies? *J. Allergy Clin. Immunol.* **2012**, *130*, 853–858. [CrossRef] [PubMed]
32. Puddu, P.; Valenti, P.; Gessani, S. Immunomodulatory effects of lactoferrin on antigen presenting cells. *Biochimie* **2009**, *91*, 11–18. [CrossRef] [PubMed]
33. Krissansen, G.W. Emerging health properties of whey proteins and their clinical implications. *J. Am. Coll. Nutr.* **2007**, *26*, 713S–723S. [CrossRef] [PubMed]
34. Munblit, D.; Peroni, D.G.; Boix-Amorós, A.; Hsu, P.S.; Van't Land, B.; Gay, M.C.L.; Kolotilina, A.; Skevaki, C.; Boyle, R.J.; Collado, M.C.; et al. Human Milk and Allergic Diseases: An Unsolved Puzzle. *Nutrients* **2017**, *17*, 9. [CrossRef] [PubMed]
35. Giansanti, F.; Panella, G.; Leboffe, L.; Antonini, G. Lactoferrin from Milk: Nutraceutical and Pharmacological Properties. *Pharmaceuticals* **2016**, *9*, 61. [CrossRef] [PubMed]
36. Sjögren, Y.M.; Jenmalm, M.C.; Böttcher, M.F.; Björkstén, B.; Sverremark-Ekström, E. Altered early infant gut microbiota in children developing allergy up to 5 years of age. *Clin. Exp. Allergy* **2009**, *39*, 518–526. [CrossRef] [PubMed]
37. Kruzel, M.L.; Bacsi, A.; Choudhury, B.; Sur, S.; Boldogh, I. Lactoferrin decreases pollen antigen-induced allergic airway inflammation in a murine model of asthma. *Immunology* **2006**, *119*, 159–166. [CrossRef] [PubMed]
38. Liao, Y.; Jiang, R.; Lönnerdal, B. Biochemical and molecular impacts of lactoferrin on small intestinal growth and development during early life. *Biochem. Cell Biol.* **2012**, *90*, 476–484. [CrossRef] [PubMed]
39. Oddy, W.H.; Rosales, F. A systematic review of the importance of milk TGF-beta on immunological outcomes in the infant and young child. *Pediatr. Allergy Immunol.* **2010**, *21*, 47–59. [CrossRef] [PubMed]
40. Kotler, B.M.; Kerstetter, J.E.; Insogna, K.L. Claudins, dietary milk proteins, and intestinal barrier regulation. *Nutr. Rev.* **2013**, *71*, 60–65. [CrossRef] [PubMed]
41. Verhasselt, V.; Milcent, V.; Cazareth, J.; Kanda, A.; Fleury, S.; Dombrowicz, D.; Glaichenhaus, N.; Julia, V. Breast milk-mediated transfer of an antigen induces tolerance and protection from allergic asthma. *Nat. Med.* **2008**, *14*, 170–175. [CrossRef] [PubMed]
42. Penttila, I. Effects of transforming growth factor-beta and formula feeding on systemic immune responses to dietary beta-lactoglobulin in allergy-prone rats. *Pediatr. Res.* **2006**, *59*, 650–655. [CrossRef] [PubMed]
43. Hawrylowicz, C.M.; O'Garra, A. Potential role of interleukin-10-secreting regulatory T cells in allergy and asthma. *Nat. Rev. Immunol.* **2005**, *5*, 271–283. [CrossRef] [PubMed]
44. Borish, L.; Aarons, A.; Rumbyrt, J.; Cvietusa, P.; Negri, J.; Wenzel, S. Interleukin-10 regulation in normal subjects and patients with asthma. *J. Allergy Clin. Immunol.* **1996**, *97*, 1288–1296. [CrossRef]
45. Pletinckx, K.; Döhler, A.; Pavlovic, V.; Lutz, M.B. Role of dendritic cell maturity/costimulation for generation, homeostasis, and suppressive activity of regulatory T cells. *Front. Immunol.* **2011**, *2*, 39. [CrossRef] [PubMed]
46. Järvinen, K.M.; Laine, S.T.; Järvenpää, A.L.; Suomalainen, H.K. Does low IgA in human milk predispose the infant to development of cow's milk allergy? *Pediatr. Res.* **2000**, *48*, 457–462. [CrossRef] [PubMed]
47. Mosconi, E.; Rekima, A.; Seitz-Polski, B.; Kanda, A.; Fleury, S.; Tissandie, E.; Monteiro, R.; Dombrowicz, D.D.; Julia, V.; Glaichenhaus, N.; et al. Breast milk immune complexes are potent inducers of oral tolerance in neonates and prevent asthma development. *Mucosal Immunol.* **2010**, *3*, 461–474. [CrossRef] [PubMed]
48. van Splunter, M.; van Osch, T.L.J.; Brugman, S.; Savelkoul, H.F.J.; Joosten, L.A.B.; Netea, M.G.; van Neerven, R.J.J. Induction of Trained Innate Immunity in Human Monocytes by Bovine Milk and Milk-Derived Immunoglobulin G. *Nutrients* **2018**, *27*, 10. [CrossRef] [PubMed]
49. Trompette, A.; Gollwitzer, E.S.; Yadava, K.; Sichelstiel, A.K.; Sprenger, N.; Ngom-Bru, C.; Blanchard, C.; Junt, T.; Nicod, L.P.; Harris, N.L.; et al. Gut microbiota metabolism of dietary fiber influences allergic airway disease and hematopoiesis. *Nat. Med.* **2014**, *20*, 159–166. [CrossRef] [PubMed]
50. Byndloss, M.X.; Olsan, E.E.; Rivera-Chávez, F.; Tiffany, C.R.; Cevallos, S.A.; Lokken, K.L.; Torres, T.P.; Byndloss, A.J.; Faber, F.; Gao, Y.; et al. Microbiota-activated PPAR-γ signaling inhibits dysbiotic Enterobacteriaceae expansion. *Science* **2017**, *357*, 570–575. [CrossRef] [PubMed]
51. Roduit, C.; Frei, R.; Ferstl, R.; Loeliger, S.; Westermann, P.; Rhyner, C.; Schiavi, E.; Barcik, W.; Rodriguez-Perez, N.; Wawrzyniak, M.; et al. High levels of butyrate and propionate in early life are associated with protection against atopy. *Allergy* **2018**. [CrossRef] [PubMed]

52. Matsumoto, M.; Aranami, A.; Ishige, A.; Watanabe, K.; Benno, Y. LKM512 yogurt consumption improves the intestinal environment and induces the T-helper type 1 cytokine in adult patients with intractable atopic dermatitis. *Clin. Exp. Allergy* **2007**, *37*, 358–370. [CrossRef] [PubMed]

53. Shoda, T.; Futamura, M.; Yang, L.; Narita, M.; Saito, H.; Ohya, Y. Yogurt consumption in infancy is inversely associated with atopic dermatitis and food sensitization at 5 years of age: A hospital-based birth cohort study. *J. Dermatol. Sci.* **2017**, *86*, 90–96. [CrossRef] [PubMed]

54. Crane, J.; Barthow, C.; Mitchell, E.A.; Stanley, T.V.; Purdie, G.; Rowden, J.; Kang, J.; Hood, F.; Barnes, P.; Fitzharris, P.; et al. Is yoghurt an acceptable alternative to raw milk for reducing eczema and allergy in infancy? *Clin. Exp. Allergy* **2018**, *48*, 604–606. [CrossRef] [PubMed]

55. Perkin, M.R. Unpasteurized milk: health or hazard? *Clin. Exp. Allergy* **2007**, *37*, 627–630. [CrossRef] [PubMed]

56. Gehring, U.; Spithoven, J.; Schmid, S.; Bitter, S.; Braun-Fahrländer, C.; Dalphin, J.C.; Hyvärinen, A.; Pekkanen, J.; Riedler, J.; Weiland, S.K.; et al. Endotoxin levels in cow's milk samples from farming and non-farming families—The PASTURE study. *Environ. Int.* **2008**, *34*, 1132–1136. [CrossRef] [PubMed]

57. Sipka, S.; Béres, A.; Bertók, L.; Varga, T.; Bruckner, G. Comparison of endotoxin levels in cow's milk samples derived from farms and shops. *Innate Immune* **2015**, *21*, 531–536. [CrossRef] [PubMed]

58. Schuijs, M.J.; Willart, M.A.; Vergote, K.; Gras, D.; Deswarte, K.; Ege, M.J.; Madeira, F.B.; Beyaert, R.; van Loo, G.; Bracher, F.; et al. Farm dust and endotoxin protect against allergy through A20 induction in lung epithelial cells. *Science* **2015**, *349*, 1106–1110. [CrossRef] [PubMed]

59. Smit, L.A.; Siroux, V.; Bouzigon, E.; Oryszczyn, M.P.; Lathrop, M.; Demenais, F. CD-14 and toll-like receptor gene polymorphisms, country living and asthma in adults. *Am. J. Respir. Crit. Care Med.* **2009**, *179*, 363–368. [CrossRef] [PubMed]

60. Roduit, C.; Wohlgensinger, J.; Frei, R.; Bitter, S.; Bieli, C.; Loeliger, S.; Büchele, G.; Riedler, J.; Dalphin, J.C.; Remes, S.; et al. Prenatal animal contact and gene expression of innate immunity receptors at birth are associated with atopic dermatitis. *J. Allergy Clin. Immunol.* **2011**, *127*, 179–185. [CrossRef] [PubMed]

61. Bieli, C.; Eder, W.; Frei, R.; Klimecki, W.; Waser, M.; Riedler, J.; von Mutius, E.; Scheynius, A.; Pershagen, G.; Doekes, G.; et al. A polymorphism in CD14 modifies the effect of farm milk consumption on allergic diseases and CD14 gene expression. *J. Allergy Clin. Immunol.* **2007**, *120*, 1308–1315. [CrossRef] [PubMed]

62. Penders, J.; Thijs, C.; van den Brandt, P.A.; Kummeling, I.; Snijders, B.; Stelma, F.; Adams, H.; van Ree, R.; Stobberingh, E.E. Gut microbiota composition and development of atopic manifestations in infancy: The KOALA Birth Cohort Study. *Gut* **2007**, *56*, 661–667. [CrossRef] [PubMed]

63. Van Nimwegen, F.A.; Penders, J.; Stobberingh, E.E.; Postma, D.S.; Koppelman, G.H.; Kerkhof, M.; Reijmerink, N.E.; Dompeling, E.; van den Brandt, P.A.; Ferreira, I. Mode and place of delivery, gastrointestinal microbiota, and their influence on asthma and atopy. *J. Allergy Clin. Immunol.* **2011**, *128*, 948–955. [CrossRef] [PubMed]

64. Bunyavanich, S.; Shen, N.; Grishin, A.; Wood, R.; Burks, W.; Dawson, P.; Jones, S.M.; Leung, D.Y.M.; Sampson, H.; Sicherer, S.; et al. Early-life gut microbiome composition and milk allergy resolution. *J. Allergy Clin. Immunol.* **2016**, *138*, 1122–1130. [CrossRef] [PubMed]

65. Melnik, B.C.; Schmitz, G. Exosomes of pasteurized milk: Potential pathogens of Western diseases. *J. Transl. Med.* **2019**, *17*, 3. [CrossRef] [PubMed]

66. Lluis, A.; Depner, M.; Gaugler, B.; Saas, P.; Casaca, V.I.; Raedler, D.; Michel, S.; Tost, J.; Liu, J.; Genuneit, J.; et al. Increased regulatory T-cell numbers are associated with farm milk exposure and lower atopic sensitization and asthma in childhood. *J. Allergy Clin. Immunol.* **2014**, *133*, 551–559. [CrossRef] [PubMed]

67. Melnik, B.C.; John, S.M.; Schmitz, G. Milk: An exosomal microRNA transmitter promoting thymic regulatory T cel maturation preventing the development of atopy? *J. Transl. Med.* **2014**, *12*, 43. [CrossRef] [PubMed]

68. Ege, M.J.; Strachan, D.P.; Cookson, W.O.; Moffatt, M.F.; Gut, I.; Lathrop, M.; Kabesch, M.; Genuneit, J.; Buchele, G.; Sozanska, B.; et al. Gene–environment interaction for childhood asthma and exposure to farming in Central Europe. *J. Allergy Clin. Immunol.* **2011**, *127*, 138–44. [CrossRef] [PubMed]

69. Loss, G.; Bitter, S.; Wohlgensinger, J.; Frei, R.; Roduit, C.; Genuneit, J.; Pekkanen, J.; Roponen, M.; Hirvonen, M.R.; Dalphin, J.C.; et al. Prenatal and early-life exposures alter expression of innate immunity genes: The PASTURE cohort study. *J. Allergy Clin. Immunol.* **2012**, *130*, 523–530. [CrossRef] [PubMed]

70. Reese, S.E.; Xu, C.J.; den Dekker, H.T.; Lee, M.K.; Sikdar, S.; Ruiz-Arenas, C.; Merid, S.K.; Rezwan, F.I.; Page, C.M.; Ullemar, V.; et al. Epigenome-wide meta-analysis of DNA methylation and childhood asthma. *J. Allergy Clin. Immunol.* **2018**. [CrossRef] [PubMed]

71. Xu, C.J.; Söderhäll, C.; Bustamante, M.; Baïz, N.; Gruzieva, O.; Gehring, U.; Mason, D.; Chatzi, L.; Basterrechea, M.; Llop, S.; et al. DNA methylation in childhood asthma: An epigenome-wide meta-analysis. *Lancet Respir. Med.* **2018**, *6*, 379–388. [CrossRef]

72. Michel, S.; Busato, F.; Genuneit, J.; Pekkanen, J.; Dalphin, J.C.; Riedler, J.; Mazaleyrat, N.; Weber, J.; Karvonen, A.M.; Hirvonen, M.R. Farm exposure and time trends in early childhood may influence DNA methylation in genes related to asthma and allergy. *Allergy* **2013**, *68*, 355–364. [CrossRef] [PubMed]

73. Perdijk, O.; van Splunter, M.; Savelkoul, H.F.J.; Brugman, S.; van Neerven, R.J.J. Cow's Milk and Immune Function in the Respiratory Tract: Potential Mechanisms. *Front. Immunol.* **2018**, *12*, 143. [CrossRef] [PubMed]

![nutrients](nutrients logo) MDPI

Article

Individual Sensitization Pattern Recognition to Cow's Milk and Human Milk Differs for Various Clinical Manifestations of Milk Allergy

Frauke Schocker [1,*], Skadi Kull [1], Christian Schwager [1], Jochen Behrends [2] and Uta Jappe [1,3]

[1] Division of Clinical and Molecular Allergology, Research Center Borstel - Leibniz Lung Center, Priority Area Asthma and Allergy, Airway Research Center North (ARCN), German Center for Lung Research (DZL), 23845 Borstel, Germany; skull@fz-borstel.de (S.K.); cschwager@fz-borstel.de (C.S.); ujappe@fz-borstel.de (U.J.)

[2] Core Facility Fluorescence Cytometry, Research Center Borstel, 23845 Borstel, Germany; jbehrends@fz-borstel.de

[3] Interdisciplinary Allergy Outpatient Clinic, MK III, University of Lübeck, Airway Research Center North (ARCN), German Center for Lung Research (DZL), 23538 Lübeck, Germany

* Correspondence: fschocker@fz-borstel.de; Tel.: +49-4537-188-4970

Received: 15 April 2019; Accepted: 10 June 2019; Published: 14 June 2019

Abstract: Cow's milk allergy (CMA) belongs to one of the most common food allergies in early childhood affecting 2–3% of children under 3 years of age. However, approximately 1% of adults remain allergic to cow's milk, often showing severe reactions even to traces of milk. In our study, we recruited patients with different clinical manifestations of CMA, including patients with anaphylaxis and less severe symptoms. We assessed the sensitization patterns and allergic responses of these subgroups through different immunological and cell-based methods. Sera of patients were investigated for IgE against whole cow's milk and its single allergens by CAP- FEIA. In a newly developed in-house multiplex dot assay and a basophil activation test (BAT), cow's milk allergens, in addition to human breast milk and single allergens from cow's and human milk were analyzed for IgE recognition and severity of CMA in the included patients. Both the CAP-FEIA routine diagnostic and the multiplex dot test could differentiate CMA with severe from milder allergic reactions by means of the patients' casein sensitization. The BAT, which mirrors the clinical response in vitro, confirmed that basophils from patients with severe reactions were more reactive to caseins in contrast to the basophils from more moderate CMA patients. By means of this improved component-resolved diagnosis of CMA, individual sensitization patterns could be assessed, also taking sensitization against human milk into consideration.

Keywords: Cow's milk allergy (CMA), anaphylaxis; sensitization pattern; cow's milk allergens; CAP-FEIA (Fluorescence Enzyme Immunoassay); multiplex dot test; basophil activation test (BAT), human breast milk

---

1. Introduction

IgE-mediated cow's milk allergy (CMA) is a common food allergy affecting 2–3% of young children under 3 years of age, involving the skin, the gastrointestinal tract, the respiratory tract or the cardio-vascular system. A high proportion of young children, approximately 85%, develop a natural tolerance. However, approximately 1% of adults have persisting allergic reactions, often severe and life threatening [1,2].

Cow's milk consists of caseins accounting for approximately 80% and whey proteins accounting for approximately 20% of the total protein content. The major cow milk allergens are the caseins (Bos d 8) yielding α-, ß- and κ-caseins. Whey proteins consist of α-lactalbumin (Bos d 4) and ß-lactoglobulin

(Bos d 5) among other proteins, e.g., bovine serum albumin (Bos d 6), immunoglobulin (Bos d 7), and lactoferrin [3]. According to the literature, the human IgE response to cow's milk is highly variable and no single allergen component alone accounts for cow's milk allergenicity [4]. However, looking for prognostic markers, low casein and ß-lactoglobulin-specific IgE-antibody concentrations were found to be predictive for the resolution of CMA [5], and whole cow's milk specific IgE above 50 kU$_A$/L were associated with persistent CMA when studying a population of children with CMA, in terms of the prognosis for developing tolerance [6]. Garcıa-Ara et al. [7] described caseins to best discriminate between transient and persistent CMA. Ito et al. [8] and D'Urbano et al. [9] confirmed these data. Likewise, recent studies have revealed differences in the IgE- and IgG$_4$- binding to epitopes of caseins, α-lactalbumin and ß-lactoglobulin as predictive markers for oral tolerance or the persistence of CMA, respectively [10].

Component-resolved diagnosis (CRD) allows the determination of specific IgE not only against whole cow's milk, but also against single allergens (α-, ß- and κ-caseins, α-lactalbumin, ß-lactoglobulin). Because allergy as such is a highly individual immune response necessitating an individual diagnostic approach, CRD may be helpful for monitoring the degree of severity of CMA. Currently, for routine allergy diagnostic tests, total serum IgE, specific IgE to milk, cheese, and milk from other species as well as single milk allergens are accessible by means of CAP- FEIA (ImmunoCAP).

In our recent study, we were able to monitor a patient with anaphylactic reactions to traces of cow's milk by means of allergen-specific IgE and basophil activation test (BAT), before and under omalizumab therapy on the molecular level, also addressing sensitization to human milk proteins [11,12].

The objective of the present study is to investigate the combination of diagnostic tests—extended by IgE recognition of whole human milk and human α-lactalbumin—for their capacity to distinguish between different clinical manifestations of CMA. The antibody-based methods such as CAP-FEIA, a newly developed multiplex dot test for the detection of specific IgE (sIgE) against whole milk and single milk allergens, as well as a highly specific and sensitive BAT, were applied in order to differentiate between the degrees of severity of CMA.

2. Materials and Methods

2.1. Study Group

Patients were recruited in the allergy outpatient clinics in Borstel and Lübeck and serum samples as well as heparinized whole blood were collected from six patients with a clear history of adverse reactions after milk ingestion in the past, with milk as the only food implicated in the episode. Specific IgE to milk was determined by in vivo and in vitro tests. Total IgE and specific IgE to milk, cheese, milk from other mammalian species and single milk allergens were determined by means of CAP-FEIA (ImmunoCAP, ThermoFisher Scientific, Freiburg, Germany). Challenge tests to milk were not performed in patients reporting life-threatening episodes (ID1–ID2); patient ID3 had a positive challenge in early childhood. The other cases refused the provocation tests as they feared adverse reactions (ID4–ID6). Skin prick tests (SPT)—if not contraindicated—were performed according to the standard procedure, with the prick to prick technique with fresh milk. Histamine dihydrochloride and phosphate-buffered saline (PBS) solution served as positive and negative controls, respectively. The local ethics committee of the University of Lübeck approved this study (approval numbers 10–126, 13–086 and 13–136). Serum and heparinized whole blood samples were additionally obtained from a non-atopic, non-sensitized healthy individual and served as controls. Allergic disease was ruled out by history, determination of total and specific IgE against pollens, house dust mite (HDM) and a panel of food allergens including milk, cheese and milk from other mammalian species. Written informed consents were obtained from all subjects included.

2.2. Dot Blot Test

For the dot blot test, 1 µL aliquots of the allergens (bovine and human milk, bovine α-, ß-,and κ-caseins, bovine ß-lactoglobulin, bovine and human α-lactalbumin) were dotted in a concentration of 1 µg/µl and 2.5 µg/µl onto nitrocellulose membrane strips (Amersham protean 0.45 µm, GE Healthcare, Freiburg, Germany) and dried carefully. Thereafter, the membranes were blocked for 1 h with SynBlock (ImmunoChemistry, Bloomington, MN, USA) (1:1 diluted with TBS-T (tris-buffered saline including 0.05% Tween), pH 7.4). In the next step, sera of milk-allergic patients and serum of a non-allergic control were diluted 1:20 with the exception of ID1 (1:50) in TBS-T and incubated on the strips overnight at room temperature. For detection of bound IgE antibodies, the strips were incubated with a horseradish peroxidase (HRP)-conjugated mouse anti-human IgE (Fc) antibody (Southern Biotech, Birmingham, AL, USA), in a 1:5,000 dilution in TBS-T for 2 h. Immunostaining was performed by means of chemiluminescent Western blot detection using Clarity Western ECL Substrate (Bio-Rad, Hercules, CA, USA). The detection of stained dots was visualized using the ChemiDoc MP System (Bio-Rad, Munich, Germany).

2.3. Basophil Activation Test (BAT)

We performed the BAT according to Schwager et al. [13] using an extensive and specific read-out. BAT was conducted with heparinized whole blood stimulated for 30 minutes at 37 °C. Cow's milk (purchased from a local food store), human breast milk (recruited in a study on peanut allergen transfer into breast milk [14,15]; ethics approval numbers 08–122 and Az19-114) and the single bovine, as well as human milk allergens bovine α-casein, ß-casein, κ-caseins, bovine ß-lactoglobulin, bovine and human α-lactalbumin (Sigma-Aldrich, Steinheim, Germany) were used in serial 10 fold dilutions (10 µg/mL; 1 µg/mL; 100 ng/mL; 10 ng/mL; 1 ng/mL).

Formyl-methionyl-leucyl phenylalanine (fMLP; 100 nM, Sigma-Aldrich, Steinheim, Germany), polyclonal goat anti-human IgE (1 mg/mL, abcam, Cambridge, UK), or PBS buffer were chosen as controls.

Basophils were analyzed using an LSR II flow cytometer. Data analysis was conducted with the FCS Express 6 program. For data visualization, Prism graphics software 6.03 (GraphPad Prism Inc., San Diego, CA, USA) was used.

3. Results

3.1. Patients with Different Complexity of Allergic Reactions to Cow's Milk

We recruited six patients with a history of allergic reactions to milk, revealing different types of clinical reactions as well as degrees of severity. The subjects included had positive responses to milk in SPT with the fresh food, if not contraindicated.

Of these patients, three patients—two adults (ID1, ID2) and one 9-year-old child (ID3)—suffered from severe allergic reactions to milk. In contrast, three patients (ID4, ID5, ID6) displayed milder allergic symptoms. The clinical data are comprised in Supplemental Table S1, summarizing the patients with severe reactions upon milk contact first (ID1, ID2, ID3), followed by the CMA patients with a milder clinical picture (ID4, ID5, ID6). The patients' history including comorbidities such as asthma, atopic dermatitis and other sensitizations/allergies are included in Supplemental Table S1.

With the exception of ID1, two blood samples were analyzed of each CMA patient named IDx (first blood sample) and IDx-1 (second blood sample, investigated in parallel to the BAT).

ID7 was a non-allergic individual serving as a control for the CAP-FEIA, the multiplex dot test and the BAT.

ID1 has suffered from severe reactions since early childhood, even under cow's milk avoidance of the nursing mother, and experienced cardioplegia at the age of thirteen in more than one episode after accidental contact with milk. Due to the severe reactions upon accidental contact to milk traces, the patient started an oral immunotherapy (OIT) in 2016, as previously reported in [11].

ID2 had already displayed massive adverse skin reactions as a breast-fed child, also under cow's milk elimination diet of the nursing mother. In 2008/2009 the patient started a self-guided desensitization with milk products. During that phase, reactions to dietary products occurred primarily exercise-induced with increasing complexity of symptoms. After having avoided milk strictly since 2012, the reactions to accidental contact with milk became more severe (angioedema, flush, urticaria and asthmatic reactions, e.g., after a kiss of her boyfriend who had consumed coffee with milk). Shortly before delivery of her first child and five months thereafter the patient presented to our clinic with the question regarding a potential sensitization to human milk and biological relevance upon contact to her own breast milk while breast feeding.

ID3, a nine-year-old girl, has had a confirmed milk allergy since 2010. The skin and respiratory tract (cough, dyspnea) have been severely affected. Milk and milk products have been strictly avoided since then.

The adult patient ID4 developed reactions to milk at the age of 23. According to the history, the reactions to milk are dose-dependent (different amounts of milk and milk products can be tolerated) and occur exercise-induced (nausea, feeling of swelling throat and troubles with swallowing).

Patient ID5, 25 years old, first observed allergic reactions to milk at the age of 21 and reported nausea and feeling of tightness in esophagus/trachea following milk ingestion.

Patient ID6 reported CMA at the age of 35. Besides gastrointestinal symptoms the patient reported the feeling of dysphagia and of chest pain (dyspnea?) and has avoided milk strictly since 2018.

3.2. Patients with Severe Reactions

Patients ID1, ID2 and ID3 with severe and anaphylactic reactions to cow's milk depict a similar sensitization pattern in the CAP-FEIA, the multiplex dot test and BAT (comprised in Figure 1).

As shown in Figure 1 (ID1-ImmunoCAP data, at the top left), the milk anaphylactic patient ID1 displayed a total IgE concentration of 85 IU/mL, a specific IgE against milk protein of 34.9 kU/L and IgE concentrations for caseins of 31.3 kU/L which were higher compared to the whey proteins α-lactalbumin and ß-lactoglobulin of 8.85 kU/L and 8.86 kU/L, respectively.

In the component-resolved dot test (ID1-Multiplex Dot Test) with IgE recognition not only against bovine but also human milk as well as single milk components (bovine α-, ß-, and κ-caseins, bovine ß-lactoglobulin, bovine and human α-lactalbumin), the patient's IgE reacted intensively against cow's milk and the single caseins. IgE reactivity was less pronounced against human milk, however existing. Dotted bovine α-lactalbumin and ß-lactoglobulin also bound antibodies of patient ID1, but less intensively than the caseins.

Using the BAT as a functional assay to assess the biologic relevance of IgE reactivity, not only cow's milk but also human milk, and the single human and bovine allergens were tested (ID1-BAT). Patient ID1 was reactive to all extracts and single components: to bovine and human milk, as well as to the caseins and whey proteins.

Patient ID2 with an increasing complexity of reactions up to anaphylactic symptoms after a self-guided trial of desensitization, displayed a sensitization pattern as shown in Figure 1 (ID2).

ID2, with total and specific IgE determinations before delivery (ID2; ImmunoCAP data with filled bars), showed a total IgE of 115.7 IU/mL and a low specific IgE against milk protein of only 1.32 kU/L. The casein-specific IgE concentration was 1.27 kU/L in contrast to negative α-lactalbumin- and ß-lactoglobulin-specific IgE. Five months after delivery (ID2–1; ImmunoCAP data with hatched bars), with a total IgE of 78.5 IU/mL; the specific IgE against milk protein was 0.79 kU/L, the casein-specific IgE 0.72 kU/L and the specific IgE against whey proteins were negative.

Dotted cow's milk and the single caseins bound antibodies of both blood samples of ID2 intensively, dotted ß-lactoglobulin only slightly. No IgE was detected by dotted human milk proteins, bovine and human α-lactalbumin (Multiplex Dot test; both boxes with a solid line and the dotted line).

Analyzing the BAT, the patient showed—before (ID2) (depicted in the upper BAT) and 5 months after delivery of her first child (ID2–1) (depicted in the lower BAT) —higher basophil activity against bovine milk, the single bovine caseins and ß-lactoglobulin, compared to human milk as well as human and bovine α-lactalbumin.

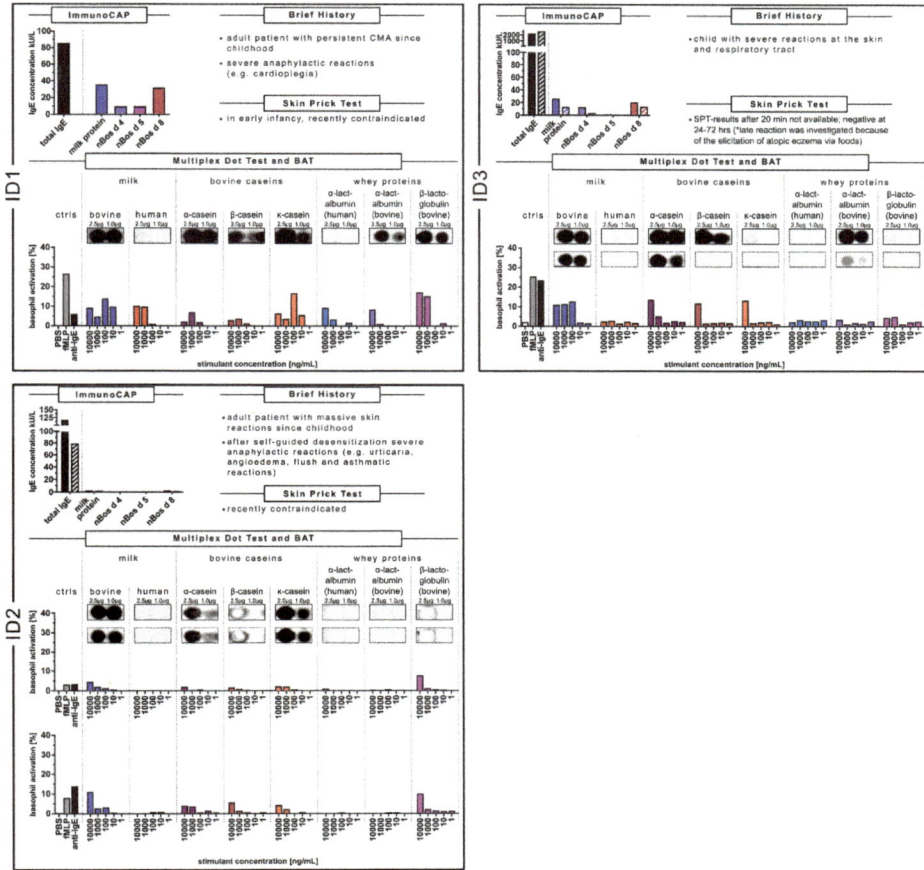

Figure 1. Synopsis of data of the sensitization patterns of ID1, ID2 and ID3 with severe reactions in the ImmunoCAP, multiplex dot test and basophil activation test (BAT). Caseins were associated with this clinical response. CAP data of the cow's milk allergy (CMA) patients are depicted in a bar graph. The filled bars show the data of the first blood sample, the hatched bars those of the second blood sample. Determination of IgE recognition to whole and single allergens of cow's milk and human milk in the multiplex dot test. The concentration of the dot-blotted analytes as indicated was 1.0 µg/µL and 2.5 µg/µL, respectively. The IgE recognition of the first blood sample is given in a box with a solid line, of the second blood sample with a box with a dotted line. The BAT was performed with blood of the patients using whole cow's milk and human milk as well as single bovine and human milk allergens. Percentages represent CD63 positive basophils determined by flow cytometric analysis. Blood sample stimulation was conducted with the analytes as indicated. fMLP, anti-IgE were run as positive controls, PBS as negative control.

For the severely allergic child ID3, with a positive oral challenge in early childhood, serum concentrations of total IgE and specific IgE to milk proteins were tested twice (ID3 in 2016) (CAP data with filled bars) and 2019 in parallel to the BAT (ID3–1) (CAP data with hatched bars). Whereas total IgE even increased from 2096 to >2500 IU/mL, specific IgE against milk protein decreased from 25.10 kU/L to 12.0 kU/L, as did IgE to bovine casein (from 19.5 kU/L to 12.8 kU/L), to bovine α-lactalbumin (from 11.8 kU/L to 2.34 kU/L), and to ß-lactoglobulin (from 0.61 kU/L to 0.31 kU/L). However, the sIgE concentration was highest against caseins (Figure 1-ID3).

In the dot test, the patient had intensive IgE reactivity against cow's milk and α- and ß-casein and with less extent to κ-casein (Multiplex Dot test; boxes with a solid line). At the second sampling, the sensitization appeared to have become less strong. Also, the intensity of IgE against dotted bovine α-lactalbumin decreased (boxes with the dotted line).

The BAT showed biological activity against all analytes exhibiting the highest basophil activity, when the basophils were incubated with caseins (ID3-BAT).

3.3. Patients with Milder Reactions

The subgroup of CMA patients with milder reactions, ID4, ID5 and ID6, showed a similar sensitization pattern in the CAP-FEIA, the multiplex dot test and BAT (comprised in Figure 2),

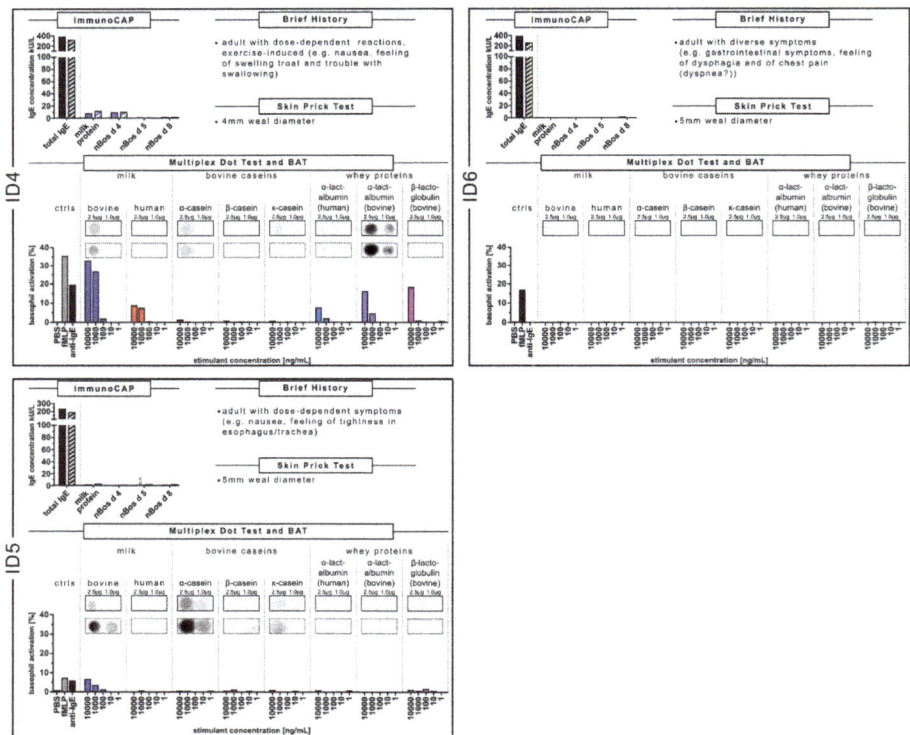

Figure 2. Synopsis of data of the sensitization pattern of ID4, ID5 and ID6 with milder allergic reactions in the ImmunoCAP, multiplex dot test and BAT. Whey proteins were more pronounced with this clinical response. CAP data of the CMA patients were depicted in a bar graph. The filled bars show the data of the first blood sample, the hatched bars are those of the second blood sample. Determination of IgE recognition to whole and single allergens of cow's milk and human milk in the multiplex dot test. The concentration of the dot-blotted analytes as indicated was 1.0 μg/μL and 2.5 μg/μL, respectively. The IgE recognition of the first blood sample was given in a box with a solid line, of the second blood sample with a box with a dotted line. The BAT was performed with blood of the patients using whole cow's milk and human milk and single bovine and human milk allergens. Percentages represent CD63 positive basophils determined by flow cytometry. Blood sample stimulation was conducted with the analytes as indicated. fMLP, anti-IgE were run as positive controls, PBS as negative control.

ID4, with dose-dependent and exercise-induced symptoms after milk ingestion, showed the following CAP-FEIA sensitization after being tested on two different visits in our allergy outpatient clinic (in 2016 (ID4)) and 2019, parallel to the BAT (ID4–1) (Figure 2). Total IgE decreased slightly from 379 to 317 IU/mL, with minor changes in the ß-lactoglobulin-IgE-concentration from 0.34 kU/L to 0.56 kU/L. At any rate, this patient showed higher IgE-concentrations against the whey protein α-lactalbumin in comparison to casein (ID4-ImmunoCAP data, at the top left).

In the dot test, the IgE reactivity against cow's milk was less pronounced. The IgE reactivity was lower against the single caseins tested, compared to the stronger IgE reactivity against bovine α-lactalbumin (Multiplex Dot Test; both boxes with a solid line and the dotted line). Accordingly, the BAT revealed the highest reactivity against the whey proteins bovine ß-lactoglobulin, bovine and human α-lactalbumin and none against caseins (ID4-BAT).

For patient ID5, with nausea and feeling of tightness in the throat after milk ingestion, we measured a decrease of total IgE from 229 IU/mL to 185 IU/mL and an increase both in the IgE-concentration to caseins (from 0.70 kU/L to 1.57 kU/mL) and to α-lactalbumin (from 0.20 kU/L to 0.56 kU/L) at two visits, where blood samples were taken (ID5 and ID5–1 parallel to the BAT; Figure 2, ID5, ImmunoCAP data). In the dot test the IgE-reactions of the patient were less intense to the dotted milk allergens, but with an increasing IgE reactivity against α- and κ-casein (ID5 in comparison to ID 5–1; ID5-Multiplex Dot Test; boxes with a solid line compared to the dotted line). In the BAT, the basophils responded merely to cow's milk and to a lower level to ß-lactoglobulin in comparison to the other stimulants (ID5-BAT).

In the CAP-FEIA analysis, ID6, with milder reactions to milk, showed higher concentrations of total IgE (385.0 IU/L) and specific IgE against casein (0.82 kU/L, next to negative specific IgE against the whey proteins α-lactalbumin and ß-lactoglobulin) within the first determination (ID6 in 2017) compared to less total IgE (251 IU/mL) and negative specific IgE against caseins in the second blood sample, parallel to the BAT (ID6–1; Figure 2; ID6-ImmunoCAP data with filled and hatched bars). Actually, all specific IgE concentrations against milk, milk products and single milk components became <0.01 kU/L between 2017 and 2019. No IgE reactivity could be determined in the dot test with the patient's serum sample parallel to the BAT. Unfortunately, the first blood sample from 2017 was no longer available. In the BAT, ID6, however, was a non-responder (Figure 2, ID6-BAT).

ID7, as a non-allergic individual, did not show any specific IgE against milk and milk allergens in the ImmunoCAP, nor any IgE reactivity against the dotted milk allergens. In the BAT, the basophil activation was very low in comparison to the anti-IgE control (unlike ID1–ID5, in which at least one stimulant was equal or higher than the anti-IgE control) (data are shown in Supplemental Figure S1; the data of the Multiplex Dot Test of all patients with CMA, in comparison to the non-allergic individual, are additionally shown in Supplemental Figure S2).

4. Discussion

In this study, we present comprehensive data of different diagnostic tests to characterize sensitization patterns of patients with different clinical pictures of CMA. The BAT and a newly developed multiplex dot test are novel tools beyond the current component-resolved diagnosis of CMA, evaluating the sensitization patterns for both cow's milk and human milk and their single allergens. We found that diagnostic measures could be improved by this molecular characterization in our study group of six patients with different clinical patterns of CMA.

Our CMA patients reported severe and even life-threatening reactions already to traces of milk, so that the challenge tests for ID1 and ID2 were contra-indicated. Even skin prick testing would have been too risky for patients—such as patient ID1—with anaphylactic reactions to food allergens.

The determination of allergen-specific IgE in CMA patients' serum is an approved test to identify cow's milk-sensitized patients. Accurate diagnosis of IgE-mediated CMA was improved by the introduction of the allergenic milk molecules caseins (Bos d 8), α-lactalbumin (Bos d 4) and ß-lactoglobulin (Bos d 5) [3]. Hence, for routine allergy diagnostic tests, total serum IgE, specific IgE to milk, cheese, milk from other species and these single milk allergens are available for CAP-FEIA.

For our CMA patients we found that the anti-casein-IgE concentrations in patients ID1, ID2 and ID3 with a history of anaphylaxis or severe respiratory reactions were higher compared to the patients with less severe clinical reactions upon milk contact (ID4, ID5, ID6). In this respect, our study could discriminate between those patients who experienced severe symptoms that have persisted since childhood (ID1, ID2) and those who have had less severe clinical responses including exercise-induced symptoms (ID4). However, patient ID5 had an increase in specific IgE against casein, probably as a prognostic marker for an upcoming increasing clinical response to cow's milk. On the contrary, ID3 showed a decrease of specific IgE against caseins between two visits, which might be an indication for a better tolerance to milk. Among others, Garcia-Ara [7] and Ito et al. [8] described that casein is a predictor for a prolonged CMA. These findings are in line with a recent study of Chatchatee et al. [16] confirming that IgE directed against the sequential casein epitopes predict persistent CMA. For the CMA patients of our study with less severe symptoms, our data were able to show lower IgE concentrations against caseins, which again support the above-mentioned literature.

In earlier studies, IgE concentrations to the whey proteins α-lactalbumin (Bos d 4) and ß-lactoglobulin (Bos d 5) did not show a clear association with a certain degree of severity of CMA. As such, the assumption that higher specific IgE concentrations to whey allergens are characteristic for persistent CMA it is not clearly defined in the literature. In our study, however, we found that higher IgE antibody concentrations against whey proteins compared to caseins helped to identify those patients who were less severely affected by CMA.

To date, the allergen-specific IgE detection via CAP-FEIA (namely sIgE against milk, against Bos d 4, Bos d 5 and Bos d 8) lacks allergens that became important for our CMA patients. In particular, patient ID2 with severe reactions upon contact to cow's milk was concerned about a possible sensitization to human milk proteins and their clinical implications for her during breastfeeding. Apart from this peculiar clinical case, IgE-reactivity to human milk was described in patients with CMA [17,18]. Hence, we developed a component-resolved dot test for a more refined diagnosis of cow's milk and human milk allergens. By this, with respect to cow's milk, IgE reactivity was able to be differentiated into a broader sensitization pattern against whole cow's milk, bovine α-, ß-, and κ-caseins, bovine ß-lactoglobulin and bovine α-lactalbumin. For human milk, IgE sensitization against whole human milk and human α-lactalbumin could be analyzed. To the best of our knowledge, this is the first component-resolved test for CMA diagnosis identifying milk allergen profiles thus detailed, as well as taking human milk proteins into consideration.

Our data could show that the dot test in a more refined analysis was a suitable test to differentiate between CMA with a more severe clinical outcome (ID1, D2, ID3) derived from the binding intensity to the panel of tested allergens compared to the patients with less pronounced symptoms upon milk contact (ID4, ID5, ID6). Compared to the anaphylactic patients with CMA, patient ID5 (ID5–1) yielded a weaker IgE reactivity to cow's milk, presumably representing a developing response to cow's milk. Therefore, this patient should be closely monitored. Moreover, we found that patients ID1 and ID2 were highly reactive to the panel of bovine α-, ß-, and κ-caseins, whereas the patient ID3 depicted less IgE against κ-caseins, which might—discussed here with due caution—indicate that the patient becomes more tolerant to milk. Again interestingly, patient ID5 recognized bovine α-casein (ID5–1), evidently indicating an upcoming response to cow's milk.

Remarkably, the human milk dot test proved to be slightly positive for patient ID1 with anaphylactic reactions to traces of cow's milk, who reported on severe reactions as a breast-fed baby, despite strict cow's milk avoidance of his nursing mother. On the other hand, the observation that no reactivity was evident for patient ID2 was particularly important for the anaphylactically reacting mother, who feared clinical reactions upon contact to human milk while breast feeding.

For patient ID2, these findings were supported by negative IgE-concentration against human α-lactalbumin. With respect to the bovine whey proteins α-lactalbumin and ß-lactoglobulin, patients ID1 and ID3 with severe reactions to cow's milk had positive IgE against α-lactalbumin and to a lesser extent also ID4. In contrast, ß-lactoglobulin appeared clearly predominant only for ID1.

Whether the developed dot assay revealed superior diagnostic accuracy to the routine CAP-FEIA test when looking at the same allergens is difficult to assess. In fact, both serologic tests reveal comparable IgE sensitization patterns. Therefore, the multiplex dot test improves CMA diagnosis as the sensitization towards human milk proteins can be mirrored, which has to be kept in mind also in CMA patients, when studying patient ID2 [17,18].

Yet, both allergy diagnostic tests primarily identify the sensitization by means of the presence of allergen-specific IgE, not the clinical response. However, to assess the current clinical response to cow's milk and also human milk allergens, we performed BAT with the blood of our CMA patients. Consistent with literature [19–22], the BAT pinpointed the clinical response of CMA best. Our results showed that the basophil activity is correlated with the degree of symptoms. In this respect, it is important that the allergen-induced basophil activation always has to be compared to the anti-IgE activation of the individual patient. In particular for our patients with a more alarming CMA history, the BAT revealed that the basophils were more reactive to allergen stimulation depicting the biological activity against the allergens tested. For ID1, basophil activation before OIT with adjunctive treatment with omalizumab was clearly shown upon stimulation with cow's milk, in particular κ-casein and ß-lactoglobulin and lower against all the other allergens. Evidently, cow's milk rather than human milk is more pronounced in the allergic response in our patient with severe cow's milk allergy since early infancy.

Remarkably, for patient ID2, our data could elucidate that the allergens tested induced basophil activation at lower allergen concentrations, meaning that less allergen is mandatory for CD63 activation. This is described by Hoffmann et al. [23] reporting on the basophil sensitivity (CDsens), defined as the allergen concentration, at which half of all reactive basophils respond. Also, patient ID3 has a high added value by means of the BAT to the sensitization profiles by dot test and CAP-FEIA determination depicting the activation of basophils after stimulation with all analytes and in low concentrations. In contrast, the BAT distinguished for ID4 and ID5 less severe sensitization status, for ID4, interestingly, a less symptomatic clinical outcome, as the patient merely represented a whey-sensitization. For ID5, our study revealed that the BAT is a suitable tool to monitor and detect changes in the in vitro immune response of his clinical milk reactivity, which is to be expected from the data of the CAP-FEIA and dot test.

However, our study group also demonstrated the limitations of BAT for CMA diagnosis. Patient ID6 represents a non-responder who did not show any CD63 activation through anti-IgE stimulation. Among others, Lötsch et al. [24] reported on 3.25–6.5% of non-responders dependent on the readout when studying BAT for hazelnut allergic patients. In such cases, the BAT was of no value; hence, the classical CAP-FEIA test and the multiplex dot test have merit and may provide reliable results.

Our study group assesses only a small number of CMA patients. However, we were able to differentiate between the various clinical pictures of CMA patients from severe anaphylactic forms identified as casein allergics to milder forms of CMA identified as whey protein allergics.

The strength of this study is that both our multiplex dot assay and the BAT are superior to IgE recognition in routine diagnostic tests, and we were the first to show that both tests make more cow's milk and human milk allergens accessible for CMA diagnosis. Thus, they enable us to show both the sensitization to cow's milk as well as to human milk proteins. Finally, our BAT with its extensive and specific readout [13] helps to improve the diagnostic accuracy: (1) It mirrors the acute degree of CMA against cow's milk and human milk allergens, and (2) provides monitoring of the development of CMA.

Supplementary Materials: The following are available online at http://www.mdpi.com/2072-6643/11/6/1331/s1, Figure S1: Synopsis of data of the non-allergic individual ID7 in the ImmunoCAP, Multiplex Dot Test and BAT. Figure S2: Determination of IgE recognition of patients with CMA and a non-allergic individual in a Multiplex Dot Test. Table S1: Summary of data for patients with milk allergy.

Author Contributions: Conceptualization, F.S. and U.J.; Data curation, F.S., S.K., C.S. and U.J.; Formal analysis, F.S., S.K., C.S., J.B. and U.J.; Funding acquisition, U.J.; Investigation, F.S., S.K., C.S. and U.J.; Methodology, F.S., S.K., C.S., J.B. and U.J.; Project administration, F.S. and U.J.; Supervision, U.J.; Validation, F.S., S.K., J.B. and U.J.; Visualization, F.S., S.K. and C.S.; Writing—original draft, F.S.; Writing—review and editing, F.S., S.K., C.S., J.B. and U.J. Recruitment of patients, U.J.; Proposal for the local ethics committee, U.J.; Supervision of all ethical procedures, U.J.

Funding: This research was funded by Research Center Borstel, Precision Medicine Fund.

Acknowledgments: We thank Marisa Böttger, Maren Hohn and Carolin Murawski for excellent technical assistance and PD Andreas Recke, Department of Dermatology, University of Lübeck, Lübeck, Germany, and Mareike de Vries, Interdisciplinary Allergy Outpatient Clinic, MK III, University of Lübeck, Airway Research Center North (ARCN), German Center for Lung Research (DZL), Lübeck, for collecting patients' data and sera.

Conflicts of Interest: The authors declare no conflict of interest.

References

1. Sampson, H.A. Update on food allergy. *J. Allergy Clin. Immunol.* **2004**, *113*, 805–819. [CrossRef] [PubMed]
2. Järvinen, K.M.; Chatchatee, P. Mammalian milk allergy: Clinical suspicion, cross-reactivities and diagnosis. *Curr. Opin. Allergy Clin. Immunol.* **2009**, *9*, 251–258. [CrossRef] [PubMed]
3. Matsuoa, H.; Yokoojib, T.; Taogoshi, T. Common food allergens and their IgE-binding epitopes. *Allergol. Int.* **2015**, *64*, 332–343. [CrossRef]
4. Wal, J.M. Bovine milk allergenicity. *Ann. Allergy Asthma Immunol.* **2004**, *93*, 2–11. [CrossRef]
5. Sicherer, S.H.; Sampson, H.A. Cow's milk protein-specific IgE concentrations in two age groups of milk-allergic children and in children achieving clinical tolerance. *Clin. Exp. Allergy* **1999**, *29*, 507–512. [CrossRef] [PubMed]
6. Skripak, J.M.; Matsui, E.C.; Mudd, K.; Wood, R.A. The natural history of IgE-mediated cow's milk allergy. *J. Allergy Clin. Immunol.* **2007**, *120*, 1172–1177. [CrossRef] [PubMed]
7. Garcıa-Ara, M.C.; Boyano-Martınez, M.T.; Dıaz-Pena, J.M.; Martın-Munoz, M.F.; Martın-Esteban, M. Cow's milk-specific immunoglobulin E levels as predictors of clinical reactivity in the follow-up of the cow's milk allergy infants. *Clin. Exp. Allergy* **2004**, *34*, 866–870. [CrossRef]
8. Ito, K.; Futamura, M.; Movérare, R.; Tanaka, A.; Kawabe, T.; Sakamoto, T.; Borres, M.P. The usefulness of casein-specific IgE and IgG4 antibodies in cow's milk allergic children. *Clin. Mol. Allergy* **2012**, *10*. [CrossRef]
9. D'Urbano, L.E.; Pellegrino, K.; Artesani, M.C.; Donnanno, S.; Luciano, R.; Riccardi, C.; Tozzi, A.E.; Rava, L.; De Benedetti, F.; Cavagni, G.; et al. Performance of a component-based allergen-microarray in the diagnosis of cow's milk and hen's egg allergy. *Clin. Exp. Allergy* **2010**, *40*, 1561–1570. [CrossRef]
10. Järvinen, K.M.; Beyer, K.; Vila, L.; Chatchatee, P.; Busse, P.J.; Sampson, H.A. B-cell epitopes as a screening instrument for persistent cow's milk allergy. *J. Allergy Clin. Immunol.* **2002**, *110*, 293–297. [CrossRef]
11. Schocker, F.; Recke, A.; Kull, S.; Worm, M.; Jappe, U. Persistent cow's milk anaphylaxis from early childhood monitored by IgE and BAT to cow's and human milk under therapy. *Pediatric Allergy Immunol.* **2018**, *29*, 210–214. [CrossRef] [PubMed]
12. Schocker, F.; Recke, A.; Kull, S.; Worm, M.; Behrends, J.; Jappe, U. Reply to Chirumbolo et al. *Pediatric Allergy Immunol.* **2018**, *29*, 461–462. [CrossRef]
13. Schwager, C.; Kull, S.; Behrends, J.; Röckendorf, N.R.; Schocker, F.; Frey, A.; Homann, A.; Becker, W.-M.; Jappe, U. Peanut oleosins associated with severe peanut allergy—Importance of lipophilic allergens for comprehensive allergy diagnostics. *J. Allergy Clin. Immunol.* **2017**, *14*, 1331–1338. [CrossRef] [PubMed]
14. Schocker, F.; Baumert, J.; Kull, S.; Petersen, A.; Becker, W.-M.; Jappe, U. Prospective investigation on the transfer of Ara h 2, the most potent peanut allergen, in human breast milk. *Pediatric Allergy Immunol.* **2016**, *27*, 348–355. [CrossRef] [PubMed]
15. Schocker, F.; Scharf, A.; Kull, S.; Jappe, U. Detection of the Peanut Allergens Ara h 2 and Ara h 6 in Human Breast Milk: Development of 2 Sensitive and Specific Sandwich ELISA Assays. *Int. Arch. Allergy Immunol.* **2017**, *174*, 17–25. [CrossRef] [PubMed]
16. Chatchatee, P.; Järvinen, K.M.; Bardina, L.; Beyer, K.; Sampson, H.A. Identification of IgE- and IgG-binding epitopes on alpha(s1)-casein: Differences in patients with persistent and transient cow's milk allergy. *J. Allergy Clin. Immunol.* **2001**, *107*, 379–383. [CrossRef] [PubMed]

17. Schulmeister, U.; Swoboda, I.; Quirce, S.; de la Hoz, B.; Ollert, M.; Pauli, G.; Valenta, R.; Spitzauer, S. Sensitization to human milk. *Clin. Exp. Allergy* **2008**, *38*, 60–68. [CrossRef]
18. Mäkinen-Kiljunen, S.; Plosila, M. A father's IgE mediated contact urticaria from mother's milk. *J. Allergy Clin. Immunol.* **2004**, *113*, 353–354. [CrossRef]
19. Santos, A.F.; Lack, G. Basophil activation test: Food challenge in a test tube or specialist research tool? *Clin. Transl. Allergy* **2016**, *6*, 10. [CrossRef]
20. Ford, L.S.; Bloom, K.A.; Nowak-Węgrzyn, A.H.; Shreffler, W.G.; Masilamani, M.; Sampson, H.A. Basophil Reactivity, Wheal Size and Immunoglobulin Levels Distinguish Degree of Cow's Milk Tolerance. *J. Allergy Clin. Immunol.* **2013**, *131*, 180–186. [CrossRef]
21. Nilsson, C.; Nordvall, L.; Johansson, G.O.; Nopp, A. Successful management of severe cow's milk allergy with omalizumab treatment and CD-sens monitoring. *Asia Pac. Allergy* **2014**, *4*, 257–260. [CrossRef] [PubMed]
22. Santos, A.F.; Brough, H.A. Making the Most of In Vitro Tests to Diagnose Food Allergy. *J. Allergy Clin. Immunol. Pract.* **2017**, *5*, 237–248. [CrossRef] [PubMed]
23. Hoffmann, H.J.; Santos, A.F.; Mayorga, C.; Nopp, A.; Eberlein, B.; Ferrer, M.; Rouzaire, P.; Ebo, D.G.; Sabato, V.; Sanz, M.L.; et al. The clinical utility of basophil activation testing in diagnosis and monitoring of allergic disease. *Allergy* **2015**, *70*, 1393–1405. [CrossRef] [PubMed]
24. Lötzsch, B.; Dölle, S.; Vieths, S.; Worm, M. Exploratory analysis of CD63 and CD203c expression in basophils from hazelnut sensitized and allergic individuals. *Clin. Transl. Allergy* **2016**, *6*, 45. [CrossRef] [PubMed]

nutrients

MDPI

Article

Comparison of the Allergenicity and Immunogenicity of Camel and Cow's Milk—A Study in Brown Norway Rats

Natalia Zofia Maryniak, Egon Bech Hansen, Anne-Sofie Ravn Ballegaard, Ana Isabel Sancho and Katrine Lindholm Bøgh *

Division of Diet, Disease Prevention and Toxicology, National Food Institute, Technical University of Denmark, 2800 Kgs. Lyngby, Denmark; nazoma@food.dtu.dk (N.Z.M.); egbh@food.dtu.dk (E.B.H.); anravn@food.dtu.dk (A.-S.R.B.), anasa@food.dtu.dk (A.I.S.)
* Correspondence: kalb@food.dtu.dk; Tel.: +45-3588-7092

Received: 2 November 2018; Accepted: 29 November 2018; Published: 4 December 2018

Abstract: Background: When breastfeeding is impossible or insufficient, the use of cow's milk-based hypoallergenic infant formulas is an option for infants suffering from or at risk of developing cow's milk allergy. As the Camelidae family has a large evolutionary distance to the Bovidae family and as camel milk differs from cow's milk protein composition, there is a growing interest in investigating the suitability of camel milk as an alternative to cow's milk-based hypoallergenic infant formulas. Methods: The aim of the study was to compare the allergenicity and immunogenicity of camel and cow's milk as well as investigating their cross-reactivity using a Brown Norway rat model. Rats were immunised intraperitoneally with one of four products: camel milk, cow's milk, cow's milk casein or cow's milk whey fraction. Immunogenicity, sensitising capacity, antibody avidity and cross-reactivity were evaluated by means of different ELISAs. The eliciting capacity was evaluated by an ear swelling test. Results: Camel and cow's milk showed similarity in their inherent immunogenicity, sensitising and eliciting capacity. Results show that there was a lower cross-reactivity between caseins than between whey proteins from camel and cow's milk. Conclusions: The study showed that camel and cow's milk have a low cross-reactivity, indicating a low protein similarity. Results demonstrate that camel milk could be a promising alternative to cow's milk-based hypoallergenic infant formulas.

Keywords: food allergy; cow's milk; camel milk; infant formula; animal models

1. Introduction

Cow's milk allergy (CMA) is the most prevalent food allergy in infants and small children [1], affecting around 2.5% [2,3], although differences are observed between studies and countries [4]. Although most CMA children outgrow their allergy, some keep it for life [5]. Originally, it was though that most children did outgrow their CMA before the age of three years, but there seems to be a tendency that more and more children outgrow their CMA later in life and for some it may even last for lifetime [6,7]. Breastfeeding is the most suited source of nutrition for a newborn infant [8]. However, in some situations, breastfeeding is impossible or insufficient and a substitute such as an infant formula is needed [9]. Infant formulas are usually based on cow's milk, as this is the most easily accessible milk source globally [10]. When an infant suffers from or is at risk of developing CMA, alternatives to conventional infant formulas are recommended such as hypoallergenic infant formulas, based on extensively or partially hydrolysed cow's milk proteins [11]. In addition to cow's milk-based hypoallergenic infant formulas, additional alternatives to conventional infant formulas are found on the market, such as amino acid-based infant formulas, plant-based infant formulas (e.g., soya-based) and infant formulas based on other mammalian milk (e.g., goat or sheep) [8,12,13].

Extensively and partially hydrolysed infant formulas as well as amino acid-based infant formulas are poor in flavour, and, thus, some newborns may refuse them [5,14]. On the other hand, it has been reported that sheep and goat milk-based infant formulas may only be an alternative for some newborns due to a high cross-reactivity between cow's milk proteins and proteins from goat and sheep milk [13,15]. In addition, plant-based infant formulas are seldom recommended due to their low nutritional value [16,17]. For those reasons, new or improved alternatives to conventional infant formulas are still of interest.

Due to the large evolutionary distance between Camelus dromedaries (Camelidae family) and the Bovidae family animals, camel milk is quite different in its composition compared to cow's milk. Equivalent to human milk, the allergenic milk protein β-lactoglobulin (BLG) is also absent in camel milk [18]. Moreover, similar to human milk, camel milk has approximately double the amount of β-casein and approximately five times the amount of immunoglobulins in comparison to cow's milk [19]. Rastani et al. [13] showed that CMA patients did not recognise camel milk by immunoblotting and concluded that camel milk is a promising alternative to cow's milk for infant formula manufacture. Further, based on double-blind, placebo-controlled food challenges, Navarre-Rodriguez et al. [20] concluded that camel milk is a safe and tolerable alternative for CMA patients above the age of one year. Camel milk is already commercially available in the Middle East, Australia, United Kingdom and the Netherlands [21–24]. In other regions such as in African countries, it is a traditionally consumed product, although without a control on its quality and safety [25]. There are a number of studies showing that camel milk is nutritionally suitable for human consumption [21,26]. For those reasons, camel milk is an exciting and suitable product with the potential to be a future alternative to hypoallergenic cow's milk-based infant formulas in prevention, treatment and management of CMA in infants and small children.

The purpose of this study was to investigate the immunogenicity and allergenicity of camel and cow's milk as well as studying cross-reactivity between proteins from the two sources. To do this, Brown Norway (BN) rats were immunised intraperitoneally (i.p.) with either camel milk, cow's milk, cow's milk casein fraction or cow's milk whey fraction and antibody responses were evaluated for level, specificity, avidity, functionality and cross-reactivity by means of different enzyme-linked immunosorbent assays (ELISAs), immunoblotting and in vivo test. This should allow for an overview of the usability of camel milk as an alternative to hypoallergenic infant formulas.

2. Materials and Methods

2.1. Products

Powders of cow's milk, cow's milk casein fraction and cow's milk whey fraction were kindly provided by Arla Foods Ingredients Videbæk, Denmark. Powder of camel milk was kindly provided by Dairy Farm Smits, Berlicum, the Netherlands. Products were tested by Pierce™ LAL Chromogenic Endotoxin Quantitation Kit (88282, Thermo Fisher, Waltham, MA, USA) in accordance with the instruction given by the manufacturer. Whereas camel milk, cow's milk and cow's milk whey fraction had an endotoxin level <2 endotoxin units (EU) per mg of protein, cow's milk casein fraction had an endotoxin level of approximately 66 EU per mg of protein.

2.2. In Silico Protein Analyses

CLC Main Workbench 8.0 (Redwood City, CA, USA) was used to compare selected protein amino acid sequences from cow's milk with those of goat, sheep, camel and human milk. Protein sequences were downloaded from UniProt (http://www.uniprot.org).

2.3. Denaturation of Products

Camel milk and cow's milk were denatured to obtain unfolded structures of proteins. Denaturation was performed by reduction and alkylation, as previously described by Madsen et al. [27].

2.4. SDS-PAGE Electrophoresis

Sodium dodecyl sulphate-polyacrylamide gel electrophoresis (SDS-PAGE) with camel milk, denatured camel milk, cow's milk, denatured cow's milk, cow's milk casein fraction and cow's milk whey fraction was performed using 5 µg of each product dissolved in Laemmli buffer (65.8 mM Tris-HCl, pH 6.8, 26.3% (*w/v*) glycerol and 2.1% (*w/v*) SDS, 161-0737, Bio-Rad, Hercules, CA, USA) with addition of β-mercaptoethanol (14.2 M, 161-0710, Bio-Rad). Samples were incubated for 5 min at 95 °C and afterwards loaded onto a 4–20% gel (Mini-Protean TGX Stain-Free gel, 456-8093, Bio-Rad). SDS-PAGE was performed in running buffer (25 mM Tris and 192 mM Glycine and with addition of 0.1% (*w/v*) SDS, pH 8.3, 161-0732, Bio-Rad). Additionally, 10 µL of the molecular weight Precision Plus Protein™ Unstained Standard (161-0363, Bio-Rad) was loaded onto the gel. Gel electrophoresis was run at 200 V with constant current at room temperature (RT). Afterwards, the gel was stained with Bio Safe™ Coomassie (161-0786, Bio-Rad) for 1 h at RT and photographed using Imager ChemiDoc XRS+ (Bio-Rad).

2.5. Animals

BN rats were from the in-house breeding colony, at the National Food Institute, Technical University of Denmark, Denmark, and kept in macrolon cages at 22 °C ± 1 °C with 55 ± 5% relative humidity at a 12-h light–dark cycle. Air exchange was applied 8–10 times per hour with overpressure. BN rats were inspected twice a day and weighted once per week. Rats were kept on a diet free from milk and soy allergens for ≥10 generations. Feed containing rice flour and fish was given ad libitum as well as was acidified tap water.

2.6. Animal Sensitisation Studies

To sensitise animals and raise antibodies against camel milk, cow's milk, cow's milk casein fraction or cow's milk whey fraction, BN rats 4–7 weeks of age, were divided into five groups of eight rats (*n* = four/gender), and housed two per cage. Groups of rats were immunised i.p. three times with 200 µg of product dissolved in phosphate buffer saline (PBS) (137 mM NaCl, 3 mM KCl, 8 mM Na_2HPO_4, 1 mM KH_2PO_4, pH 7.2) without the use of adjuvant one time at Day 0, 14 and 28 (Figure 1). One group of rats was not immunised to act as a control group (naïve animals) for an ear swelling test. At Day 35, rats were sacrificed and blood collected. The animal experiment was carried out at the National Food Institute, Technical University of Denmark under ethical approval given by the Danish Animal Experiments Inspectorate and the authorisation number 2015-15-0201-00553-C1. The experiment was overseen by the National Food Institute's in-house Animal Welfare Committee for animal care and use.

2.7. Ear Swelling Test

To investigate the eliciting capacity of camel and cow's milk, at Day 33 of the experiment, an ear swelling test was performed. Rats were anesthetised with hypnorm-dormicum and baseline ear thickness was measured. Subsequently, 20 µL of PBS with 10 µg of camel milk or 10 µg of cow's milk were injected into the right or left ear, respectively, and ear thicknesses were measured again one hour after injections. Naïve rats were included to see unspecific ear swelling and irritation capacity after camel and cow's milk protein ear injection. Delta ear swelling was calculated.

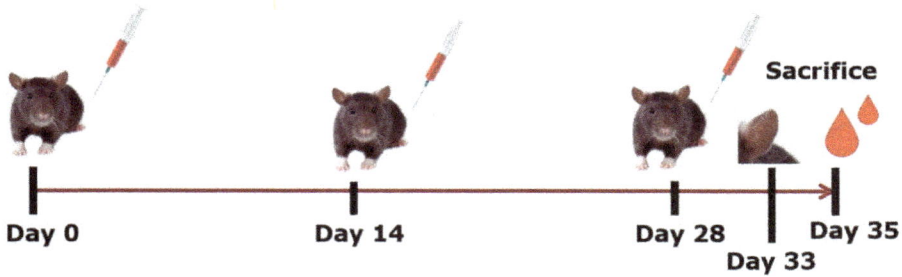

Figure 1. Animal experimental design. Brown Norway rats were immunised i.p. with 200 µg of camel milk, cow's milk, cow's milk casein fraction or cow's milk whey fraction three times, at Days 0, 14 and 28. At Day 33 an ear swelling test was performed and at Day 35 rats were sacrificed and blood collected. Pictures were purchased from https://www.colourbox.com.

2.8. Indirect ELISA for Specific IgG1 Detection

To detect IgG1 antibodies specific for camel milk, denatured camel milk, cow's milk and denatured cow's milk, indirect ELISAs were performed using Maxisorp microtitre plates (96-well, Nunc, Roskilde, Denmark). Plates were coated with 100 µL/well of 10 µg/mL of camel milk, denatured camel milk, cow's milk or denatured cow's milk, in coating buffer (15 mM Na_2CO_3, 35 mM $NaHCO_3$, pH 9.6), and incubated overnight at 4 °C. Between each step, plates were washed five times in PBS with 0.01% (w/v) Tween 20 (PBS-T). For all steps that required incubation, plates were incubated for one hour in the dark at RT, with gentle agitation. First, plates were incubated with 50 µL/well of two-fold serial dilution of serum samples (v/v) in PBS-T. In each plate, positive and negative control serum samples were included in order to identify potential plate-to-plate variance. For antibody detection, 50 µL/well of secondary antibody (horse radish peroxidase (HRP)-labelled-mouse-anti-rat IgG1, 3060-05, Southern Biotech, Birmingham, AL, USA) diluted 1:20,000 (v/v) in PBS-T was added to the plates. After incubation plates were additionally washed twice with tap water. To visualise specific antibody detection, 100 µL/well of TMB-one (3,3′,5,5′-tetramethylobenzidine, 4380A, Kementec Diagnosis, Taastrup, Denmark) was added and incubated for 12 min at RT. The reaction was stopped with 100 µL/well 0.2 M H_2SO_4 and the absorbance was measured at 450 nm with a reference wavelength of 630 nm using a microtitre reader (Gen5, BioTek, EL800 Instrument, Winooski, VT, USA). The cut-off values were set to be higher than the mean absorbance of negative control plus three times the standard deviation (SD). Results were expressed in log2 titre values with a cut-off at the optical density (OD) of 0.1 for IgG1 specific for camel milk, cow's milk and denatured cow's milk and 0.15 for IgG1 specific for denatured camel milk.

2.9. Antibody Capture ELISA to Detect Specific IgE

To detect IgE specific for camel milk, denatured camel milk, cow's milk and denatured cow's milk, antibody capture ELISAs were performed using Maxisorp microtitre plates (96-well, Nunc) coated with 100 µL/well of mouse anti-rat IgE (HDMAB-123, Hydri-Domus, Nottingham, UK) diluted 1:2000 in coating buffer and incubated overnight at 4 °C. Between each step, plates were washed five times with PBS-T. For all steps that required incubation, plates were incubated for one hour in the dark at RT, with gentle agitation. For camel and cow's milk specific IgE detection, antibody capture ELISA was optimised to use proper blocking for each product. Plates were blocked at 37 °C, 200 µL/well, with 3% (v/v) horse serum for camel milk specific IgE detection and 5% (v/v) rabbit serum for cow's milk specific IgE detection, diluted in PBS-T. Subsequently, plates were incubated for one hour with 50 µL/well of two-fold serial dilution of serum samples (v/v) in PBS-T. In each plate, positive and negative control serum samples were included. Afterwards, 50 µL/well of 0.05 µg/mL of 10:1 digoxigenin (DIG)-coupled camel milk or 0.1 µg/mL of 10:1 DIG-coupled cow's milk in PBS-T were added, to detect specific IgE. Next, plates were incubated with 100 µL/well of HRP-labelled sheep-anti-DIG-POD (11633716001, Roche, Diagnostics GmbH, Mannheim, Germany) diluted 1:1000 (v/v) in PBS-T. After this step, plates were additionally washed twice with tap water and incubated for 12 min with 100 µL/well of TMB-one (Kementec Diagnosis). The reaction was stopped with 100 µL/well of 0.2 M H_2SO_4 and the absorbance was measured. Results were expressed as log2 titre value with an individual cut-off of plates at an OD of 0.145–0.2 for IgE specific for camel milk and of 0.125–0.175 for IgE specific for cow's milk.

2.10. Avidity Measurements

To measure binding strength between antigens and IgG1 antibodies from serum samples, avidity ELISA was performed as previously described by Bøgh et al. [28]. Serum samples from rats that reached an OD of at least 0.5 were included.

2.11. Inhibitory ELISA

To examine the cross-reactivity between proteins from camel and cow's milk, inhibitory ELISA was performed. The procedure was as described for the indirect IgG1 ELISA with few exceptions. Serum samples for each group of animals were pooled and diluted in PBS-T to reach an OD of approximately 2.0. Serum pools were then pre-incubated for one hour with ten-fold serial dilutions of camel and cow's milk. After pre-incubation, samples were added to the plates in duplicates and incubated for one hour. The assay was performed twice. The results were expressed in percentage inhibition against the concentration of the inhibitor.

2.12. Immunoblotting

To do immunoblotting, SDS-PAGE was performed with 5 µg of camel and cow's milk as described previously. In addition, SDS-PAGE with an eight-time higher load of proteins (40 µg) was performed to visualise cross-reactivity. After SDS-PAGE, proteins were transferred onto polyvinylidene difluoride membranes (Trans-Blot® Turbo™ Mini PVDF Transfer Pack, 1704156, Bio-Rad) by semidry blotting (Trans-Blot® Turbo™ Transfer System, 170-4150, Bio-Rad) at constant 200 V. Membranes were washed three times for 5 min in PBS-T (0.05% v/v Tween 20) and each blocked with 20 mL of 5% ovalbumin (OVA, egg whites from chicken, Sigma Aldrich, St. Louis, MO, USA) diluted in PBS-T (0.1% v/v Tween 20) and incubated for one hour in the dark at RT, on a shaking table. The 5% OVA solution was used during the whole experiment as a blocking solution. After blocking, membrane was divided into two pieces, both pieces with 5 µg of camel and cow's milk. Next, 10 mL of serum pooled from rats immunised with cow's milk diluted 1:3000 (v/v) in blocking solution or serum pooled from rats immunised with camel milk diluted 1:8000 (v/v) in blocking solution were added separately to each half of the membrane containing 5 µg of camel and cow's milk and incubated for one hour in the dark at RT, on a shaking table. Half of the membrane with 40 µg of cow's milk was incubated with serum pooled from rats immunised with camel milk diluted 1:500 (v/v) in blocking solution, while the other half of the membrane with 40 µg of camel milk was incubated with serum pooled from rats immunised with cow's milk diluted 1:500 (v/v) in blocking solution. Afterwards, membranes were washed three times for 5 min in PBS-T (0.05% v/v Tween 20) and 10 mL of the secondary antibody diluted 1:15,000 together with StrepTacin-HRP conjugate (Bio-Rad) for Precision Plus Protein™ Unstained Standard detection, diluted 1:15,000 in blocking solution were added to each half of the membrane. Membranes were incubated for one hour in the dark at RT, on a shaking table. Subsequently, membranes were washed three times for 5 min in PBS-T (0.05% v/v Tween 20) followed by PBS washing two times for 5 min to remove the detergent. Membranes were incubated with peroxidase substrate (Clarity™ Western ECL Substrate, 1705060, Bio-Rad) for 5 min. After incubation, membranes were developed and photographed using Imager ChemiDoc XRS+ (Bio-Rad).

2.13. Statistical Analysis of Data

Graphs and statistical analyses of the data were performed using GraphPrism version 7.0 (San Diego, CA, USA). Results from indirect and antibody-capture ELISAs were expressed as log2 antibody titre values.

ELISA results expressed as log2 antibody titres were tested for normality distribution. Based on the results, either parametric or non-parametric *t*-tests were performed. Differences were regarded as statistically significant when $p \leq 0.05$. Asterisks indicate statistically significant differences between two given groups: $* = p \leq 0.05$, $** = p \leq 0.01$, $*** = p \leq 0.001$, $**** p \leq 0.0001$.

Inhibition curves resulting from avidity and inhibitory ELISA were examined with one-way repeated-measurements ANOVA test. Analyses showed no statistically significant differences between curves, thus IC_{50} calculations were performed. IC_{50} was calculated using sigmoidal dose response with non-linear regression.

3. Results

3.1. Protein Characterisation

The primary sequence from selected cow's milk proteins were aligned to their counterpart proteins in milk from goat, sheep, camel and human to investigate the amino acid sequence identity between the different species and to predict the potential cross-reactivity between proteins of interest. As shown in Table 1, goat and sheep milk protein sequences show a very high percentage identity to cow's milk proteins, ranging from a protein sequence identity of 85–95% for goat and sheep. A much lower protein sequence identity was evidenced between camel and cow's milk proteins, where the protein identity ranged from 47% to 81%. This is very similar to the protein sequence identity of human and cow's milk proteins ranging from 33% to 76% and human and camel milk proteins ranking from 40% to 76%. In addition, neither camel nor human milk contains BLG [18]. Similarities between camel and cow's milk caseins sequences were shown to be slightly lower than between the whey proteins.

Table 1. Amino acid sequence identity between cow's and goat, sheep, camel and human milk proteins.

		Goat	Sheep	Camel	Human [c]
Casein	β-casein	91	91	67	55 (60)
	αs1-casein	88	88	47	33 (40)
	αs2-casein	88	89	56	NA [a]
	κ-casein	85	85	58	52 (60)
Whey	α-lactalbumin	95	95	60	74 (62)
	β-lactoglobulin	93	93	NA [b]	NA [b]
	serum albumin	88	92	81	76 (76)
	lactoferrin	92	92	75	70 (74)

Sequence identity (%) between selected cow's milk proteins and their counterpart milk proteins from goat, camel and human expressed in percentage. Sequence alignments were performed using CLC Main Workbench 8.0 and UniProt and NCBI database. NA: not available. (a) αs2-casein not identified in human milk [29,30]. (b) β-lactoglobulin not available in camel and cow's milk [18]. (c) Numbers in brackets represents sequence identity between human and camel milk. Accession number: β-casein: Cow: AAA30431; Goat: AAA30906; Sheep: CAA56139; Camel: CDO50354; Human: AAC82978. αs1-casein: Cow: AAA30429; Goat: CAA51022; Sheep: AEN84772; Camel: O97943; Human: CAA55185. αs2-casein: Cow: NP_776953; Goat: CAC21704; Sheep: CAA26983; Camel: O97944. κ-casein: Cow: CAA33034; Goat: CAA43174; Sheep: NP_001009378; Camel: CCI79378; Human: CAA47048. α-lactalbumin: Cow: CAA29664; Goat: CAA28797; Sheep: CAA29665; Camel: P00710; Human: AAA60345. β-lactoglobulin: Cow: CAA32835; Goat: CAA79623; Sheep: CAA31305. serum albumin: Cow: CAA41735; Goat: XP_005681801; Sheep: CAA34903; Camel: XP_010981066; Human: AAN17825; lactoferrin: Cow: AAA30610; Goat: AAA97958; Sheep: ACT76166; Camel: CAB53387; Human AAA59511.

SDS-PAGE electrophoresis was performed to display the protein profile of the products used in this study. Caseins run as thick bands between 25 and 37 kilodalton (kDa) in both camel and cow's milk (Lanes 1–4, Figure 2) as well as in the casein fraction of cow's milk (Lane 5) [31]. The band corresponding to a molecular weight (MW) of around 30 kDa represents β-casein while the band immediately above represents α-caseins with a MW of around 35 kDa [19,31]. In cow's milk as well as in the whey fraction of cow's milk (Lanes 3 and 6), a clear band representing BLG is evident (~18 kDa) [1], which is not present in camel milk. In all lanes except for the lane corresponding to the casein fraction of cow's milk (Lane 6), the lower band represents α-lactalbumin (ALA) (~14 kDa), while the two upper bands most likely represent lactoferrin (LF) (~75 kDa) and serum albumin (SA) (~66 kDa) [19]. Immunoglobulins (~150 kDa) are only hardly seen due to their low amount in the milk products. LF and SA are slightly more visible in the denatured version of the milk products (Lanes 2 and 4) than their native counterparts. Another difference between the native and denatured version of the milk products are a lower mobility of proteins in the denatured versions compared to the native versions.

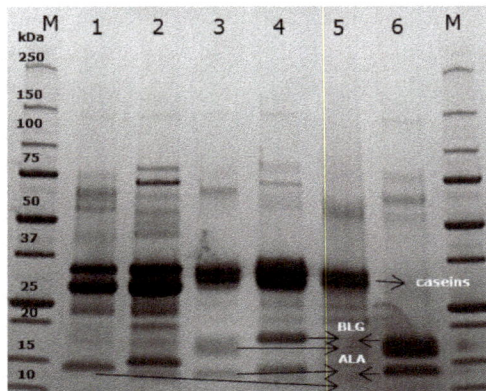

Figure 2. SDS-PAGE electrophoresis. Gel electrophoresis, with native and denatured camel and cow's milk as well as with native cow's milk casein fraction and native cow's milk whey fraction, was performed to display protein profiles. M, protein standard (kDa); 1, camel milk; 2, denatured camel milk; 3, cow's milk; 4, denatured cow's milk; 5, cow's milk casein fraction; 6, cow's milk whey fraction. BLG, β-lactoglobulin; ALA, α-lactalbumin.

3.2. Camel and Cow's Milk Immunogenicity and Cross-Reactivity

Serum samples from individual BN rats immunised with camel milk, cow's milk, cow's milk casein fraction or cow's milk whey fraction were assessed for specific IgG1 by means of indirect ELISAs. Figure 3A shows the IgG1 responses against both the native and denatured version of camel as well as cow's milk proteins.

The immunogenicity of camel and cow's milk appears to be very similar as there is no statistically significant difference between the level of camel milk specific IgG1 raised against camel milk and the level of cow's milk specific IgG1 raised against cow's milk (Figure 3A). For both antibodies raised against camel or cow's milk proteins, there is a statistically significant difference between the IgG1 reactivity towards camel milk and cow's milk proteins, indicating a low cross-reactivity between camel and cow's milk proteins. For antibodies raised against camel milk, the IgG1 reactivity against cow's milk proteins was ~30 fold lower than the reactivity against camel milk proteins, measured by the amount of specific antibodies. Opposite the IgG1 reactivity against camel milk proteins was ~50-fold lower than the reactivity against cow's milk proteins for sera raised against cow's milk proteins. This was shown irrespectively of responses that were measured against the native or denatured version of the milk proteins.

The IgG1 responses in rats immunised with either the casein or the whey fraction of cow's milk, are shown in Figure 3B. For both groups of animals, the IgG1 responses against native and denatured camel milk were statistically significantly lower than the responses against native and denatured cow's milk, stressing a low cross-reactivity for both the casein and the whey fraction of camel and cow's milk proteins. For antibodies raised against casein, the IgG1 reactivity against native camel milk proteins was ~250-fold lower than the reactivity against cow's milk proteins, while for antibodies raised against whey, the IgG1 reactivity against camel milk proteins was ~15-fold lower than the reactivity against cow's milk proteins. This indicates a lower cross-reactivity between camel and cow's milk caseins than whey proteins.

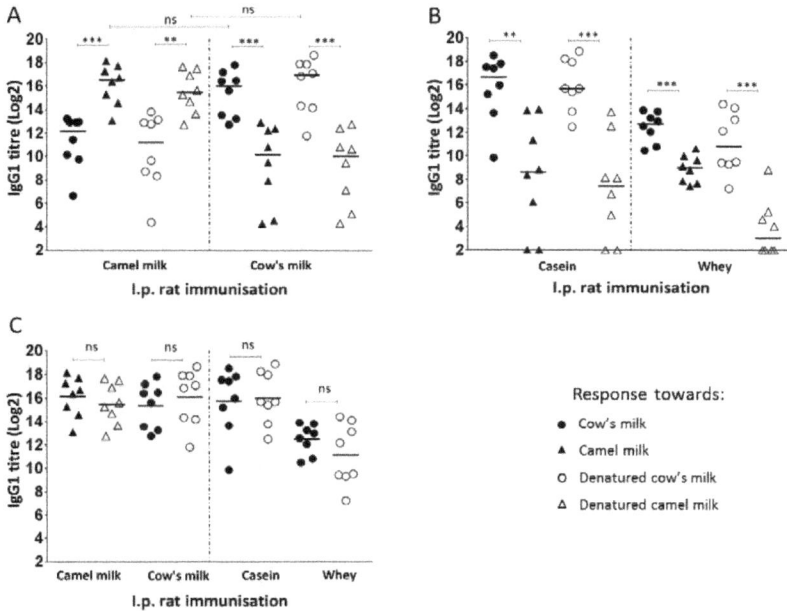

Figure 3. Specific IgG1 antibody responses. Comparison of specific IgG1 antibody responses toward cow's milk (●), camel milk (▲), denatured cow's milk (○) and denatured camel milk (△) raised in rats immunised with camel milk, cow's milk, cow's milk casein fraction or cow's milk whey fraction. Each symbol represents the specific IgG1 titre value for an individual rat. (**A**) Comparison of native and denatured camel milk and cow's milk specific IgG1 antibody responses in rats immunised with camel or cow's milk, respectively. Horizontal lines display the median values for each group of rats. Statistically significant difference between two groups was determined using the non-parametric Mann–Whitney test. Asterisks indicate statistically significant differences between the two given groups when: $* = p \leq 0.05$, $** = p \leq 0.01$, $*** = p \leq 0.001$, $**** p \leq 0.0001$. (**B**) Comparison of native and denatured camel milk and cow's milk specific IgG1 antibody responses in rats immunised with cow's milk casein or whey fraction. Horizontal lines display the median values for each group of rats. Statistically significant difference between two groups was determined using the non-parametric Mann–Whitney test. Asterisks indicate statistically significant differences between the two given groups when: $* = p \leq 0.05$, $** = p \leq 0.01$, $*** = p \leq 0.001$, $**** p \leq 0.0001$. (**C**) Comparison of IgG1 antibody reactivity against native vs. denatured camel and cow's milk. Horizontal lines display the mean values for each group of rats. Statistically significant difference between two groups was determined using the parametric *t*-test. Asterisks indicate statistically significant differences between the two given groups when: $* = p \leq 0.05$, $** = p \leq 0.01$, $*** = p \leq 0.001$, $**** p \leq 0.0001$.

3.3. Linear and Conformational Epitope Recognition

The study showed that there were no statistically significant differences between the IgG1 responses against the native and denatured versions of milk proteins for rats immunised with neither camel milk nor cow's milk, indicating that linear epitopes are dominating both responses (Figure 3C). In addition, the IgG1 raised against cow's milk caseins showed no statistically significant difference in their reactivity against the native or denatured version of milk proteins with an approximate ratio of 1:1. In contrast, although IgG1 raised against cow's milk whey showed no statistically significant difference in their reactivity against the native and denatured version of milk proteins, the ratio between IgG1 specific for native vs. denatured cow's milk was 4:1. This demonstrates that, while caseins primarily induce antibodies against linear epitopes, whey primarily induces antibodies against conformational epitopes.

3.4. Inhibitory ELISA

Inhibitory ELISA was performed with sera pools from groups of rats immunised with camel milk, cow's milk, cow's milk casein or whey fraction in order to evaluate the competitive capacity of native as well as denatured camel and cow's milk.

3.4.1. IgG1 Antibody Competition of Native Camel and Cow's Milk

While native camel milk was able to fully inhibit antibodies raised against camel milk, native cow's milk was only able to inhibit ~50% of the antibodies raised against camel milk (Figure 4A). On the other hand, while native cow's milk was fully capable of inhibiting the antibodies raised against cow's milk, native camel milk was only capable of inhibiting ~35% of antibodies raised against cow's milk (Figure 4B). While native cow's milk was able to fully inhibit antibodies raised against both the casein and the whey fraction of cow's milk, native camel milk was only able to inhibit ~30% of antibodies raised against cow's milk casein (Figure 4C) and ~45% of antibodies raised against cow's milk whey (Figure 4D). This confirms previous results, showing a lower cross-reactivity between casein compared to the whey fraction of camel and cow's milk.

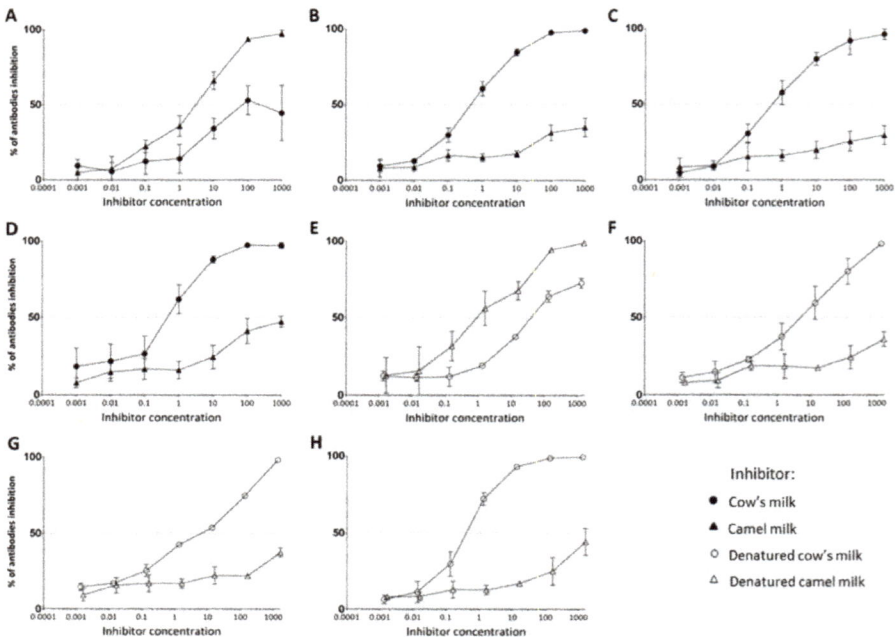

Figure 4. IgG1 antibody binding competition. Inhibitory ELISA with native (●) or denatured (○) cow's milk or native (▲) or denatured (△) camel milk as inhibitors was performed using serum pools from rats immunised with camel milk, cow's milk, cow's milk casein fraction or cow's milk whey fraction.

Each symbol represents the percent inhibition of IgG1 specific antibodies at different inhibitor concentrations. Error bars in the inhibition curves represent ± standard deviation (SD). (**A**) Inhibition curve for sera raised against camel milk. (**B**) Inhibition curve for sera raised against cow's milk. (**C**) Inhibition curve for sera raised against cow's milk casein fraction. (**D**) Inhibition curve for sera raised against cow's milk whey fraction. (**E**) Inhibition curve for sera raised against linear epitopes of camel milk. (**F**) Inhibition curve for sera raised against linear epitopes of cow's milk. (**G**) Inhibition curve for sera raised against linear epitopes of cow's milk casein fraction. (**H**) Inhibition curve for sera raised against linear epitopes of cow's milk whey fraction.

3.4.2. IgG1 Antibody Competition Towards Denatured Camel and Cow's Milk

By performing inhibitory ELISA with the use of denatured versions of camel and cow's milk, we could only study the cross-reactivity as a measure of antibodies raised against linear epitopes. While denatured camel milk was able to inhibit fully antibodies raised against linear epitopes of camel milk, denatured cow's milk was able to inhibit ~70% of the antibodies raised against linear epitopes of camel milk (Figure 4E). On the other hand, while denatured cow's milk was fully capable of inhibiting the antibodies raised against linear epitopes of cow's milk, denatured camel milk was only capable of inhibiting ~35% of antibodies raised against cow's milk (Figure 4F). While denatured cow's milk was able to fully inhibit antibodies raised against both linear epitopes of the casein and the whey fraction of cow's milk, denatured camel milk was only able to inhibit ~35% of antibodies raised against linear epitopes of cow's milk casein (Figure 4G) and ~45% of antibodies raised against linear epitopes of cow's milk whey (Figure 4H). This indicated a slightly higher cross-reactivity between linear epitopes compared to conformational epitopes of camel and cow's milk.

3.5. Specific IgG1 Antibody Avidity

Avidity ELISAs were performed to evaluate binding strength between specific IgG1 antibodies and the milk proteins. Figure 5 displays the amount of potassium thiocyanate (KSCN) needed to inhibit 50% of the antibody–antigen binding. Results indicated that there were no statistically significant differences in binding strength between IgG1 raised against camel milk and camel milk or cow's milk. Similar results were shown for IgG1 raised against cow's milk and their binding strength towards cow's milk and camel milk, although slightly higher avidity was shown between antibodies raised against cow's milk and cow's milk compared to the avidity between antibodies raised against cow's milk and camel milk.

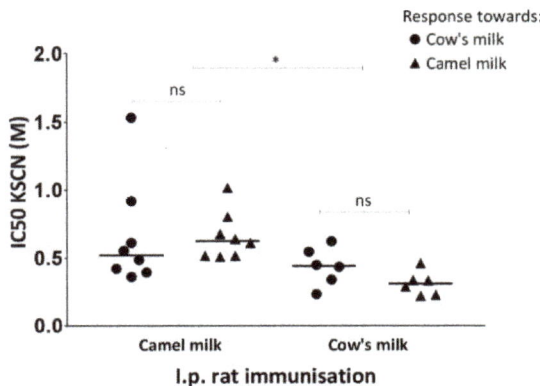

Figure 5. Avidity of IgG1 specific for cow's milk (●) or camel milk (▲). Serum samples from rats immunised with camel milk or cow's milk were evaluated to compare specific IgG1 antibody binding strength towards camel or cow's milk. Each symbol represents an individual rat. The avidity is expressed

as potassium thiocyanate concentration needed to inhibit 50% of the IgG1 response towards camel or cow's milk for groups of rats immunised with camel milk or cow's milk. Horizontal lines display the median values for each group of rats. Statistically significant difference between two groups was determined using the non-parametric Mann–Whitney test. Asterisks indicate statistically significant differences between two given groups when: $* = p \leq 0.05$, $** = p \leq 0.01$, $*** = p \leq 0.001$, $**** p \leq 0.0001$.

3.6. Sensitising Capacity of Camel and Cow's Milk

As IgE is the main player in food allergies [32], specific IgE titres were determined by the use of antibody-capture ELISAs. The results showed no obvious differences in the sensitising capacity of camel and cow's milk proteins, both products containing the capacity to induce high levels of specific IgE antibodies (Figure 6). No statistical analysis could be performed as the camel and cow's milk assays cannot be directly compared because of their potential different sensitivity. In line with the specific IgG1 responses, also for the specific IgE responses a low cross-reactivity between camel and cow's milk proteins was identified. Furthermore, in accordance with the IgG1 results, also for the IgE results a lower cross-reactivity could be observed for the casein fraction compared to the whey fraction of camel and cow's milk proteins.

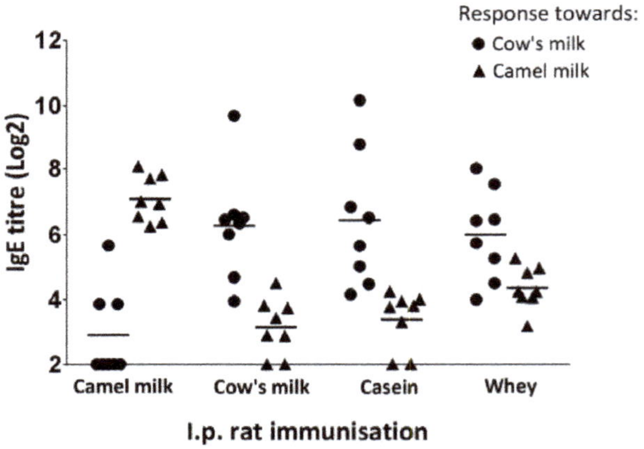

Figure 6. Specific IgE antibody responses. Comparison of specific IgE responses towards cow's (●) and camel milk (▲) in rats immunised with camel milk, cow's milk, cow's milk casein fraction or cow's milk whey fraction. Each symbol represents a specific IgE titre value for an individual rat. Horizontal lines on the graph display the median values for each group of rats.

3.7. Eliciting Capacity of Camel and Cow's Milk

The ability of camel and cow's milk to elicit allergic reactions was determined by an ear swelling test (Figure 7). Rats sensitised to camel milk showed a larger reaction towards camel milk than towards cow's milk, and opposite rats sensitised to cow's milk showed a larger reaction against cow's milk than camel milk, which correlates very well with the specific IgE responses (Figure 6), and confirms the low cross-reactivity between camel and cow's milk proteins. While a statistically significant difference was obtained for rats sensitised with camel milk, no statistically significant difference was obtained for rats sensitised with cow's milk. This may be explained by the fact that only seven animals are included in the cow's milk sensitised group compared to eight animals in the camel milk sensitised group, as one cow's milk sensitised animal died during the ear swelling test due to anaphylaxis. This has biased the results as this animal would probably be the one that would have responded with the greatest ear swelling. The groups immunised with cow's milk casein fraction or cow's milk whey fraction both showed a significantly larger response towards cow's milk compared to camel milk; however, in line with the antibody responses, the casein proteins were shown to have a lower cross-reactivity than the whey proteins.

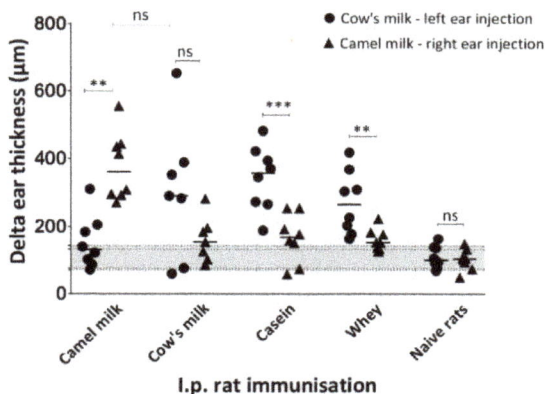

Figure 7. IgE functionality. Comparison of eliciting capacity of camel and cow's milk measured by an ear swelling test in rats immunised with camel milk, cow's milk, cow's milk casein fraction or cow's milk whey fraction. Delta ear thicknesses was calculated based on differences in ear thickness before and one hour after the ear injection of cow's milk solution to the left ear (●) and camel milk solution to the right ear (▲) at Day 33. Each symbol represents the delta ear thickness for an individual rat. Horizontal lines on the graph display the median values for each group of rats. Naïve rats correspond to the control group and define the median delta ear thickness with SD (grey coloured area) indicating no elicitation but an ear swelling caused by the injection volume. Statistically significant difference between two groups was determined using the non-parametric Mann–Whitney test. Asterisks indicate statistically significant differences between two given groups when: $* = p \leq 0.05$, $** = p \leq 0.01$, $*** = p \leq 0.001$, $**** p \leq 0.0001$.

3.8. Immunoblot

Immunoblotting was performed to investigate the specificity of the responses towards camel and cow's milk. Figure 8A shows the specificity of antibodies raised against cow's milk. BLG (~18 kDa) and the two casein fractions between 25 and 37 kDa were the proteins that antibodies specific for cow's milk reacted most pronounced to. Moreover, a hardly visible band was seen between 50 and 75 kDa indicating a weak reactivity towards SA (~66 kDa). There was no visible reaction of antibodies specific for cow's milk for camel milk proteins. The opposite situation is shown in Figure 8B where the specificity of antibodies raised against camel milk was evaluated. Here, antibodies reacted most pronounced with the camel milk β-casein fraction seen between 25 and 37 kDa standard marker bands, while there was no detectable reaction towards cow's milk proteins. However, the pooled serum dilutions used for the immunoblots were high, with a dilution of 1:3000 for sera raised against cow's milk and 1:8000 for sera raised against camel milk, for which reasons only the proteins with the strongest IgG1 binding capacity were visualised. As the ELISA assay showed very low cross-reactivity between camel and cow's milk proteins, we decided to use eight times higher protein concentration and lower serum pools dilution in order to visualise the proteins responsible for the cross-reactivity. The dilution used for both camel and cow's milk raised sera was 1:500. Figure 8C,D show cross-reactivity between camel and cow's milk proteins. Antibodies specific for cow's milk were able to cross-react exclusively with camel milk whey proteins. There were visible bands between 50 and 75 kDa indicating the most pronounced cross-reactivity with camel milk SA (~66 kDa) and LF (~75 kDa). Another weak but visible band was detected around 15 kDa indicating a very low cross-reactivity with camel milk ALA (Figure 8C). Another very weakly detectable band was at approximately 150 kDa. This probably corresponded to immunoglobulins [33]. Antibodies specific for camel milk showed a very weak reaction with cow's milk caseins between the 25 and 37 kDa standard marker bands, and with a whey protein appeared between the 50 and 75 kDa standard marker bands (Figure 8D).

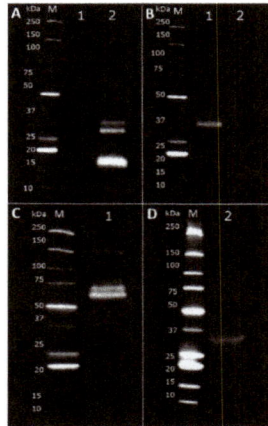

Figure 8. Immonublotting with camel and cow's milk. M, 2 μL of Protein Standard (kDa); 1, camel milk; 2, cow's milk. (**A**) Comparison of the reactivity of IgG1 antibodies specific for cow's milk diluted 1:3000 (v/v) towards 5 μg of camel milk and cow's milk (**B**) Comparison of the reactivity of IgG1 antibodies specific for camel milk diluted 1:8000 (v/v) towards 5 μg of camel milk and cow's milk. (**C**) Cross-reactivity between IgG1 antibodies raised towards cow's milk diluted 1:500 (v/v) and 40 μg of camel milk proteins. (**D**) Cross-reactivity between IgG1 antibodies raised towards camel milk diluted 1:500 (v/v) and 40 μg of cow's milk proteins.

4. Discussion

CMA is a major health issue of growing concern, for which reason the World Health Organisation (WHO) has created a guideline for diagnosis and rationale action [4]. Special hypoallergenic infant formulas for CMA infants as well as for infants in risk of developing CMA are available. These infant formulas are based on hydrolysed cow's milk proteins and designated extensively and partially hydrolysed infant formulas, respectively, depending on the degree of hydrolysis and peptide size distribution profile. Additional formulas, based on plants or milk from other mammalians have also been suggested for CMA infants [8]. However, for example, goat and sheep milk cannot be recommended for all CMA infants due to the high protein homology and consequently high cross-reactivity with cow's milk proteins [2,4,12]. One-humped camel—*Camelus dromedaries* (Camelidae family)—has a great evolutionary distance to animals from the Bovidae family [34]. Evolutionary distance directly influences milk protein composition variances, suggesting great differences between camel and cow's milk. Having a different protein composition, camel milk is anticipated to be a suitable alternative to hypoallergenic cow's milk-based infant formulas in the near future. To confirm a role for camel milk in management, primary prevention, and treatment of CMA, a combination of animal and human studies is needed. Using a BN rat model, we have compared immunogenicity, allergenicity, and cross-reactivity of camel and cow's milk proteins.

The present study showed that camel and cow's milk contain similar immunogenicity as well as allergenicity, being able to induce comparable levels of specific IgG1 and IgE antibodies with similar avidity. In addition, the eliciting capacity of the two milk products was shown to be similar. However, evaluation of the specific antibody reactivity towards cross-reactive proteins was low.

Whereas antibody responses raised against caseins were dominated by epitopes of the linear type, antibody responses raised against whey proteins were dominated by conformational epitopes. This is in line with a previous study showing that while caseins primarily raised antibodies towards linear epitopes, BLG and ALA primarily induced antibodies towards conformational epitopes, irrespectively of animals were dosed i.p. or orally [21]. I.p. dosing enables the immune system to recognise proteins in their native, undigested state. These results correlate very well to the structural folding of the proteins within the casein and whey fraction of milk, where caseins possess a flexible unstructured folding [21,35,36], while the predominant proteins within whey, BLG and ALA are globular proteins containing two and four disulphide bonds, respectively [19,21,37].

Camel and cow's milk proteins were in general shown to have a very low cross-reactivity. While approximately only 1 in 30 IgG1 antibodies raised against camel milk could react with cow's milk, only approximately 1 in 50 IgG1 antibodies raised against cow's milk could react with camel milk. The low cross-reactivity was confirmed by inhibitory ELISA where camel milk could only inhibit approximately 35% of the response against cow's milk and cow's milk could only inhibit approximately 50% of the response against camel milk. Similar results were observed for the IgE responses. Low cross-reactivity may reflect differences in the epitope pattern between camel and cow's milk proteins directly correlated with a fairly low protein sequence identities.

The present study demonstrates that camel milk may be a suitable alternative to hypoallergenic infant formulas for CMA infant, as the low cross-reactivity should confer the camel milk low risk of inducing reactions. This is consistent with human studies showing that the introduction of camel milk to children with confirmed CMA, who did not respond to the conventional management, had a positive, rapid and long-lasting effect on their health [20,38]. Other studies have shown that neither camel milk caseins nor whey proteins could inhibit or bind to sera antibodies from patients with confirmed CMA [8,39]. In contrast to camel milk, both goat and sheep milk show a large cross-reactivity to cow's milk [2,14,40], which is also reflected by the high protein identity, causing a similar epitope pattern. Human studies also showed that children with confirmed CMA reacted with goat milk due to IgE antibody cross-reactions [2,15]. In general, goat milk is not recommended for CMA patients without restrictive supervision of specialists [2,14].

The study showed that cow's milk was more efficient in inhibiting binding to antibodies raised against camel milk than camel milk was in inhibiting binding to antibodies raised against cow's milk. Certainly, the lack of BLG, one of the major allergenic proteins in cow's milk [28], may at least partly explain this difference. This indicates that camel milk in general is a more suitable infant formula for CMA infant, than is cow's milk for potential camel milk allergic infants.

The cross-reactivity between camel and cow's milk caseins was found to be less than the cross-reactivity between camel and cow's milk whey proteins, indicating that camel milk would be a more suitable alternative to hypoallergenic infant formulas for casein allergic infants than for whey allergic infants. This corresponds very well to the protein identity within the casein fraction compared to the whey fraction. In addition, immunoblot confirmed that antibodies specific for cow's milk were able to exclusively react with camel milk whey proteins, confirming a predominance of whey proteins cross-reactivity. The reactivity was mostly towards camel milk SA, which is a protein that is rarely detected to independently cause cow's milk allergy, and mostly sensitise together with other milk allergens [41,42].

Small differences were seen between the cross-reactivity accounted for by linear epitopes in comparison to cross-reactivity accounted for by conformational epitopes, where this study indicated that there is a tendency to a lower cross-reactivity between conformational epitopes compared to linear epitopes.

The difference in titre values in each group of immunised rats could reflect weaker antibody binding due to imperfect matching epitopes or be due to a low a low amount of shared epitopes. It can be stressed that the second option is the most likely, as the avidity of the cross-reacting antibodies was equal to the avidity of total population of antibodies.

Nutrients **2018**, *10*, 1903

Overall, it is suggested that approximately 35–40% sequence identity between allergens is adequate to induce IgE cross-reactive binding [43]. However, cross-reactions are unusual below 50% identity and mostly requires more than 70% identity [44]. We can therefore conclude that the low level of cross-reactivity found in the present study is at the expected level for proteins of an evolutionary distance around 60%. A lower cross-reactivity would probably require an even lower sequence homology, which again would require milk from an animal with even larger evolutionary distance to cows. An alternative approach would be to look for milk from animals with a shorter evolutionary distance to humans.

5. Conclusions

This study showed that, although camel and cow's milk display similar immunogenicity and allergenicity, cross-reactivity between their proteins is low. Moreover, selected protein sequence alignments showed lower protein sequence identity between camel and cow's milk proteins in comparison to other mammalian milk proteins such as goat or sheep. With this study, we showed that camel milk is a promising alternative to hypoallergenic cow's milk-based infant formulas. For further evaluation of camel milk and its usefulness as a suitable alternative for hypoallergenic cow's milk-based infant formulas in prevention, treatment and management of CMA, studies including oral animal sensitisation, primary prevention and treatment should be performed. In addition, mechanistic studies, including in vivo analyses of IgE functionality after oral challenge as well as evaluation of cellular changes in the gastrointestinal tract, would be of a great importance.

Author Contributions: N.Z.M. performed all lab work including denaturation of camel and cow's milk, ELISAs and SDS-PAGE electrophoresis and immunoblots. N.Z.M. participated in the lab work and result discussion during the entire study. N.Z.M. did statistical analyses and converted the student report to a paper manuscript. E.B.H. participated in the protein structure analyses as well as immunoblots. E.B.H. was involved in the lab work and result discussion during the entire study. E.B.H. reviewed the manuscript. A.-S.R.B. and A.I.S. participated in SDS-PAGE electrophoresis and immunoblots. A.I.S. especially contributed to the immunoblots optimisation. A.-S.R.B. and A.I.S. reviewed the manuscript. K.L.B. designed the animal experiment and led the animal study. K.L.B. performed the ear swelling test and participated in the denaturation of camel and cow's milk. K.L.B. participated in the lab work and current issues and results discussion during the entire study. K.L.B. reviewed the manuscript.

Funding: This research received no external funding.

Acknowledgments: The authors thank Karsten Bruun Qvist for inspiration and stimulating discussions prior to the initiation of this study. We also thank Juliane Margrethe Gregersen and Sarah Grundt Simonsen for assistance in the laboratory. Special thanks to Anne Ørngreen, Maja Danielsen, Olav Dahlgaard, Elise Navntoft and Kenneth Worm for great assistance in the animal facility.

Conflicts of Interest: The authors declare no conflict of interest.

Abbreviations

ALA	α-lactalbumin
BLG	β-lactoglobulin
BN	Brown Norway
CMA	Cow's milk allergy
DIG	dioxigenin
ELISA	Enzyme-linked immunosorbent assay
EU	endotoxin units
HRP	horse-radish peroxidase
IC$_{50}$	half minimum inhibitory concentration
i.p.	intraperitoneally
kDa	kilodalton
KSCN	potassium thiocyanate
LF	actoferrin
LP	lactoperoxidase
MW	molecular weight

NA   not available
OD   optical density
OVA   ovalbumin
PBS   phosphate buffered saline
PBT-T   phosphate buffered saline-tween
PAGE   polyacrylamide gel electrophoresis
PVDF   polyvinylidene difluoride
RT   room temperature
SA   serum albumin
SD   standard deviation
SDS   sodium dodecyl sulphate
TMB   3,3′,5,5′-tetramethylbenzidine
Tris   2-amino-2-(hydroxymethyl)-1,3-propenediol
WHO   World Health Organisation

References

1. Hochwallner, H.; Schulmeister, U.; Swoboda, I.; Spitzauer, S.; Valenta, R. Cow's milk allergy: From allergens to new forms of diagnosis, therapy and prevention. *Methods* **2014**, *66*, 22–33. [CrossRef] [PubMed]
2. Bellioni-Businco, B.; Paganelli, R.; Lucenti, P.; Giampietro, P.G.; Perborn, H.; Businco, L. Allergenicity of goat's milk in children with cow's milk allergy. *J. Allergy Clin. Immunol.* **1999**, *103*, 1191–1194. [CrossRef]
3. Carrard, A.; Rizzuti, D.; Sokollik, C. Update on food allergy. *Eur. J. Allergy Clin. Immunol.* **2015**, *70*, 1511–1520. [CrossRef] [PubMed]
4. Fiocchi, A.; Brozek, J.; Schünemann, H. World Allergy Organization (WAO) Diagnosis and Rationale for Action against Cow's Milk Allergy (DRACMA) Guidelines. *Pediatric. Allergy Immunol.* **2010**, *21*, 1–125. [CrossRef]
5. El-Agamy, E.I. The challenge of cow milk protein allergy. *Small Rumin. Res.* **2007**, *68*, 64–72. [CrossRef]
6. Many Kids May Not Outgrow Cow's Milk Allergy. Reuters. Available online: https://www.reuters.com/article/us-milk-allergy/many-kids-may-not-outrow-cows-milk-allergy-idUSCOL87231020071218 (accessed on 25 November 2018).
7. Milk and Dairy Allergy. ACAAI. Available online: https://acaai.org/allergies/types-allergies/food-allergy/types-food-allergy/milk-dairy-allergy (accessed on 25 November 2018).
8. El-Agamy, E.I.; Nawar, M.; Shamsia, S.M.; Awad, S.; Haenlein, G.F.W. Are camel milk proteins convenient to the nutrition of cow milk allergic children? *Small Rumin. Res.* **2009**, *82*, 1–6. [CrossRef]
9. Von Berg, A.; Koletzko, S.; Grübl, A.; Filipiak-Pittroff, B.; Wichmann, H.E.; Bauer, C.P.; Reinhardt, D.; Berdel, D.; German Infant Nutritional Intervention Study Group. The effect of hydrolyzed cow's milk formula for allergy prevention in the first year of life: The German Infant Nutritional Intervention Study, a randomized double-blind trial. *J. Allergy Clin. Immunol.* **2003**, *111*, 533–540. [CrossRef] [PubMed]
10. Gateway to Dairy Production and Products. Food Agric. Organ. United Nations. Available online: http://www.fao.org/dairy-production-products/production/dairy-animals/en/ (accessed on 1 January 2018).
11. Järvinen, K.-M.; Chatchatee, P.; Bardina, L.; Beyer, K.; Sampson, H.A. IgE and IgG binding epitopes on α-lactalbumin and β-lactoglobulin in cow's milk allergy. *Int. Arch. Allergy Immunol.* **2001**, *126*, 111–118. [CrossRef]
12. Dean, T.P.; Adler, B.R.; Ruge, F. In vitro allergenicity of cow's milk substitutes. *Clin. Exp. Allergy* **1993**, *23*, 205–210. [CrossRef]
13. Restani, P.; Gaiaschi, A.; Plebani, A.; Beretta, B.; Cavagni, G.; Fiocchi, A.; Poiesi, C.; Velona, A.; Ugazio, G.; Galli, C.L. Cross-reactivity between milk proteins from different animal species. *Clin. Exp. Allergy* **1999**, *29*, 997–1004. [CrossRef] [PubMed]
14. Ehlayel, M.; Bener, A.; Abu Hazeima, K.; Al-Mesaifri, F. Camel Milk Is a Safer Choice than Goat Milk for Feeding Children with Cow Milk Allergy. *ISRN Allergy* **2011**, *2011*, 1–5. [CrossRef] [PubMed]
15. Spuergin, P.; Walter, M.; Schiltz, E.; Deichmann, K.; Forster, J.; Mueller, H. Allergenicity of α-caseins from cow, sheep, and goat. *Allergy* **1997**, *52*, 293–298. [CrossRef] [PubMed]

16. Fiocchi, A.; Dahda, L.; Dupont, C.; Campoy, C.; Fierro, V.; Nieto, A. Cow's milk allergy: Towards an update of DRACMA guidelines. *World Allergy Organ. J.* **2016**, *9*, 1–11. [CrossRef] [PubMed]

17. Lonnerdal, B. Nutritional aspects of soy formula. *Acta Pediatr Suppl.* **1994**, *83*, 105–108. [CrossRef]

18. Hinz, K.; O'Connor, P.M.; Huppertz, T.; Ross, R.P.; Kelly, A.L. Comparison of the principal proteins in bovine, caprine, buffalo, equine and camel milk. *J. Dairy Res.* **2012**, *79*, 185–191. [CrossRef] [PubMed]

19. Hailu, Y.; Hansen, E.B.; Seifu, E.; Eshetu, M.; Ipsen, R.; Kappeler, S. Functional and technological properties of camel milk proteins: A review. *J. Dairy Res.* **2016**, *83*, 422–429. [CrossRef] [PubMed]

20. Navarrete-Rodríguez, E.M.; Ríos-Villalobos, L.A.; Alcocer-Arreguín, C.R.; Del-Rio-Navarro, B.E.; Del Rio-Chivardi, J.M.; Saucedo-Ramírez, O.J.; Sienra-Monge, J.J.L.; Frias, R.V. Cross-over clinical trial for evaluating the safety of camel's milk intake in patients who are allergic to cow's milk protein. *Allergol. Immunopathol.* **2017**, *4*, 149–154. [CrossRef]

21. Nagy, P.; Thomas, S.; Marko, O.; Juhasz, J. Milk production, raw milk quality and fertility of Dromedary camels (Camelus Dromedarius) under intensive management. *Acta Weterinaria Hungarica.* **2013**, *61*, 71–84. [CrossRef]

22. Spencer, P.B.S.; Woolnough, A.P. Assessment and genetic characterisation of Australian camels using microsatellite polymorphisms. *Live Stock Sci.* **2010**, *129*, 241–245. [CrossRef]

23. Nutrition. Available online: https://desertfarms.co.uk/pages/nutrition (accessed on 23 November 2018).

24. Fresh and Frozen Camel Milk. Available online: https://www.oasismilk.com/en/camel-milk/ (accessed on 23 November 2018).

25. Wolkaro, T.; Yusuf Kurtu, M.; Eshetu, M.; Ketema, M. Analysis of camel milk value chain in the pastoral areas of eastern Ethiopia. *J. Camelid Sci.* **2017**, *10*, 1–16.

26. Sawaya, W.N.; Khalil, J.K.; Al-Shalhat, A.; Al-Mohammad, H. Chemical composition and nutritional quality of camel milk. *J. Food Sci.* **1984**, *49*, 744–747. [CrossRef]

27. Madsen, J.; Kroghsbo, S.; Madsen, C.; Pozdnyakova, I.; Barkholt, V.; Bøgh, K.L. The impact of structural integrity and route of administration on the antibody specificity against three cow's milk allergens—A study in Brown Norway rats. *Clin. Transl. Allergy* **2014**, *4*, 25. [CrossRef] [PubMed]

28. Bøgh, K.L.; Barkholt, V.; Madsen, C.B. The sensitising capacity of intact β-lactoglobulin is reduced by Co-administration with digested β-lactoglobulin. *Int. Arch. Allergy Immunol.* **2013**, *161*, 21–36. [CrossRef] [PubMed]

29. Bertino, E.; Gastaldi, D.; Monti, G.; Baro, C.; Fortunato, D.; Garoffo, L.P. Detailed proteomic analysis on DM: Insight into its hypoallergenicity. *Front. Biosci.* **2010**, *2*, 526–536. [CrossRef]

30. Rijnkels, M.; Elnitski, L.; Miller, W.; Rosen, J.M. Multispecies comparative analysis of a mammalian-specific genomic domain encoding secretory proteins. *Genomics* **2003**, *82*, 417–432. [CrossRef]

31. Omar, A.; Harbourne, N.; Oruna-Concha, M.J. Quantification of major camel milk proteins by capillary electrophoresis. *Int. Dairy J.* **2016**, *58*, 31–35. [CrossRef]

32. Broekman, H.C.H.; Eiwegger, T.; Upton, J.; Bøgh, K.L. IgE–the main player of food allergy. *Drug Discov. Today Dis. Model.* **2015**, *17–18*, 37–44. [CrossRef]

33. Levieux, D.; Levieux, A.; El-Hatmi, H.; Rigaudière, J.P. Immunochemical quantification of heat denaturation of camel (Camelus dromedarius) whey proteins. *J. Dairy Res.* **2006**, *73*, 1–9. [CrossRef]

34. Youcef, N.; Saidi, D.; Mezemaze, F.; Kaddouri, H.; Negaoui, H.; Chekroun, A. Cross Reactivity between Dromedary Whey Proteins and IgG Anti Bovine α-Lactalbumin and Anti Bovine β-Lactoglobulin. *Am. J. Appl. Sci.* **2009**, *6*, 1448–1452. [CrossRef]

35. O'Mahony, J.A.; Fox, P.F. Milk Proteins: Introduction and Historical Aspects. *Adv. Dairy Chem.* **2013**, *1*, 43–85. [CrossRef]

36. McMahon, D.J.; Oommen, B.S. Supramolecular Structure of the Casein Micelle. *J. Dairy Sci. Elsevier* **2008**, *91*, 1709–1721. [CrossRef] [PubMed]

37. Beg, O.U.; Bahr-Lindstrom, H.; Zaidi, Z.H.; Jornvall, H. The primary structure of α-lactalbumin from camel milk. *Eur. J. Biochem.* **1985**, *147*, 233–239. [CrossRef] [PubMed]

38. Shabo, Y.; Barzel, R.; Margoulis, M.; Yagil, R. Camel milk for food allergies in children. *Isr. Med. Assoc. J.* **2005**, *7*, 796–798. [PubMed]

39. Restani, P.; Beretta, B.; Fiocchi, A.; Ballabio, C.; Galli, C.L. Cross-reactivity between mammalian proteins. *Ann. Allergy Asthma Immunol.* **2002**, *89*, 11–15. [CrossRef]

40. Wolpe, L.Z.; Silverstone, P.C. A series of substitutes for milk in the treatment of allergies. *J. Pediatr.* **1942**, *21*, 635–658. [CrossRef]
41. Järvinen, K.M.; Chatchatee, P. Mammalian milk allergy: Clinical suspicion, cross-reactivities and diagnosis. *Curr. Opin. Allergy Clin. Immunol.* **2009**, *9*, 251–258. [CrossRef]
42. Natale, M.; Bisson, C.; Monti, G.; Peltran, A.; Garoffo, L.P.; Valentini, S.; Fabris, C.; Bertino, E.; Coscia, A.; Conti, A. Cow's milk allergens identification by two-dimensional immunoblotting and mass spectrometry. *Mol. Nutr. Food Res.* **2004**, *48*, 363–369. [CrossRef]
43. Kroghsbo, S.; Bøgh, K.L.; Rigby, N.M.; Mills, E.N.C.; Rogers, A.; Madsen, C.B. Sensitization with 7S globulins from peanut, hazelnut, soy or pea induces ige with different biological activities which are modified by soy tolerance. *Int. Arch. Allergy Immunol.* **2011**, *155*, 212–224. [CrossRef]
44. Aalberse, R.C. Structural biology of allergens. *J. Allergy Clin. Immunol.* **2000**, *106*, 228–238. [CrossRef]

nutrities

MDPI

Article

Delayed-Type Hypersensitivity Underlying Casein Allergy Is Suppressed by Extracellular Vesicles Carrying miRNA-150

Magdalena Wąsik [1], Katarzyna Nazimek [1,2], Bernadeta Nowak [1], Philip W. Askenase [2] and Krzysztof Bryniarski [1,2,*]

[1] Department of Immunology, Jagiellonian University Medical College, 31-121 Krakow, Poland; magdalena.wasik@interia.pl (M.W.); katarzyna.nazimek@uj.edu.pl (K.N.); bernadeta.nowak@uj.edu.pl (B.N.)

[2] Section of Rheumatology, Allergy and Clinical Immunology, Yale University School of Medicine, New Haven, CT 06520, USA; philip.askenase@yale.edu

* Correspondence: mmbrynia@cyf-kr.edu.pl; Tel.: +48-12-632-58-65

Received: 28 March 2019; Accepted: 18 April 2019; Published: 23 April 2019

Abstract: In patients with non-IgE-mediated milk allergy, a cellular mechanism of delayed-type hypersensitivity (DTH) is considered. Recent findings prove that cell-mediated reactions can be antigen-specifically inhibited by extracellular vesicles (EVs) carrying miRNA-150. We sought to establish a new mouse model of DTH to casein and test the possibility of antigen-specific suppression of the inflammatory reaction. To produce soluble antigenic peptides, casein was subjected to alkaline hydrolysis. DTH reaction to casein was induced in CBA, C57BL/6, and BALB/c mice by intradermal (id) injection of the antigen. Cells collected from spleens and lymph nodes were positively or negatively selected and transferred to naive recipients intravenously (iv). CBA mice were tolerized by iv injection of mouse erythrocytes conjugated with casein antigen and following id immunization with the same antigen. Suppressive EVs were harvested from cell cultures and serum of tolerized donors by means of ultrafiltration and ultracentrifugation for further therapeutic utilization. The newly established mouse model of DTH to casein was mediated by CD4+ Th1 cells and macrophages, while EVs produced by casein-tolerized animals effectively suppressed effector cell response, in an miRNA-150-dependent manner. Altogether, our observations contribute to the current understanding of non-IgE-mediated allergy to casein and of the possibilities to downregulate this reaction.

Keywords: cow's milk allergy; casein; cell-mediated reactions; delayed-type hypersensitivity; extracellular vesicles; miRNA-150

1. Introduction

Cow's milk allergy is one of the most common food allergies in children. It is hard to precisely estimate the prevalence of milk allergy, because its symptoms are often mistaken with lactose intolerance, a non-immunological reaction [1]. Allergy to cow's milk usually manifests itself as a type I hypersensitivity reaction that is IgE-mediated and occurs immediately after contact with the allergen in sensitized individuals. There is also a group of patients with onset delayed more than 20 h after the exposure to the allergen and where the reaction does not depend on IgE level [2–4]. This suggests involvement of another mechanism likely cell-mediated.

Delayed-type hypersensitivity (DTH) to common allergens is far more difficult to diagnose than type I allergy and thus remains a challenge for clinicians. Symptoms of non-IgE-mediated allergy to food proteins are mostly gastro-intestinal and include malabsorption, bloody diarrhea, emesis, pallor, lethargy, and weight loss [1,5].

DTH, classified by Gell and Coombs as a type IV allergic reaction, has been widely studied in drug-induced models, and a new subclassification has been established according to the manifestation of the allergy and the type of cells involved in the reaction. Subtype IVa is mediated by Th1 lymphocytes and macrophages (manifested by bullous exanthem), while IVb by Th2 lymphocytes and eosinophils is accompanied by maculopapular exanthem. Subtype IVc is mediated by cytotoxic T lymphocytes, and either maculopapular or bullous exanthem can be observed as the clinical symptoms. Pustular exanthem is a hallmark of the IVd subtype, to which reaction depends on T lymphocytes producing IL-8 and GM-CSF that activate neutrophils [6]. However, it still remains an open question if DTH subtypes underlie the non-IgE-mediated allergy to casein.

Cow's milk consists of around 30 different proteins, from which caseins constitute approximately 80% [7]. Currently, the most effective treatment of non-IgE-dependent allergy to cow's milk is the elimination of the offending food and all products that may contain its proteins from child's diet. If an infant with a cow's milk allergy is breastfed, then it should also be eliminated from the mother's diet. This method is a great burden for parents, since milk proteins, such as casein, can be found not only in dairy products but also in many processed foods [1,8]. Total avoidance of specific food may even predispose infants to develop an allergy by disturbing IgA-mediated oral tolerance [9,10].

Recently, it was proved that cell-mediated allergic reactions to haptens can be inhibited by exosome-like extracellular vesicles (EVs) carrying miRNA-150 [11]. Exosomes representing a population of small EVs have recently become the subject of increasing interest as a new means of intercellular communication. They are produced intracellularly in multivesicular bodies and are released by exocytosis. EVs can contain a "cargo" of miRNA that may be delivered to the acceptor cell and affect its physiology [12]. In a murine model of contact sensitivity to haptens, EVs carrying miRNA-150 had the capacity to antigen-specifically target effector T cells and suppress the inflammation [11]. These EVs were produced by a distinct population of CD8+ suppressor T (Ts) lymphocytes, but not from FoxP3+ regulatory T cells, after intravenous administration of syngeneic erythrocytes conjugated with hapten to naive mice. Moreover, they became antigen-specific when coated with IgM light chains, produced by B1 lymphocytes after contact immunization with the same hapten. The suppressive Ts cell-derived miRNA-150-carrying EVs and B1 cell-produced antibody light chains were described to act together as soluble T suppressor factor (TsF) [13]. Our preliminary data suggest that analogous tolerance mechanism mediated by Ts cell EV-contained miRNA-150 could also be induced in DTH to ovalbumin.

Our study aimed to describe the effector mechanisms and to standardize a new, mouse model of DTH to casein antigen. For this purpose, we established a method of obtaining a soluble casein (Cas) antigen that appears to be safe to use in an animal model of active immunization and to induce cellular inflammatory response. The next aim of our research was to test the possibility of antigen-specific inhibition of DTH reactions to protein antigen, such as Cas, in adoptive transfer and in active immunization.

2. Materials and Methods

2.1. Mice

Mice of CBA, BALB/c, and C57BL/6 inbred strains were obtained from the breeding unit of the Faculty of Medicine, Jagiellonian University Medical College in Krakow, Poland. Animals were fed with casein-free chow supplied by Labofeed H (Kcynia, Poland). All experiments were conducted according to the guidelines of the Animal Care and Use Committee of the Jagiellonian University Medical College and Yale School of Medicine, New Haven, CT (Permit Number 07381) and under ethical approval of the 1st Local Ethics Committee in Krakow (approval no. 243/2015). Control and treatment groups consisted of 4–5 animals.

2.2. Casein Alkaline Hydrolysis

Insoluble bovine casein (Sigma-Aldrich, St Louis, MO, USA), containing all casein fractions, i.e., α-s1, α-s2, β, and κ-casein, was suspended in $NaHCO_3$ (0.5 M) in a 1:8 ratio (w/v) and hydrolyzed in 37 °C for 5–7 h with occasional mixing. The solution was then kept at room temperature for subsequent 48 h. Next, all probes were centrifuged (3300 g, 15 min), and the soluble fraction was dialyzed to DPBS. The remaining pellet was resuspended in NaOH (1 M), adjusted to pH = 10 with HCl solution, mixed, and kept at room temperature for the next 24 h. Again, the resulting soluble fraction was processed as above. Apart from immunogenicity testing, soluble Cas fractions after hydrolysis with $NaHCO_3$ or NaOH were mixed and then dialyzed to either DPBS (for coupling with erythrocytes) or 0.9% NaCl (for immunization). Yielded soluble Cas antigen was filtrated, and the protein concentration was then assessed with a UV spectrophotometer at 280 nm.

2.3. Active Immunization

Mice under light anesthesia were immunized with the Cas antigen by intradermal administration of a total of 200 μL of soluble Cas peptides (0.5 mg/ml) administered into 4 separate sides of the abdomen on Days 0 and 1. On Day 5, mice were challenged on both ears by intradermal administration of 10 μL of the same Cas solution. The increase in ear thickness was measured by a blinded observer using an engineer's micrometer (Mitutoyo, Kawasaki, Japan) from 24 to 120 h later [14]. Background earlobe thickness was subtracted to obtain a value of ear thickness increase for each mouse, while nonspecific increase in earlobe thickness in non-immunized, but challenged littermates was subtracted from experimental groups to yield a net swelling value expressed as delta ± standard error (SE) $[U \times 10^{-2}$ mm].

2.4. Adoptive Transfer of DTH Effector Cells

Cas-immunized mice (see above) were sacrificed on Day 5 by means of cervical dislocation under deep anesthesia. Effector cells were obtained from spleens and lymph nodes, and their single cell suspensions in DPBS were transferred intravenously to naive recipients (7×10^7 cells per mouse). The next day, animals were ear-challenged by intradermal administration of 10 μL of soluble Cas, and subsequent earlobe thickness was measured as above [14].

2.5. Negative Selection Assay

DTH effector cell suspensions from spleens and lymph nodes of Cas-immunized mice were incubated with monoclonal antibodies specific for CD4 (rat IgG2b antibody, GK1.5 clone) or CD8 (rat IgG2b antibody, TIB-105 clone, both cell lines were cultured and produced monoclonal antibodies were chromatographically purified at Department of Immunology, Jagiellonian University Medical College, Krakow, Poland) markers. The rabbit complement (BIOMED, Lublin, Poland) was then added and suspensions were incubated for 1 h in a 37 °C water-bath. DTH effector cells were depleted of macrophages by triple filtration through a 70 μm nylon mesh. Selection was followed by Ficoll centrifugation to remove dead cells. Selected viable cells were suspended in DPBS and administered to naive mice intravenously in adoptive transfer (see above) [15].

2.6. Positive Selection Assay

Cell suspensions from spleens and lymph nodes of mice immunized or tolerized with Cas antigen or Cas-coupled erythrocytes, respectively, were positively selected on MiniMACS MS columns with microbeads coated with monoclonal antibodies anti-CD4 or anti-CD8 according to manufacturer's procedures (Miltenyi Biotec, Bergisch Gladbach, Germany). Selected viable cells were suspended in DPBS and intravenously administered to naive mice in adoptive transfer (see above).

2.7. Induction and Separation of Ts cell EVs from Cell Supernatants

Mice, on Days 0 and 4, were injected intravenously with 10% suspension of syngeneic erythrocytes conjugated with Cas antigen in the presence of 1-ethyl-3-(3-dimethylaminopropyl) carbodiimide (EDC) for activation of coupling process. This induced a generation of miRNA-150-carrying EVs by Ts cells. On Days 7 and 8, Cas antigen without an adjuvant was administered intradermally into 4 separate sites of the abdomen (100 µg per mouse, see above) in order to induce production of specific IgM light chains by B1 lymphocytes. Spleens and lymph nodes of tolerized mice were harvested on Day 10 and their single cell suspensions, containing Ts cells and B1 lymphocytes, were cultured for 48 h in protein-free Mishell-Dutton medium (MDM, 2×10^7 cells per ml, 37 °C, 5% CO_2). Supernatants from cell cultures were then centrifuged twice ($300\times g$ and $3000\times g$, for 15 min) and filtered through 0.45 and 0.22 µm syringe filters (Merck Millipore, Burlington, MA, USA). Finally, EVs were concentrated by double ultracentrifugation in a Beckman L870M ultracentrifuge ($100,000\times g$, 70 min, 4 °C) and resuspended in DPBS [15,16] for further experimental usage as Ts cell EVs. In some cases, EVs were incubated with anti-miR-150 (miRIDIAN Hairpin Inhibitor of Mouse mmu-miR150, Dharmacon, GE Healthcare, Lafayette, CO, USA) at 37 °C for 1 h, in a dose of 3 µg per eventual recipient mouse, to block the biological activity of miRNA-150, and then washed to remove excessive anti-miR-150 molecules [15]. Where indicated, cells of spleens and lymph nodes of tolerized mice were subjected to a positive selection assay, as described above.

2.8. Induction and Separation of B1 Cell-Produced EVs

Mice were injected intradermally on Days 0 and 1 with Cas antigen without adjuvant into 4 separate sites of the abdomen (100 µg per mouse, see above) in order to induce production of B1 cell EVs, coated with specific IgM light chains, but devoid of miRNA-150 [11]. Spleens and lymph nodes of immunized mice were harvested on Day 3 and single cell suspensions were cultured for 48 h in protein-free MDM (2×10^7 cells per ml, 37 °C, 5% CO_2). Supernatants from cell cultures were then centrifuged twice ($300\times g$ and $3000\times g$, for 15 min) and filtered through 0.45 and 0.22 µm syringe filters. B1 cell-produced EVs were concentrated by ultracentrifugation in a Beckman L870M ultracentrifuge ($100,000\times g$, 70 min, 4 °C) and resuspended in DPBS. In some cases, B1 cell EVs were incubated overnight on ice with miRNA-150 (miRIDIAN Mimic Mouse mmu-miR-150, Dharmacon, GE Healthcare, Lafayette, CO, USA), in a dose of 3 µg per eventual mouse recipient, and then washed to remove excessive miRNA-150 molecules.

2.9. Antigen-Affinity Chromatographic Separation of Cas-Specific EVs

Soluble Cas antigen or purified anti-CD9 monoclonal antibodies (BD Biosciences, San Diego, CA, USA) were linked to cyanogen bromide-activated Sepharose 4FF (fast flow; Pharmacia, Uppsala, Sweden) according to the manufacturer's procedure. Supernatant from tolerized mouse lymph node and spleen suppressive cell culture was then applied onto either Cas- or anti-CD9- Sepharose-filled columns. The fraction of nanovesicles that first passed through the column was collected as flow through (FT, i.e., Cas-non-binding or CD9 negative EVs), and, after column washing with DPBS, the fraction of nanovesicles that was eluted with 5 M guanidine HCl (pH = 4.7) was collected as eluate (i.e., Cas-binding or CD9 positive EVs). Both fractions were filtered through 0.45 and 0.22 µm filters and ultracentrifuged twice in DPBS ($100,000\times g$, 70 min, 4 °C) [11].

2.10. Suppression of Effector Cells Tested in Adoptive Transfer

DTH effector cells collected as above from spleens and lymph nodes of Cas-immunized mice were incubated with Ts cell EVs for 30 min in 37 °C, which was followed by washing in DPBS and 70 µm nylon mesh filtration. Cell suspensions in DPBS were transferred intravenously to naive, anesthetized recipients in a ratio of 7×10^7 cells per mice. The next day, animals were ear-challenged by intradermal administration of 10 µL of Cas and ear thickness was measured after 24 h, as described above.

Approximately 1.3×10^{10} EVs were used either in adoptive transfer per 7×10^7 effector cells per mouse or as a single dose administered to actively immunized mouse. The exact quantity of EVs was estimated using nanoparticle tracking analysis as described in detail previously [11].

2.11. Cytometric Analysis of EVs and Antibody Light Chains

Aldehyde/sulfate latex beads of 4 μm size (Life Technologies, Thermo-Fisher Scientific, Carlsbad, CA, USA) were incubated in DPBS with Ts cell EVs in a total volume of 1 mL of DPBS for 2 h at room temperature with gentle agitation. Afterwards, EV-coated beads were blocked with 100 mM glycine for 30 min at room temperature with gentle agitation. After washing and resuspending in DPBS, EV-coated beads were stained with fluorescein isothiocyanate (FITC)-conjugated monoclonal antibodies against mouse kappa light chains (BD Biosciences, San Diego, CA) and/or phycoerythrin (PE)-conjugated monoclonal antibodies against mouse CD9, CD63, or CD81 tetraspanins (BD Biosciences, San Diego, CA, USA) and acquired by a BD FACSCalibur (BD Bioscience, San Jose, CA, USA).

2.12. Test of Different Routes of Therapeutic EV Administration and Active Tolerance Induction

Ts cell-derived EVs or B1 cell-produced EVs were administered to the Cas-immunized animals intradermally, intraperitoneally, intravenously or per os in equal doses (see above) at the peak of the allergic response, i.e., 24 h post-challenge. Ear thickness was measured up to 120 h after challenge by a blinded observer, unaware of the experimental protocol, using an engineer's micrometer (Mitutoyo, Kawasaki, Japan). Otherwise, mice, 5 days prior to active immunization, had been injected intravenously with 10% suspension of syngeneic erythrocytes conjugated with Cas antigen, and subsequent ear swelling was elicited as above 5 days after immunization.

2.13. MHC Criss-Cross Testing

DTH effector cells collected from Cas-immunized CBA, BALB/c, or C57BL/6 mice were incubated with Ts cell EVs obtained from CBA mice tolerized to Cas antigen. Next, EV-pulsed or non-pulsed effector cells were intravenously transferred to naive mice of respective strain. On the following day, mice were intradermally ear-challenged with Cas antigen, and ear thickness was measured as described above.

2.14. Statistical Analysis

Readings were done twice on both ears, and the inflammatory response induced in each mouse was expressed as a mean of the 4 measurements. After meeting of test assumptions, one-way analysis of variance (ANOVA) with post-hoc RIR Tukey test was used to evaluate the statistical significance between the groups with $p < 0.05$ taken as a minimum level of significance, which was marked in the figures as * $p < 0.05$; ** $p < 0.01$; *** $p < 0.001$; **** $p < 0.0001$.

3. Results

3.1. Soluble Casein Antigen Induces Allergic Reaction

Mice intradermally immunized and challenged with soluble Cas antigen fractions hydrolyzed in either $NaHCO_3$ or NaOH, developed a significantly greater ear swelling response, when compared to the negative control group of mice only challenged with respective Cas fraction (Figure 1). In addition, ear swelling response peaked 24 h after challenge in immunized animals, which resembles DTH reaction. Thus, we assumed that Cas antigen preserves its immunogenicity during alkaline hydrolysis and is able to induce DTH reaction in mice after intradermal administration without an adjuvant. Besides, inflammatory reactions caused by immunization with both fractions of soluble Cas antigen were comparable, so we decided to mix them in order to use in further experiments.

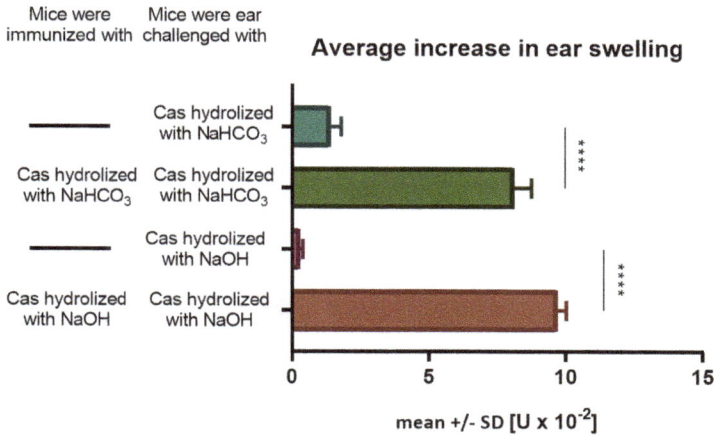

Figure 1. Immunogenicity of soluble casein (Cas) antigen obtained by alkaline hydrolysis with either $NaHCO_3$ or NaOH. Mice had been intradermally (id) immunized with a saline solution of soluble Cas antigen (100 µg per mouse) 5 days before challenging by id administration of the same Cas solution (5 µg per earlobe). Twenty-four hours later ear swelling response was measured and expressed as mean $\pm$ SD [units (U) $\times$ 10^{-2} mm] (n = 4, N = 3). **** $p < 0.0001$.

3.2. CD4+ T Cells and Macrophages Mediate the Effector Phase of DTH Reaction to Cas Antigen

Phenotype of DTH effector cells mediating allergic reaction in Cas-immunized mice was assessed using negative and positive selection assays (Figure 2). Statistically significant decrease in ear swelling in comparison to a positive control was observed in mice depleted of CD4+ T lymphocytes or macrophages prior to adoptive transfer, which indicates that both cell populations are important for induction of DTH reaction to Cas antigen. The inflammatory response in mouse recipients of macrophage-depleted DTH effector cells was noticeably higher than that in groups deprived of CD4+ T cells, which is possibly caused by the activity of a recipient's macrophages, which can still elicit inflammatory response of transferred effector CD4+ T cells.

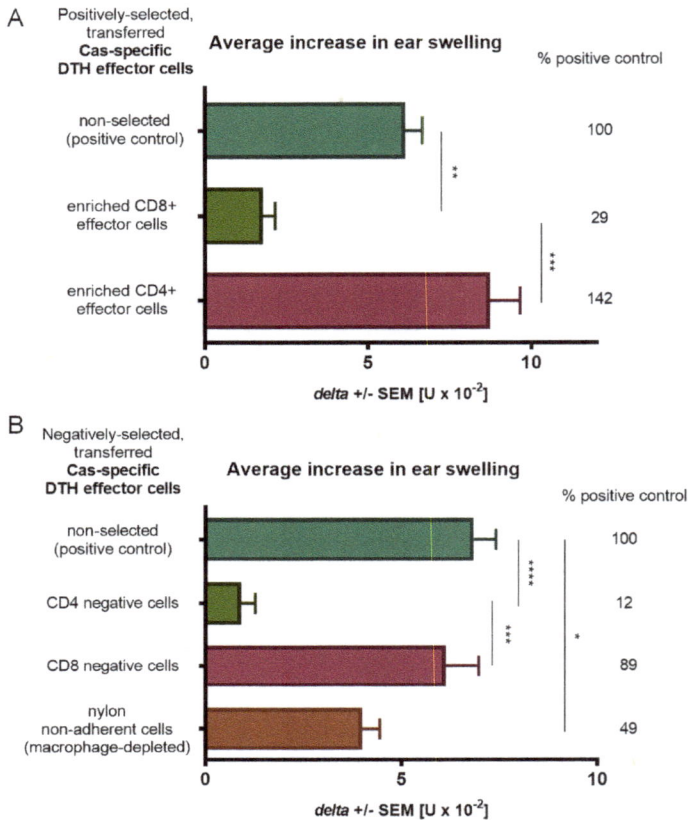

Figure 2. Phenotyping of effector cells of delayed-type hypersensitivity (DTH) to soluble casein (Cas) antigen. Mice had been intradermally (id) immunized with a saline solution of soluble Cas antigen (100 μg per mouse) 5 days before harvest of lymph nodes and spleens containing effector cells, which were then subjected to positive (**A**) or negative (**B**) selection assays by, respectively, magnetic-advanced cell sorting or depletion with either monoclonal antibodies and complement or nylon wool separation. Afterwards, selected effector cells were transferred to naive recipients, which 24 h later were challenged by id administration of the same Cas solution (5 μg per earlobe). After 24 h, ear swelling response was measured and expressed as delta $\pm$ SEM [units (U) $\times$ 10^{-2} mm] (n = 5, N = 2), * $p < 0.05$; ** $p < 0.01$; *** $p < 0.001$; **** $p < 0.0001$.

3.3. CD8+ T Cells Are Responsible for Production of Suppressive EVs that Express CD9 and CD81 Tetraspanins and Are Specific to Casein due to Expression of Antibody Light Chains

Phenotype of suppressive cells was confirmed in positive selection according to CD8 expression. Selected, CD8+ enriched and CD8 negative cells were cultured separately for 48 h, and the pelleted EVs from ultracentrifuged culture supernatant were incubated with DTH effector cells prior to adoptive transfer. Statistically significant decrease in ear swelling was observed in challenged recipient mice had been administered with DTH effector cells incubated with EVs released by either CD8+ T cell enriched or non-selected cells from tolerized mouse lymph nodes and spleens (Figure 3A). This observation confirmed that the population of CD8+ T lymphocytes is responsible for the production of suppressive, miRNA-150-carrying EVs. By means of flow cytometry, Ts cell EVs were shown to express CD9 and CD81 tetraspanins, and CD9+ EVs eluted from anti-CD9 chromatography column suppressed DTH effector cells (Figure 3B,C). Interestingly, the flow through fraction of the anti-CD9 column increased

DTH ear swelling. We assumed that this may result from the presence of Cas-specific antibody light chains non-associated with CD9+ EVs in the flow through fraction from the anti-CD9 column, as detected by flow cytometry (Figure 3B). Furthermore, Ts cell EVs expressed antibody light chains (Figure 3B), which could be responsible for their antigen specificity. To test this assumption, suppressive EVs were separated by antigen-affinity chromatography. Cas-binding EVs eluted from the Cas column, when incubated with DTH effector cells prior to adoptive transfer, significantly suppressed the elicited DTH ear swelling (Figure 3D). These results prove that suppressive EVs are Cas-specific.

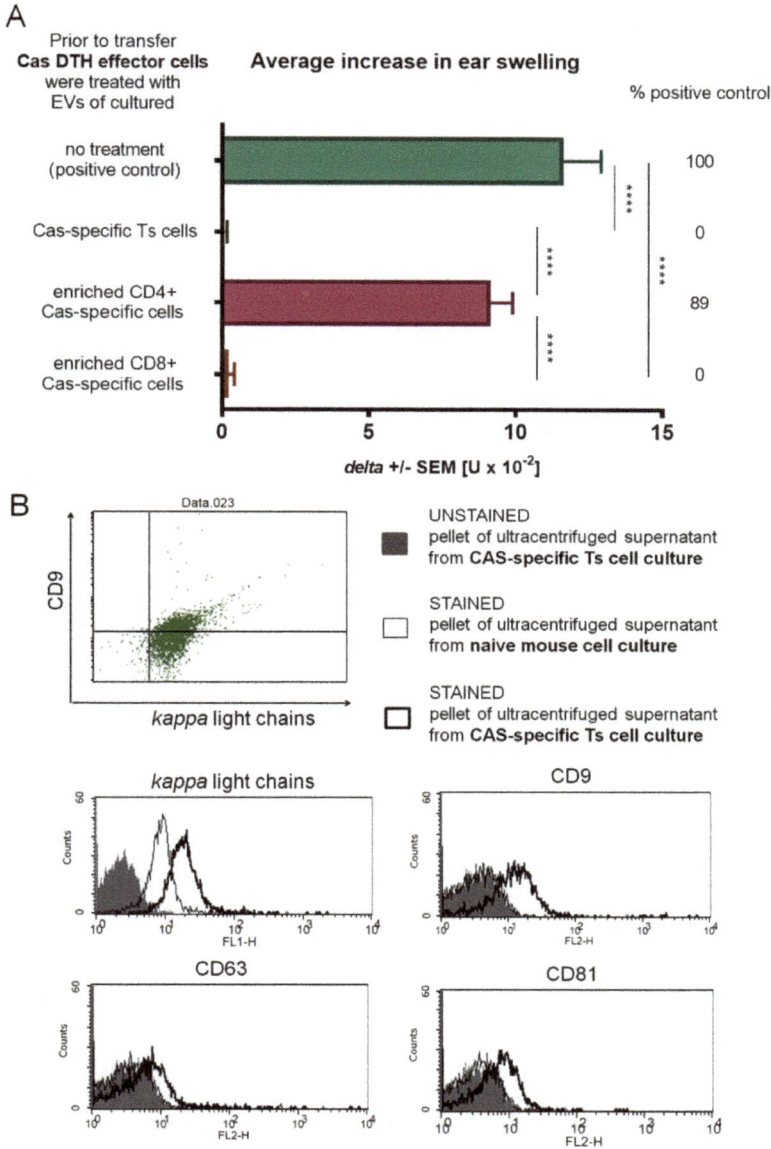

Figure 3. *Cont.*

C

Prior to transfer
Cas DTH effector cells Average increase in ear swelling
were treated with EVs of

% positive control

no treatment
(positive control) — 100

unseparated
Cas-specific Ts cell — 20
culture supernatant

flow through from — 150
anti-CD9 column

eluate from — 25
anti-CD9 column

0   2   4   6   8

delta +/- SEM [U x 10^{-2}]

D

Prior to transfer
Cas DTH effector cells Average increase in ear swelling
were treated with EVs of

% positive control

no treatment
(positive control) — 100

flow through from — 97
Cas-column

eluate from — 0
Cas-column

0   5   10   15

delta +/- SEM [U x 10^{-2}]

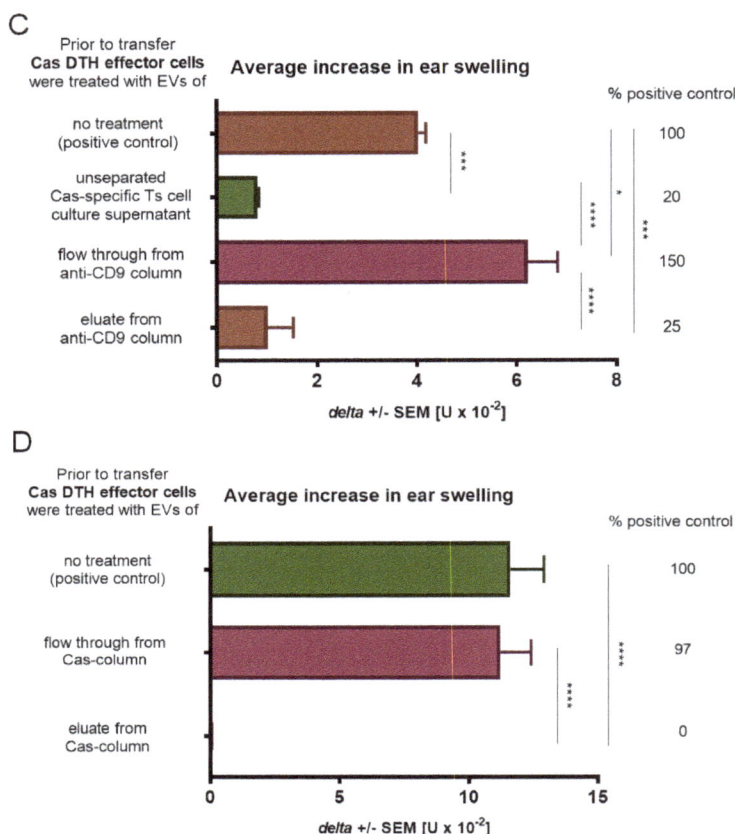

Figure 3. Suppression of delayed-type hypersensitivity (DTH) to soluble casein (Cas) antigen is mediated by CD8+ suppressor T (Ts) cell-derived extracellular vesicles (EVs) expressing CD9 and CD81 tetraspanins and Cas-specific antibody light chains. (**A**) Mice had been intravenously administered with Cas-coupled syngeneic red blood cells, followed by intradermal (id) immunization with a saline solution of soluble Cas antigen (100 µg per mouse) 3 days before harvest of lymph nodes and spleens containing Ts cells, which were then subjected to positive selection assay, i.e., magnetic-advanced cell sorting, and cultured for subsequent 48 h. The resulting supernatant was filtered and ultracentrifuged to concentrate Ts cell EVs, used to treat DTH effector cells prior to adoptive transfer to recipients, challenged 24 h later by id administration of the same Cas solution (5 µg per earlobe). After 24 h, ear swelling response was measured and expressed as delta ± SEM [units (U) × 10^{-2} mm] (n = 5, N = 2). (**B**) Ts cell EVs were coated onto latex beads, stained with monoclonal antibodies against CD9, CD63, and CD81 tetraspanins or against mouse antibody kappa light chains, and then analyzed by flow cytometry (n = 3, N = 2). (**C**) Ts cell EVs were separated by antigen affinity chromatography on column filled with Sepharose coated with anti-CD9 monoclonal antibodies, and the resulting fractions, i.e., flow through or eluate, were used to treat DTH effector cells prior to adoptive transfer to recipients, challenged 24 h later by id administration of the same Cas solution (5 µg per earlobe). After 24 h, ear swelling response was measured and expressed as delta ± SEM [units (U) × 10^{-2} mm] (n = 5, N = 2). (**D**) Ts cell EVs were separated by antigen affinity chromatography on column filled with Cas-coated Sepharose, and the resulting fractions, i.e., flow through or eluate, were used to treat DTH effector cells prior to adoptive transfer to recipients, challenged 24 h later by id administration of the same Cas solution (5 µg per earlobe). After 24 h, ear swelling response was measured and expressed as delta ± SEM [units (U) × 10^{-2} mm] (n = 5, N = 3). * $p < 0.05$; ** $p < 0.01$; *** $p < 0.001$; **** $p < 0.0001$.

3.4. Cas-Coupled Erythrocyte Intravenous Administration Induces Tolerance and the Resulting EVs Are Suppressive after Administration via Different Routes

Mice were tolerized with syngeneic erythrocytes conjugated with Cas antigen (Cas-MRBC), prior to intradermal immunization with Cas. Figure 4A shows the average inflammatory response after 24, 48, 72, 96, and 120 h after challenge with Cas antigen. Statistically significant decrease in ear swelling in the case of tolerized mice was then observed at every time point, which proves that intravenous tolerization with Cas-MRBC prevents the development of DTH to Cas, likely due to Ts cell EVs carrying miRNA-150.

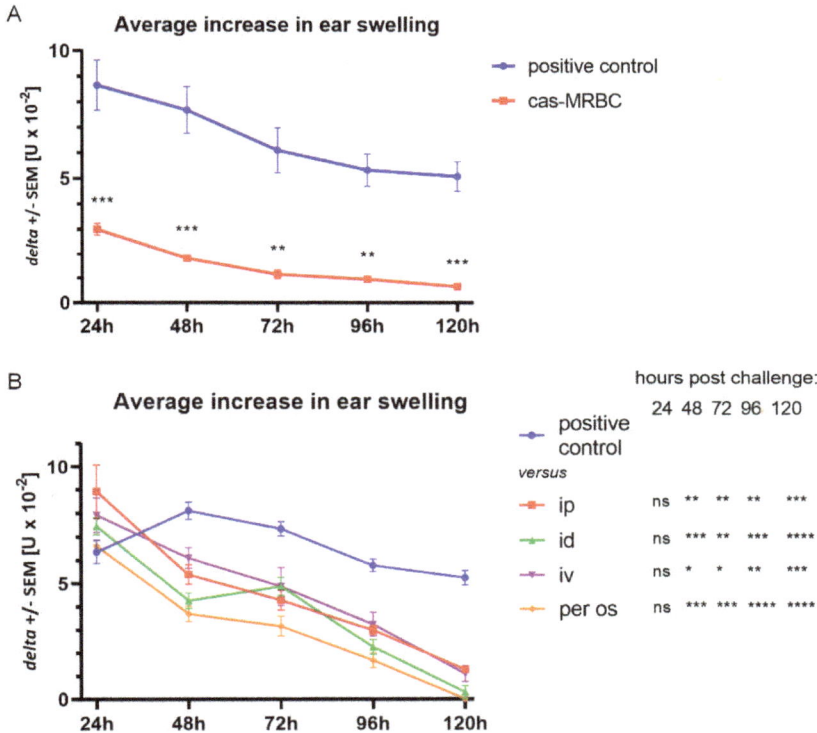

Figure 4. Suppression of delayed-type hypersensitivity (DTH) to soluble casein (Cas) antigen is induced by intravenous (iv) injection of Cas-coupled syngeneic red blood cells and by CD8+ suppressor T (Ts) cell-derived extracellular vesicles (EVs) administered via different routes. (**A**) Mice had been administered iv with Cas-coupled syngeneic red blood cells, which was followed by intradermal (id) immunization with a saline solution of soluble Cas antigen (100 μg per mouse) 5 days before challenging by id administration of the same Cas solution (5 μg per earlobe). Subsequent ear swelling response was measured daily up to 120 h after challenge and expressed as delta ± SEM [units (U) × 10^{-2} mm] (n = 5, N = 2). (**B**) Mice had been id immunized with a saline solution of soluble Cas antigen (100 μg per mouse) 5 days before challenging by id administration of the same Cas solution (5 μg per earlobe). After measurement of 24 h ear swelling, Ts cell EVs were administered intraperitoneally (ip), id, iv, or per os, and subsequent ear swelling response was measured daily up to 120 h after challenge and expressed as delta ± SEM [units (U) × 10^{-2} mm] (n = 5, N = 2). * $p < 0.05$; ** $p < 0.01$; *** $p < 0.001$; **** $p < 0.0001$.

Thus, Ts cell EVs were administered to Cas-immunized mice intravenously (iv), intradermally (id), intraperitoneally (ip), or per os, at the peak of 24 h inflammatory response. Figure 4B presents the average inflammatory response in each group, measured daily up to 120 h after challenge with Cas antigen. All administration routes induced a significant decrease in ear swelling in comparison to the positive control, with the most notable decrease observed when Ts cell EVs were administered intradermally, which is the route used for immunization of animals, and per os, which is the natural means of immunization with food protein allergens, such as Cas.

3.5. Ts Cell EVs Obtained from Tolerized CBA Mice Suppress the Adoptively Transferred DTH Reaction in Different Strains of Mice

DTH effector cells of CBA, BALB/c, or C57BL/6 mice immunized with Cas antigen were incubated with Ts cell EVs from CBA mice tolerized to Cas. Next, the effector cells were administered iv to the naive recipients of respective strain. A significant decrease in ear swelling was observed in all tested strains in groups treated with CBA Ts cell EVs, with the strongest effect detected in the case of CBA mice (Figure 5A), which suggests that EV action is rather not MHC-restricted.

3.6. The Blockade of miRNA-150 Activity in EVs Deprives Them of Suppressive Properties

To confirm, whether miRNA-150 is indeed the suppressive component of Ts cell EVs, they were incubated with antagonist of miRNA-150 (anti-miR-150) for 1 h before incubation with DTH effector cells from animals immunized with Cas antigen. Then, DTH effector cells were administered iv to naive mice that were ear-challenged with Cas antigen on the following day. Significant suppression of inflammatory response was observed in a group administered with DTH effector cells incubated with intact Ts cell EVs, while no significant suppression was observed in a group treated with inactivated miRNA-150 (Figure 5B). This confirms that miRNA-150 provides suppressive capacity of Ts cell EVs.

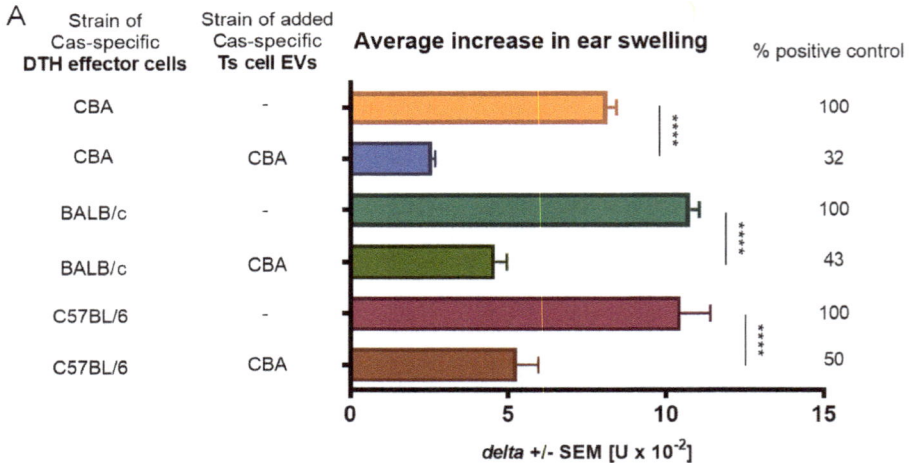

Figure 5. *Cont.*

B

Prior to transfer
Cas-specific
DTH effector cells
were treated with

Average increase in ear swelling

% positive control

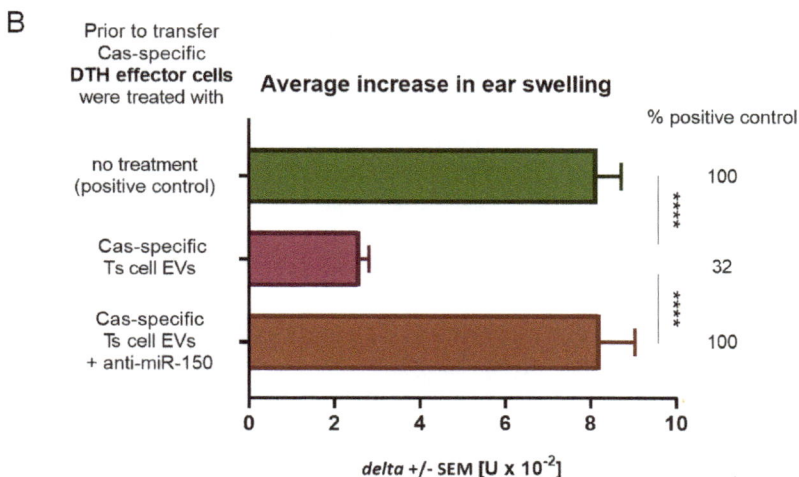

Figure 5. Major histocompatibility complex (MHC) restriction and dependence on miRNA-150 of suppression of delayed-type hypersensitivity (DTH) to soluble casein (Cas) antigen. (**A**) Mice of CBA, BALB/c, or C57BL/6 strains had been intradermally (id) immunized with a saline solution of soluble Cas antigen (100 µg per mouse) 5 days before harvest of lymph nodes and spleens containing effector cells, which were then treated with Cas-specific suppressor T (Ts) cell-derived extracellular vesicles (EVs) of tolerized CBA mice. Afterwards, effector cells were transferred to naive recipients of respective strain, which 24 h later were challenged by id administration of the same Cas solution (5 µg per earlobe). After 24 h, ear swelling response was measured and expressed as delta $\pm$ SEM [units (U) $\times 10^{-2}$ mm] (n = 4, N = 2). (**B**) Part of Ts cell EVs was incubated with miRNA-150 antagonist, i.e., anti-miR-150, prior to treatment of DTH effector cells, which were then adoptively transferred to recipients, challenged 24 h later by id administration of the same Cas solution (5 µg per earlobe). After 24 h, ear swelling response was measured and expressed as delta $\pm$ SEM [units (U) $\times 10^{-2}$ mm] (n = 5, N = 3). **** $p <$ 0.0001.

3.7. Cas-Specific B1 Cell-Produced EVs Gain Suppressive Activity after Incubation with miRNA-150 and Express Therapeutic Potential after Administration via Different Routes

B1 cell-derived EVs were proposed effective vehicles for transmission of regulatory miRNAs. To verify this hypothesis, we had incubated B1 cell-derived EVs from Cas-immunized mice with miRNA-150 and used them to treat Cas-specific DTH effector cells, which were next administered to naive recipients. No significant suppression of inflammatory response was observed in a group administered with B1 cell-produced EVs alone, while B1 cell EVs supplemented with miRNA-150 significantly suppressed DTH reaction (Figure 6A). On the other hand, ovalbumin-specific B1 cell EVs supplemented with miRNA-150 failed to suppress Cas-specific DTH effector cells (Figure 6A). These results prove that B1 cell EVs themselves are non-suppressive but can gain suppressive properties when supplemented with miRNA-150 and that they suppress DTH reaction afterwards in an antigen-specific manner.

To verify Cas-specific, B1 cell EVs' therapeutic potential after supplementation with miRNA-150, they were administered to mice intravenously (iv), intradermally (id), intraperitoneally (ip), or per os at the peak of 24 h inflammatory response to Cas antigen. The most notable decrease in ear swelling response was observed in a group of mice administered with EVs per os (Figure 6B), analogously to Ts cell EVs (see Figure 4B). Statistically significant suppression was also noted in groups administered with EVs ip and id, while there were no significant differences between the positive control and the iv treated group until 120 h (Figure 6B). These results confirm that Cas-specific B1 cell EVs can gain suppressive properties when supplied with miRNA-150 and that the most effective method of their administration is the same as the natural route of immunization with Cas antigen, i.e., the oral route, which is well tolerated by patients.

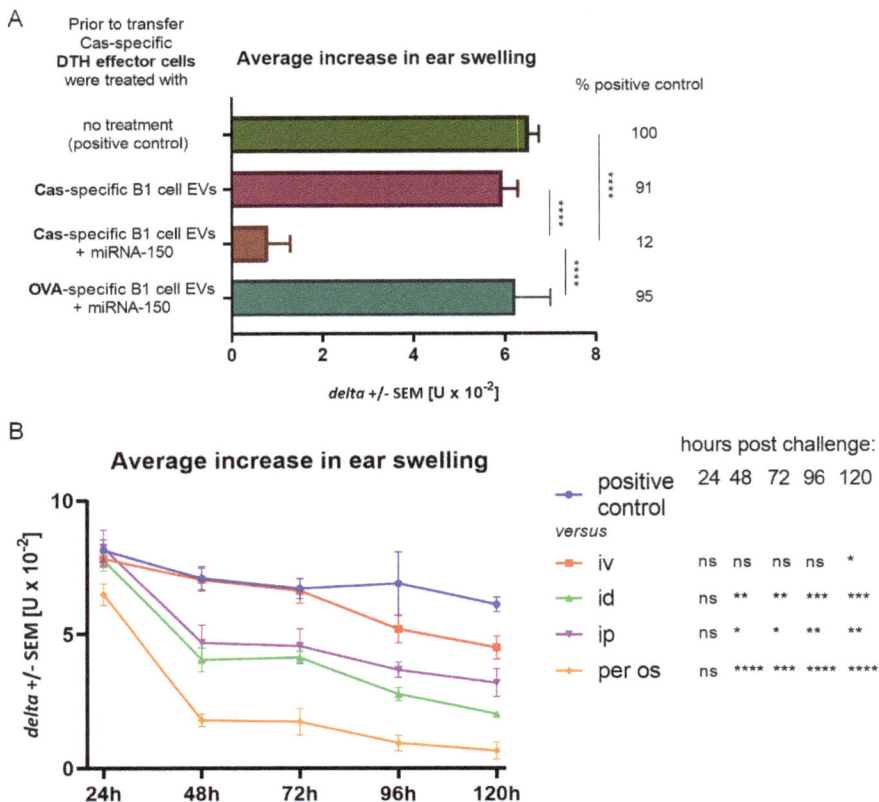

Figure 6. Suppression of delayed-type hypersensitivity (DTH) to soluble casein (Cas) antigen by miRNA-150-supplemented B1 cell-derived extracellular vesicles (EVs) of Cas-immunized mice. (**A**) Mice had been intradermally (id) immunized with a saline solution of soluble Cas antigen or ovalbumin (OVA, both 100 μg per mouse) 3 days before harvest of lymph nodes and spleens containing B1 cells, which were cultured for a subsequent 48 h. The resulting supernatant was filtered and ultracentrifuged to concentrate Cas-specific or OVA-specific B1 cell EVs, then supplemented with miRNA-150 prior to treatment of Cas-specific DTH effector cells, which were adoptively transferred to recipients, challenged 24 h later by id administration of the same Cas solution (5 μg per earlobe). After 24 h, ear swelling response was measured and expressed as delta ± SEM [units (U) × 10^{-2} mm] (n = 5, N = 2). (**B**) Mice had been id immunized with a saline solution of soluble Cas antigen (100 μg per mouse) 5 days before challenging by id administration of the same Cas solution (5 μg per earlobe). After measurement of 24 h ear swelling, miRNA-150-supplemented, Cas-specific B1 cell EVs were administered intraperitoneally (ip), id, intravenously (iv), or per os, and subsequent ear swelling response was measured daily up to 120 h after challenge and expressed as delta ± SEM [units (U) × 10^{-2} mm] (n = 4, N = 2). * $p < 0.05$; ** $p < 0.01$; *** $p < 0.001$; **** $p < 0.0001$.

4. Discussion

Food allergy is considered a consequence of a breakdown of oral tolerance. This process can be initiated by disturbances in antigen uptake caused by increased skin and gut permeability that lead to dissemination of the food allergen, presentation of its epitope by antigen-presenting cells

(APCs), and sensitization [17]. Animals in the experiments were fed with casein-free chow in order to prevent any previous contact with the allergen and to exclude the possibility of oral tolerance development. A new mouse model of DTH to Cas was established using soluble Cas antigen administered intradermally without an adjuvant, and the inflammatory response, peaking at 24 h after challenge, was also effectively transferred from immunized donors to naive recipients, as unraveled by CD4+ T lymphocytes and macrophages as effector cells. The important role of macrophages in inflammatory reactions suggests that CD4+ T lymphocytes belong to the Th1 subpopulation. According to Lerch's and Pichler's subclassification of cell-mediated hypersensitivity, DTH to Cas can be classified as a IVa allergic reaction [6].

EVs and their ability to transport bioactive molecules, such as miRNAs, to acceptor cells are a new, promising tool that may be used in diagnosis and therapy for a variety of human diseases [18,19]. Bai et al. reported that miRNA-150 may inhibit the proliferation and promote the cell cycle arrest in thyroid cancer cells [20]. Exogenous miRNA-150 was also described to inhibit proliferation and metastasis and to enhance cell apoptosis in human osteoblasts [21]. Chen et al. investigated the therapeutic effect of mesenchymal stem cell-derived EVs carrying miRNA-150-5p in potential therapy of rheumatoid arthritis. Injection of miRNA-150-loaded EVs reduced joint destruction by inhibition of hyperplasia and angiogenesis in a mouse model of collagen-induced arthritis [22]. miRNA-150 was proved to be a regulatory factor carried by Ts cell-derived exosomes [11]. Suppressive capacity of Ts cell EVs was confirmed by numerous independent groups in a mouse model of contact sensitivity to haptens [13]. We managed to separate Ts cell EVs from cell culture supernatant from animals tolerized to food protein, such as Cas, and to prove its ability to suppress inflammatory reaction in DTH to Cas antigen both in adoptive transfer and in active immunization. Cas-specific Ts cell EVs treated with miRNA-150 antagonist prior to incubation with Cas DTH effector cells proved to be inactive. Additionally, B1 cell-derived EVs did not suppress DTH reaction to Cas antigen, unless they were supplemented with miRNA-150. These results confirmed that Ts cell EVs, produced by animals tolerized to Cas antigen, owe their suppressive activity to miRNA-150 that is carried by EVs derived from CD8+ Ts cells and not B1 cells.

We confirmed that suppressive EVs in DTH to Cas, like in CS to haptens, were a product of CD8+ Ts lymphocytes by means of positive selection. Furthermore, antigen-specificity of Ts cell EVs was tested in antigen-affinity chromatography and the eluate from the column proved to be strongly suppressive in contrast to flow through. Bryniarski et al. tested antigen-specificity of EVs in CS to haptens in both antigen-affinity chromatography and in a crisscross experiment, where trinitrophenol-specific Ts cell EVs failed to suppress inflammatory reaction to oxazolone. The antigen-specificity of EVs was then demonstrated by means of flow cytometry to be the result of IgM antibody light chain (LC) coating [11]. Those LCs are produced by B1 lymphocytes activated by immunization with an antigen, which follows the tolerization with antigen-coupled syngeneic red blood cells [15]. Current results demonstrated the presence of LC on Cas-specific Ts cell EVs, which additionally confirms their antigen-specificity. Furthermore, miRNA-150-supplemented B1 cell-derived EVs from mice immunized with ovalbumin failed to suppress Cas-induced DTH, which brought more evidence for the LC-mediated specificity of EV action. IgM antibodies produced by B1 lymphocytes are usually characterized by low specificity to antigen as their secretion does not require signals from helper T cells. Lately, it was presented that a special subset of B1a cells generates high antigen-affinity IgM antibodies and free LCs as a consequence of immunoglobulin V-region mutations induced by activation-induced cytidine deaminase. B1a lymphocytes are suggested to initiate early responses in immune resistance to pneumococcal pneumonia, CS, and DTH [23,24], which was herein indirectly confirmed by the enhancement of DTH reaction caused by the flow through fraction of the anti-CD9 column, demonstrated to contain Cas-binding LC. Recent findings confirm that B1 cell-derived LCs provide the specificity of EVs in CS to haptens and enable execution of their suppressive function, as the Ts cell EVs of B-cell-deficient or immunoglobulin-deficient mice are non-suppressive [25]. Previously, it was demonstrated that injection of haptenized MRBC before contact sensitization with oxazolone and elicitation of CS ear swelling response leads to suppression of inflammatory response in wild-type mice [25]. Here, we confirmed that double injection of cas-MRBC prior to immunization with the Cas antigen protected animals from developing of DTH reaction to Cas. The results suggest that animals tolerized with Cas antigen actively produce Ts cell EVs that prevent elicitation of inflammatory response up to 120 h.

Ts cell EVs modulate inflammatory response indirectly. Transfer of effector cells incubated with peritoneal macrophages treated previously with Ts cell EVs inhibited reaction in CS effector cell recipients [26]. Lately, it was confirmed that EVs containing miRNA-150 act as mediators in communication between effector T cells and APC, since mice depleted of macrophages cannot be effectively tolerized [27]. APCs express major histocompatibility complex class II (MHC II) on their surface that was reported to be also expressed on their EVs, which enables targeting of CD4+ T cells [28]. Here, we investigated whether the suppressive effect of Ts cell EVs depends on MHC II by treatment of DTH effector cells from CBA, BALB/c, and C57BL/6 mice immunized with Cas with EVs collected from CBA mice tolerized to Cas antigen. In our experiment, CBA Cas-specific Ts cell EVs were effective in each strain. This suggests that T cell-derived EVs, in contrast to T cells themselves, cannot check MHC homology, which implies that those EVs, like miRNAs, may mediate a highly conserved mode of communication across strains.

We tested four different routes of Ts cell EV administration and all of them effectively suppressed inflammatory response in active immunization. Interestingly, the most effective routes of EV administration, i.e., id and per os, were also the routes of immunization with the antigen. In our experiments, animals were immunized intradermally and the natural route of immunization with food protein allergens would be through the gastrointestinal tract. Furthermore, the oral route is also the natural means of food tolerance induction. Intestinal epithelial cell-derived EVs, carrying $\alpha v \beta 6$ integrin and food antigens, were reported to stimulate tolerance in dendritic cells (DCs) and promote regulatory T cell development in a model of oral tolerance induction [29]. Previously, DCs were described to directly sample the content of intestine lumen through the trans-epithelial dendrites [17,30]. Food antigens could also be internalized in Peyer's patches and captured by APCs, which then migrate to lymph nodes and activate effector or regulatory T cells [31]. These mechanisms of oral tolerance create a suitable opportunity for Ts cell EVs to reach APCs and transfer the suppressive information. In a following experiment, we tested the same routes of administration of miRNA-150-supplemented B1 cell EVs, and the oral administration again happened to be the most effective. Statistically significant differences were also observed when EVs were administered ip and id, which is possibly a result of particularly easy access to resident macrophages and DCs in peritoneal cavity and skin [32,33]. Intravenous injection of the miRNA-150-supplemented B1 cell EVs to Cas-immunized mice was not effective until 120 h in contrast to previous experiments with Ts cell EVs, where all routes suppressed inflammatory reaction within 24 h of administration. Our results may indicate that miRNA-150 is not internalized by B1 cell EVs but adheres to their surface instead, where it is susceptible to enzymatic hydrolysis. In contrast, when miRNA-150 is originally sorted to EVs in MVB of Ts cells, it is protected from ribonucleases activity in serum. Thus, miRNA-150-supplemented B1 cell EVs require more time to access sufficient amount of APCs to deliver the suppressive information.

It is worth noting that EVs are resistant to harsh conditions, including very low pH and activity of digestive enzymes, which enables the protection of contained RNA cargo, as reported in the case of the transmission of dietary miRNAs associated with EVs via intestines [34,35]. This was true also for maternal milk exosome RNA cargo transferring epigenetic information to neonates after intestinal absorption [36], likely via epithelial cell endocytosis [37,38]. Interestingly, EV-associated miRNAs absorbed via intestinal barrier have been proposed to regulate immunity [39], and our current results seem to confirm this assumption.

5. Conclusions

We standardized a method to produce immunogenic, soluble Cas antigen activating DTH when administered intradermally without an adjuvant. Further, we developed a new model of DTH to a milk protein allergen and identified the phenotype of effector cells responsible for the inflammatory reaction as CD4+ Th1 lymphocytes and macrophages. The possibility of antigen-specific suppression of cell-mediated reaction in allergic response to food protein, such as Cas, was also confirmed. Suppressive EVs obtained from cell supernatants of mice tolerized to Cas were a product of CD8+ Ts lymphocytes. miRNA-150 was described as the suppressive compound of Cas-specific Ts cell EVs. Animals tolerized to Cas prior to immunization actively produced Ts cell EVs that protected them from developing DTH response to Cas. The oral route of administration was the most effective for treatment with either Ts cell EVs or miRNA-150-supplemented B1 EVs, as it is the natural route of oral tolerance development. Cas-specific Ts cell EVs derived from CBA mice suppressed inflammatory reactions in different strains of mice immunized to Cas antigen, which suggests that the intercellular communication via suppressive EVs is conserved among strains.

Author Contributions: Conceptualization: M.W., K.N., P.W.A., and K.B.; funding acquisition: P.W.A. and K.B.; investigation: M.W., K.N., B.N., and K.B.; methodology: B.N. and P.W.A.; supervision: K.B.; validation: M.W., K.N., and K.B.; visualization: M.W. and K.N.; writing—original draft: M.W.; writing—review & editing: K.N. and K.B.

Funding: This research was supported by the grant of Polish Ministry of Science and Higher Education No K/ZDS/006148 to K.B. and partly by grant No AI-1053786 from the National Institutes of Health to P.W.A.

Conflicts of Interest: The authors declare no conflict of interest.

References

1. Wąsik, M.; Nazimek, K.; Bryniarski, K. Allergic reactions to cow's milk: Pathomechanism, diagnostic and therapeutic strategies, possibilities of food tolerance induction. *Postepy Hig. Med. Dosw.* **2018**, *72*, 1–11. (In Polish) [CrossRef]
2. Hill, D.J.; Ball, G.; Hosking, C.S. Clinical manifestations of cows' milk allergy in childhood. I. Associations with in-vitro cellular immune responses. *Clin. Allergy* **1988**, *18*, 469–479. [CrossRef]
3. Hill, D.; Firer, M.; Shelton, M.; Hosking, C. Manifestations of milk allergy in infancy: Clinical and immunologic findings. *J. Pediatr.* **1986**, *109*, 270–276. [CrossRef]
4. Díaz, M.; Guadamuro, L.; Espinosa-Martos, I.; Mancabelli, L.; Jiménez, S.; Molinos-Norniella, C.; Pérez-Solis, D.; Milani, C.; Rodríguez, J.M.; Ventura, M.; et al. Microbiota and Derived Parameters in Fecal Samples of Infants with Non-IgE Cow's Milk Protein Allergy under a Restricted Diet. *Nutrients* **2018**, *10*, 1481. [CrossRef]
5. Nowak-Węgrzyn, A.; Katz, Y.; Mehr, S.S.; Koletzko, S. Non-IgE-mediated gastrointestinal food allergy. *J. Allergy Clin. Immunol.* **2015**, *135*, 1114–1124. [CrossRef] [PubMed]
6. Lerch, M.; Pichler, W.J. The immunological and clinical spectrum of delayed drug-induced exanthems. *Curr. Opin. Allergy Clin. Immunol.* **2004**, *4*, 411–419. [CrossRef]
7. Wróblewska, B.; Jędrychowski, L. Effect of technological modification on the immunoreactive properties of cow milk proteins. *Alergia Astma Immunol.* **2003**, *8*, 157–164. (In Polish)
8. Hochwallner, H.; Schulmeister, U.; Swoboda, I.; Spitzauer, S.; Valenta, R. Cow's milk allergy: From allergens to new forms of diagnosis, therapy and prevention. *Methods* **2014**, *66*, 22–33. [CrossRef]
9. Jêvinen, K.-M.; Laine, S.T.; Jêvenpêê, A.-L.; Suomalainen, H.K. Does Low IgA in Human Milk Predispose the Infant to Development of Cow's Milk Allergy? *Pediatr. Res.* **2000**, *48*, 457–462.
10. Järvinen, K.M.; Westfall, J.E.; Seppo, M.S.; James, A.K.; Tsuang, A.J.; Feustel, P.J.; Sampson, H.A.; Berin, C. Role of maternal elimination diets and human milk IgA in the development of cow's milk allergy in the infants. *Clin. Exp. Allergy* **2014**, *44*, 69–78. [CrossRef] [PubMed]
11. Bryniarski, K.; Ptak, W.; Jayakumar, A.; Püllmann, K.; Caplan, M.J.; Chairoungdua, A.; Lu, J.; Adams, B.D.; Sikora, E.; Nazimek, K.; et al. Antigen-specific, antibody-coated, exosome-like nanovesicles deliver suppressor T-cell microRNA-150 to effector T cells to inhibit contact sensitivity. *J. Allergy Clin. Immunol.* **2013**, *132*, 170–181. [CrossRef]

12. Nazimek, K.; Bryniarski, K.; Santocki, M.; Ptak, W. Exosomes as mediators of intercellular communication: clinical implications. *Pol. Arch. Intern. Med.* **2015**, *125*, 370–380. [CrossRef]

13. Ptak, W.; Nazimek, K.; Askenase, P.W.; Bryniarski, K. From Mysterious Supernatant Entity to miRNA-150 in Antigen-Specific Exosomes: A History of Hapten-Specific T Suppressor Factor. *Arch. Immunol. Ther. Exp.* **2015**, *63*, 345–356. [CrossRef] [PubMed]

14. Szczepanik, M.; Akahira-Azuma, M.; Bryniarski, K.; Tsuji, R.F.; Kawikova, I.; Ptak, W.; Kiener, C.; Campos, R.A.; Askenase, P.W. B-1 B Cells Mediate Required Early T Cell Recruitment to Elicit Protein-Induced Delayed-Type Hypersensitivity. *J. Immunol.* **2003**, *171*, 6225–6235. [CrossRef]

15. Bryniarski, K.; Ptak, W.; Martin, E.; Nazimek, K.; Szczepanik, M.; Sanak, M.; Askenase, P.W. Free Extracellular miRNA Functionally Targets Cells by Transfecting Exosomes from Their Companion Cells. *PLoS ONE* **2015**, *10*, e0122991. [CrossRef] [PubMed]

16. Stremersch, S.; Brans, T.; Braeckmans, K.; De Smedt, S.; Raemdonck, K. Nucleic acid loading and fluorescent labeling of isolated extracellular vesicles requires adequate purification. *Int. J. Pharm.* **2018**, *548*, 783–792. [CrossRef]

17. Wambre, E.; Jeong, D. Oral Tolerance Development and Maintenance. *Immunol Allergy Clin. North Am.* **2018**, *38*, 27–37. [CrossRef]

18. Hu, G.; Drescher, K.M.; Chen, X.-H.; Chen, X. Exosomal miRNAs: Biological Properties and Therapeutic Potential. *Front. Genet.* **2012**, *3*, 56. [CrossRef] [PubMed]

19. Rashed, M.H.; Bayraktar, E.; Helal, G.K.; Abd-Ellah, M.F.; Amero, P.; Chavez-Reyes, A.; Rodriguez-Aguayo, C.; Pichler, M. Exosomes: From Garbage Bins to Promising Therapeutic Targets. *Int. J. Mol. Sci.* **2017**, *18*, 538. [CrossRef]

20. Bai, D.; Sun, H.; Wang, X.; Lou, H.; Zhang, J.; Wang, X.; Jiang, L. MiR-150 Inhibits Cell Growth In Vitro and In Vivo by Restraining the RAB11A/WNT/β-Catenin Pathway in Thyroid Cancer. *Med Sci.* **2017**, *23*, 4885–4894. [CrossRef]

21. Li, X.; Chen, L.; Wang, W.; Meng, F.-B.; Zhao, R.-T.; Chen, Y. MicroRNA-150 Inhibits Cell Invasion and Migration and Is Downregulated in Human Osteosarcoma. *Cytogenet. Genome Res.* **2015**, *146*, 124–135. [CrossRef] [PubMed]

22. Chen, Z.; Wang, H.; Xia, Y.; Yan, F.; Lu, Y. Therapeutic Potential of Mesenchymal Cell–Derived miRNA-150-5p–Expressing Exosomes in Rheumatoid Arthritis Mediated by the Modulation of MMP14 and VEGF. *J. Immunol.* **2018**, *201*, 2472–2482. [CrossRef] [PubMed]

23. Askenase, P.W.; Bryniarski, K.; Paliwal, V.; Redegeld, F.; Kormelink, T.G.; Kerfoot, S.; Hutchinson, A.T.; Van Loveren, H.; Campos, R.; Itakura, A.; et al. A subset of AID-dependent B-1a cells initiates hypersensitivity and pneumococcal pneumonia resistance. *Ann. N. Y. Acad. Sci.* **2015**, *1362*, 200–214. [CrossRef] [PubMed]

24. Yamamoto, N.; Kerfoot, S.M.; Hutchinson, A.T.; Dela Cruz, C.S.; Nakazawa, N.; Szczepanik, M.; Majewska-Szczepanik, M.; Nazimek, K.; Ohana, N.; Bryniarski, K.; et al. Expression of activation-induced cytidine deaminase enhances the clearance of pneumococcal pneumonia: Evidence of a subpopulation of protective anti-pneumococcal B1a cells. *Immunology* **2016**, *147*, 97–113. [CrossRef]

25. Nazimek, K.; Askenase, P.W.; Bryniarski, K. Antibody Light Chains Dictate the Specificity of Contact Hypersensitivity Effector Cell Suppression Mediated by Exosomes. *Int. J. Mol. Sci.* **2018**, *19*, 2656. [CrossRef] [PubMed]

26. Zembala, M.; Asherson, G.L. T cell suppression of contact sensitivity in the mouse. II. The role of soluble suppressor factor and its interaction with macrophages. *Eur. J. Immunol.* **1974**, *4*, 799–804. [CrossRef]

27. Nazimek, K.; Ptak, W.; Nowak, B.; Askenase, P.W.; Bryniarski, K. Macrophages play an essential role in antigen-specific immune suppression mediated by T CD8$^+$ cell-derived exosomes. *Immunology* **2015**, *146*, 23–32. [CrossRef]

28. Raposo, G. B lymphocytes secrete antigen-presenting vesicles. *J. Exp. Med.* **1996**, *183*, 1161–1172. [CrossRef]

29. Chen, X.; Song, C.-H.; Feng, B.-S.; Li, T.-L.; Zheng, P.-Y.; Xing, Z.; Yang, P.-C. Intestinal epithelial cell-derived integrin αβ6 plays an important role in the induction of regulatory T cells and inhibits an antigen-specific Th2 response. *J. Leukoc. Boil.* **2011**, *90*, 751–759. [CrossRef]

30. Wawrzyniak, M.; O'Mahony, L.; Akdis, M. Role of Regulatory Cells in Oral Tolerance. *Allergy Asthma Immunol. Res.* **2017**, *9*, 107–115. [CrossRef]

31. Coombes, J.L.; Powrie, F. Dendritic cells in intestinal immune regulation. *Nat. Rev. Immunol.* **2008**, *8*, 435–446. [CrossRef]

32.   Davies, L.C.; Jenkins, S.J.; Allen, J.E.; Taylor, P.R. Tissue-resident macrophages. *Nat. Immunol.* **2013**, *14*, 986–995. [CrossRef] [PubMed]

33.   Romani, N.; Flacher, V.; Tripp, C.H.; Sparber, F.; Ebner, S.; Stoitzner, P. Targeting skin dendritic cells to improve intradermal vaccination. *Curr. Top. Microbiol. Immunol.* **2012**, *351*, 113–138. [CrossRef] [PubMed]

34.   Benmoussa, A.; Lee, C.H.C.; Laffont, B.; Savard, P.; Laugier, J.; Boilard, E.; Gilbert, C.; Fliss, I.; Provost, P. Commercial Dairy Cow Milk microRNAs Resist Digestion under Simulated Gastrointestinal Tract Conditions. *J. Nutr.* **2016**, *146*, 2206–2215. [CrossRef] [PubMed]

35.   Izumi, H.; Kosaka, N.; Shimizu, T.; Sekine, K.; Ochiya, T.; Takase, M. Bovine milk contains microRNA and messenger RNA that are stable under degradative conditions. *J. Dairy Sci.* **2012**, *95*, 4831–4841. [CrossRef]

36.   Irmak, M.K.; Oztas, Y.; Öztaş, E. Integration of maternal genome into the neonate genome through breast milk mRNA transcripts and reverse transcriptase. *Theor. Boil. Med. Model.* **2012**, *9*, 20. [CrossRef] [PubMed]

37.   Kusuma, R.J.; Manca, S.; Friemel, T.; Sukreet, S.; Nguyen, C.; Zempleni, J. Human vascular endothelial cells transport foreign exosomes from cow's milk by endocytosis. *Am. J. Physiol. Physiol.* **2016**, *310*, C800–C807. [CrossRef] [PubMed]

38.   Hock, A.; Miyake, H.; Li, B.; Lee, C.; Ermini, L.; Koike, Y.; Chen, Y.; Määttänen, P.; Zani, A.; Pierro, A. Breast milk-derived exosomes promote intestinal epithelial cell growth. *J. Pediatr. Surg.* **2017**, *52*, 755–759. [CrossRef]

39.   Melnik, B.C.; John, S.M.; Schmitz, G. Milk: An exosomal microRNA transmitter promoting thymic regulatory T cell maturation preventing the development of atopy? *J. Transl. Med.* **2014**, *12*, 43. [CrossRef]

![nutrients logo] *nutrients*

MDPI

Article

Suppression of Food Allergic Symptoms by Raw Cow's Milk in Mice is Retained after Skimming but Abolished after Heating the Milk—A Promising Contribution of Alkaline Phosphatase

Suzanne Abbring [1], Joseph Thomas Ryan [2], Mara A.P. Diks [1], Gert Hols [2], Johan Garssen [1,2] and Betty C.A.M. van Esch [1,2,*]

[1]   Division of Pharmacology, Utrecht Institute for Pharmaceutical Sciences, Faculty of Science, Utrecht University, 3584 CG Utrecht, The Netherlands
[2]   Danone Nutricia Research, 3584 CT Utrecht, The Netherlands
*   Correspondence: e.c.a.m.vanesch@uu.nl; Tel.: +31-625732735

Received: 19 April 2019; Accepted: 25 June 2019; Published: 30 June 2019

Abstract: Raw cow's milk was previously shown to suppress allergic symptoms in a murine model for food allergy. In the present study, we investigated the contribution of fat content and heat-sensitive milk components to this allergy-protective effect. In addition, we determined the potency of alkaline phosphatase (ALP), a heat-sensitive raw milk component, to affect the allergic response. C3H/HeOuJ mice were treated with raw milk, pasteurized milk, skimmed raw milk, pasteurized milk spiked with ALP, or phosphate-buffered saline for eight days prior to sensitization and challenge with ovalbumin (OVA). Effects of these milk types on the allergic response were subsequently assessed. Similar to raw milk, skimmed raw milk suppressed food allergic symptoms, demonstrated by a reduced acute allergic skin response and low levels of OVA-specific IgE and Th2-related cytokines. This protective effect was accompanied by an induction of $CD103^+CD11b^+$ dendritic cells and TGF-β-producing regulatory T cells in the mesenteric lymph nodes. Pasteurized milk was not protective but adding ALP restored the allergy-protective effect. Not the fat content, but the heat-sensitive components are responsible for the allergy-protective effects of raw cow's milk. Adding ALP to heat-treated milk might be an interesting alternative to raw cow's milk consumption, as spiking pasteurized milk with ALP restored the protective effects.

Keywords: alkaline phosphatase; allergic diseases; food allergy; immune regulation; milk processing; raw cow's milk

---

1. Introduction

Breastfeeding is the gold standard of infant nutrition. It is a complex matrix providing a unique combination of lipids, carbohydrates, proteins, vitamins and minerals. In addition, breast milk contains numerous components with immunomodulatory properties, such as immunoglobulins, lactoferrin, oligosaccharides, long-chain fatty acids, antioxidants and anti-inflammatory cytokines [1]. These bioactive components are potentially responsible for the allergy-protective effects associated with breastfeeding [2–4].

In analogy to breast milk, numerous epidemiological studies have shown that the consumption of raw, unprocessed, cow's milk can also reduce the risk of allergic diseases [5–9]. These epidemiological findings were recently confirmed by causal evidence, showing that raw cow's milk prevents the development of house dust mite-induced allergic asthma [10] and of OVA-induced food allergy [11] in murine animal models. However, due to the possible contamination with pathogens, raw cow's milk consumption is discouraged by regulatory authorities [12]. Even though risks from certified raw cow's milk, produced under strict hygienic and microbiological standards, are considered to be low [13], a zero-risk can never be attained. Cow's milk used for commercial purposes is therefore processed.

Milk processing, i.e., heat treatment and homogenization, ensures microbial safety and increases shelf life. Unfortunately, it also impacts the asthma- and allergy-protective effect of raw cow's milk [5,10,14]. Milk processing considerably alters raw cow's milk with most prominent effects on the fat content and heat-sensitive milk components. For both constituents, associations have been found in relation to the asthma- and allergy-protective effects. For the fat content of the milk, effects were mainly attributed to the levels of *n*-3 polyunsaturated fatty acids [14], whereas for the heat-sensitive milk components, the whey protein fraction was found to be associated with a reduced allergy risk [5]. Confirming that these raw milk constituents are indeed responsible for the observed allergy protection by showing causality is crucial. This knowledge will further support the development of mildly processed milk, or the addition of specific raw milk ingredients to heat-treated milk as an alternative to raw milk consumption.

In the current study, we investigated to which extent the fat content of the milk and the heat-sensitive milk components contribute to the allergy-protective effects of raw cow's milk by examining skimmed raw milk and pasteurized milk, respectively, in a murine ovalbumin (OVA)-induced food allergy model. In addition, we added alkaline phosphatase (ALP), one of the first bioactive raw milk components losing its activity upon heat treatment, to pasteurized milk to assess whether this restores the allergy-protective effect.

2. Materials and Methods

2.1. Mice

Three-week-old, specific pathogen-free, female C3H/HeOuJ mice, purchased from Charles River Laboratories (Sulzfeld, Germany) were housed in filter-topped makrolon cages (one cage/group, $n = 6$–8/cage) at the animal facility of the Utrecht University (Utrecht, The Netherlands) on a 12 h light/dark cycle with unlimited access to food ("Rat and Mouse Breeder and Grower Expanded"; Special Diet Services, Witham, UK) and water. Upon arrival, mice were randomly allocated to the control and experimental groups and were habituated to the laboratory conditions for one week prior to the start of the study. Animal procedures were approved by the Ethical Committee for Animal Research of the Utrecht University and conducted according to the European Directive 2010/63/EU on the protection of animals used for scientific purposes (AVD108002015346).

2.2. Milk Types

Raw cow's milk was collected from a dairy farm (Macroom, Ireland). After collection, the raw cow's milk was divided into three aliquots. Aliquot 1 was stored without any treatment at −20 °C until further use (raw milk). Aliquot 2 was heated for 15 s at 78 °C, cooled to 4 °C and then stored at −20 °C until further use (pasteurized milk). Aliquot 3 was skimmed at 55 °C to remove the milk fat, cooled to 4 °C and stored at −20 °C until further use (skimmed milk; 0.1% fat). All milk types were produced for experimental purposes only (Danone Nutricia Research, Utrecht, The Netherlands). On the days of milk treatment (Experimental Days -9 to -2; Figure 1), milks were thawed at room temperature and part of the pasteurized milk was spiked with bovine intestinal ALP (pasteurized milk + ALP; 3 units/0.5 mL pasteurized milk; 10× higher concentration than present in raw cow's milk). ALP was kindly provided by Prof. Dr. W. Seinen (Utrecht University, Utrecht, The Netherlands).

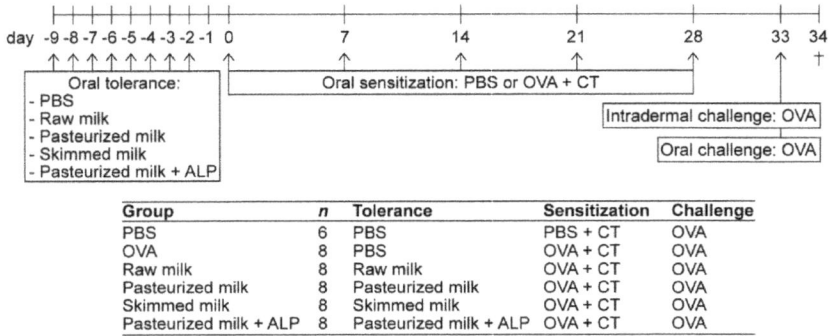

Figure 1. Schematic overview of the experimental setup. Female C3H/HeOuJ mice were randomly allocated to the control and experimental groups: PBS group (PBS-sensitized control mice; $n = 6$), OVA group (OVA-sensitized allergic mice; $n = 8$), raw milk group (raw milk-treated mice; $n = 8$), pasteurized milk group (pasteurized milk-treated mice; $n = 8$), skimmed milk group (skimmed milk-treated mice; $n = 8$) and pasteurized milk + ALP group (pasteurized milk + ALP-treated mice; $n = 8$). Mice were orally treated with 0.5 mL raw milk, pasteurized milk, skimmed milk, pasteurized milk spiked with ALP, or PBS (as a control). Following this oral tolerance induction period, mice were orally sensitized to OVA (20 mg/0.5 mL PBS) with CT as an adjuvant (10 µg/0.5 mL PBS). PBS-sensitized control mice (PBS group) received CT alone. Subsequently, all mice were intradermally and orally challenged with OVA. Mice were killed on day 34 (as indicated by †). PBS, phosphate-buffered saline; ALP, alkaline phosphatase; OVA, ovalbumin; CT, cholera toxin.

2.3. Animal Procedures

A schematic representation of the experimental design is shown in Figure 1. On Experimental Days 0, 7, 14, 21 and 28, mice ($n = 8$/group) were orally sensitized to 20 mg of the hen's egg protein OVA (grade V; Sigma-Aldrich, Zwijndrecht, The Netherlands) dissolved in 0.5 mL phosphate-buffered saline (PBS) containing 10 µg cholera toxin (CT; List Biological Laboratories, Campbell, CA, USA) as an adjuvant. The PBS-sensitized control mice ($n = 6$) received CT alone (10 µg/0.5 mL PBS). Prior to sensitization, mice were orally treated by using a blunt needle with 0.5 mL raw milk, pasteurized milk, skimmed raw milk, pasteurized milk spiked with ALP, or PBS (as a control) for eight consecutive days (Days -9 to -2). On Day 27, one day before the last sensitization, a blood sample was drawn via cheek puncture to measure basophil activation. On Day 33, five days after the last sensitization, all mice were challenged intradermally in both ears with OVA (10 µg/20 µL PBS) to determine the acute allergic skin response. On the same day, mice were challenged orally with 50 mg OVA dissolved in 0.5 mL PBS. Sixteen hours after the oral challenge (Day 34), a blood sample was taken, and mice were killed by cervical dislocation.

2.4. Evaluation of the Acute Allergic Skin Response

To assess the magnitude of the acute allergic skin response to OVA, mice were intradermally challenged in the ear pinnae of both ears with 10 µg OVA in 20 µL PBS. Ear thickness was measured in duplicate for each ear prior to and 1 h after the intradermal challenge using a digital micrometer (Mitutoyo, Veenendaal, The Netherlands). By subtracting the mean basal ear thickness from the mean ear thickness measured 1 h after the intradermal challenge, the ear swelling (expressed as Δ µm) was calculated. Isoflurane (Abbott, Breda, The Netherlands) was used for inhalation anesthesia to perform the intradermal challenge as well as the ear measurements. Measurements were performed blinded.

2.5. Basophil Activation Test

The basophil activation test was performed as described previously [15], with few alterations. Briefly, whole blood was drawn from each mouse via cheek puncture on experimental Day 27 (one day before the last sensitization). Blood samples from two mice were pooled and incubated with RPMI 1640 medium (Lonza, Verviers, Belgium), anti-mouse IgE (0.125 µg/mL; eBioscience, Breda, The Netherlands) or OVA (20 µg/mL; Sigma-Aldrich) for 90 min at 37 °C. Activation was stopped with PBS containing 5 mM EDTA (Thermo Fisher Scientific, Paisley, Scotland). After washing the cells twice with PBS, red blood cells were lysed and fixed using a whole blood lysing reagent kit (Beckman Coulter, Brea, CA, USA) according to the manufacturer's instructions. Cells were then washed again, and non-specific binding sites were blocked by incubating cells for 15 min on ice with anti-mouse CD16/CD32 (Mouse BD Fc Block; BD Biosciences, Alphen aan de Rijn, The Netherlands). Cells were subsequently stained for 30 min on ice with CD4-PE and CD45R/B220-PE to gate out T cells and B cells and with IgE-FITC and CD49b-APC to select basophils. CD200R-PerCP-eFluor® 710 was used as a marker for basophil activation. All antibodies were purchased from eBioscience. Flow cytometry was performed using FACS Canto II (BD Biosciences) and the results were analyzed using FlowLogic Software (Inivai Technologies, Mentone, Australia). Cut-off gates for positivity were established using the fluorescence-minus-one technique.

2.6. Measurement of OVA-Specific Immunoglobulins in Serum

Blood samples collected prior to sacrifice were centrifuged at 10,000 rpm for 10 min and serum was stored at −20 °C until analysis of OVA-specific immunoglobulins by means of ELISA. OVA-specific IgE levels were quantified as described previously [16], with few modifications. Briefly, high binding Costar 9018 plates (Corning Inc., New York, NY, USA) were coated overnight at 4 °C with 2 µg/mL purified rat anti-mouse IgE (BD Biosciences) in carbonate/bicarbonate buffer (0.05 M, pH 9.6; Sigma-Aldrich). The next day, plates were washed, blocked for 1 h with PBS/1% bovine serum albumin (BSA; Sigma-Aldrich) and incubated for 2 h with serum samples at room temperature. After washing, plates were incubated for 1 h with 1 µg/mL OVA coupled to digoxigenin (DIG). Plates were then washed again, followed by 1 h incubation with 300 mU/mL anti-DIG-POD Fab fragments conjugated to horseradish peroxidase (Sigma-Aldrich). After washing again, the reaction was developed using *o*-phenylenediamine (Sigma-Aldrich) and stopped by 4 M H_2SO_4. The absorbance was measured at 490 nm using a Benchmark microplate reader (Bio-Rad, Veenendaal, The Netherlands). OVA-specific IgE levels are expressed in arbitrary units, calculated based on a titration curve of pooled sera serving as an internal standard. For OVA-specific IgG1 and IgA, high binding Costar 9018 plates were coated with 20 µg/mL OVA (Sigma-Aldrich) in carbonate/bicarbonate buffer and incubated overnight at 4 °C. After overnight incubation, plates were washed and blocked for 1 h with PBS/1%BSA. Serum samples were then incubated for 2 h at room temperature and after washing, plates were incubated for 1.5 h with biotinylated rat anti–mouse IgG1 or IgA detection antibody (1 µg/mL; BD Biosciences). Plates were subsequently washed, incubated for 45 min with streptavidin–horseradish peroxidase (0.5 µg/mL; Sanquin), washed again and developed as described above for IgE. OVA-specific IgG1 and IgA levels are expressed as OD values.

2.7. Spleen, Mesenteric Lymph Nodes (MLN) and Lamina Propria (LP) Cell Isolation

Spleen and MLN single cell suspensions were obtained by crushing tissues through a 70 µm nylon cell strainer using a syringe. Splenocyte suspensions were incubated with lysis buffer (8.3 g NH_4Cl, 1 g KHC_3O and 37.2 mg EDTA dissolved in 1 L demi water, filter sterilized) to remove red blood cells. Cell suspensions were resuspended in RPMI 1640 medium (Lonza), supplemented with 10% heat-inactivated fetal bovine serum (FBS; Bodinco, Alkmaar, The Netherlands), penicillin (100 U/mL)/streptomycin (100 µg/mL; Sigma-Aldrich) and β-mercaptoethanol (20 µM; Thermo Fisher Scientific) prior to ex vivo OVA-specific restimulation assays or in PBS/1% BSA (Sigma-Aldrich) prior to cell stainings for flow cytometric analysis. For the isolation of small intestinal LP cells (n = 6/group), the small intestine was removed, cleared from fat and Peyer's patches, opened longitudinally, washed in PBS, and cut into 0.5 cm pieces. To remove epithelial cells and intraepithelial lymphocytes, these pieces were washed using Hank's Balanced Salt Solution (HBSS; Thermo Fisher Scientific) containing 15 mM HEPES (Thermo Fisher Scientific), pH 7.2, and incubated 4 × 15 min at 37 °C with HBSS/HEPES buffer supplemented with 5 mM EDTA, 10% FBS and penicillin (100 U/mL)/streptomycin (100 µg/mL), pH 7.2. After washing with RPMI 1640 medium containing 5% FBS and penicillin/streptomycin, tissue samples were digested for 2 × 45 min on a plate shaker at 37 °C with RPMI 1640 medium supplemented with 5% FBS, penicillin/streptomycin and 0.5 mg/mL collagenase type VIII (Sigma-Aldrich). To collect lamina propria cells, samples were vortexed for 10 s after each incubation and passed through a 100 µm nylon cell strainer. LP cell suspensons were subsequently washed with HBSS/HEPES and purified using a Percoll® density gradient (pH 7.2; GE Healthcare, Uppsala, Sweden). Purified LP cell suspensions were washed and resuspended in PBS/1% BSA for flow cytometric analysis.

2.8. Flow Cytometric Analysis of Immune Cells

Spleen-, MLN-, and LP-derived single cell suspensions (0.5–1 × 10^6 cells/well) were incubated for 15 min on ice with anti-mouse CD16/CD32 (Mouse BD Fc Block; BD Biosciences) in PBS/1% BSA/5% FBS buffer to block non-specific binding sites. Subsequently, cells were extracellularly stained with CD4-PerCP-Cy5.5, CD69-APC, CXCR3-PE, CD25-Alexa Fluor® 488, F4/80-APC-eFluor® 780, CD11c-PerCP-Cy5.5, CD103-APC, CD11b-PE, MHCII-FITC, CD45-PE-Cy7, CD19-PerCP-Cy5.5, CD45R/B220-FITC, latency-associated peptide (LAP)-PE-Cy7 (all purchased from eBioscience), T1ST2-FITC (MD Bioproducts, St. Paul, MN, USA) or CD138-APC (BD Biosciences) for 30 min on ice. Viable cells were distinguished using Fixable Viability Dye-eFluor® 780 (eBioscience). Cells only stained for extracellular markers were fixed using IC Fixation Buffer (eBioscience). Cells additionally stained with intracellular markers were fixed and permeabilized using the FoxP3 Transcription Factor Staining Buffer Set (eBioscience) according to the manufacturer's protocol and then stained with FoxP3-PE-Cy7 or -APC (eBioscience). Stained cells were measured on the FACS Canto II (BD Biosciences) and analyzed with FlowLogic Software (Inivai Technologies). To increase LAP expression on the surface of MLN-derived lymphocytes, cells were polyclonally stimulated with anti-CD3 (10 µg/mL)/CD28 (1 µg/mL; eBioscience) for 48 h at 37 °C, 5% CO_2 prior to staining, and boosted afterwards with leukocyte activation cocktail (BD Biosciences) for 4 h at 37 °C, 5% CO_2.

2.9. Cytokine Measurements after ex vivo OVA-Specific Stimulation of Splenocytes

Single cell splenocyte suspensions (8 × 10^5 cells/well) were cultured in U-bottom culture plates (Greiner, Frickenhausen, Germany) with either medium or OVA (50 µg/mL) for four days at 37 °C, 5% CO_2. Culture supernatant was collected and stored at −20 °C until measurements of IFNγ, IL-13 and IL-10 by means of ELISA, as described elsewhere [17].

2.10. Short-Chain Fatty Acid (SCFA) Analysis in Caecum

Caecal content was collected, snap-frozen in liquid nitrogen and stored at −80 °C until further analysis. After thawing, samples were homogenized by vortexing and diluted in cold PBS (1:10). Samples were subsequently centrifuged, the supernatant was collected and concentrations of acetic, propionic, butyric, isobutyric, valeric and isovaleric acid were determined as previously described [18] by means of a Shimadzu GC2010 gas chromatograph (Shimadzu Corporation, Kyoto, Japan), using 2-ethylbutyric acid as internal standard.

2.11. Statistical Analysis

Data are presented as mean ± SEM, including individual data points, and differences between pre-selected groups were statistically determined with one-way ANOVA followed by a Bonferroni's multiple comparisons test. For plasma cells in the MLN, log-transformed data were used to obtain normality for one-way ANOVA. For the same reason, OVA-specific IgG1 and IgA levels were square root-transformed. As OVA-specific IgE levels were not normally distributed, data were presented as individual data points in a box-and-whisker Tukey plot and analyzed using Kruskal–Wallis test followed by a Dunn's multiple comparisons test for pre-selected groups. All statistical analyses were performed using GraphPad Prism software (version 7.03; GraphPad Software, San Diego, CA, USA) and results were considered statistically significant when $P < 0.05$.

3. Results

3.1. Suppression of the Allergic Effector Response by Raw Milk is Retained after Skimming but Abolished after Heating the Milk

To determine whether milk processing affects the capacity of raw cow's milk to induce tolerance to a non-milk, food allergen, mice were orally treated with raw milk, pasteurized milk or skimmed milk before being sensitized and challenged with OVA. As expected, OVA-sensitized allergic mice showed an increased acute allergic skin response upon intradermal challenge compared to PBS-sensitized control mice (Figure 2A). Exposing mice to raw milk before OVA-sensitization significantly reduced the acute allergic skin response compared to PBS-treated allergic mice (Figure 2A). This protective effect was retained after skimming but abolished after pasteurization of the milk (Figure 2A). Since ALP is one of the first bioactive raw milk components losing activity upon heat treatment, we investigated whether spiking pasteurized milk with ALP would restore the allergy-protective effect. Interestingly, addition of ALP to pasteurized milk significantly lowered the acute allergic skin response compared to PBS-treated allergic mice and pasteurized milk-treated mice (Figure 2A). To study the extent of basophil activation, basophil surface expression of CD200R after stimulation of whole blood with OVA was determined. Even though no difference was observed in CD200R expression on basophils of OVA-sensitized allergic mice compared to PBS-sensitized control mice, CD200R expression was significantly reduced on basophils of mice treated with pasteurized milk + ALP compared to mice treated with pasteurized milk alone (Figure 2B), which is in line with the effects observed on the acute allergic skin response (Figure 2A). OVA-specific IgE levels and plasma cells were not significantly affected by exposure to the different milk types, but they did follow a similar pattern as the acute allergic skin response, with low levels in the raw milk, skimmed milk and pasteurized milk + ALP group and higher levels in the pasteurized milk group (Figure 2C,D). Unfortunately, OVA-specific IgE levels were not significantly increased in OVA-sensitized allergic mice compared to PBS-sensitized control mice (Figure 2C). However, OVA-specific IgG1 and IgA levels did (tend to) increase in these mice, demonstrating an immune response to OVA and supporting sensitization (Figure 2E,F). Functionality of IgE antibodies was furthermore confirmed using a murine bone marrow-derived mast cell degranulation assay (data not shown). For OVA-specific IgG1 and IgA, no differences between milk groups were observed (Figure 2E,F).

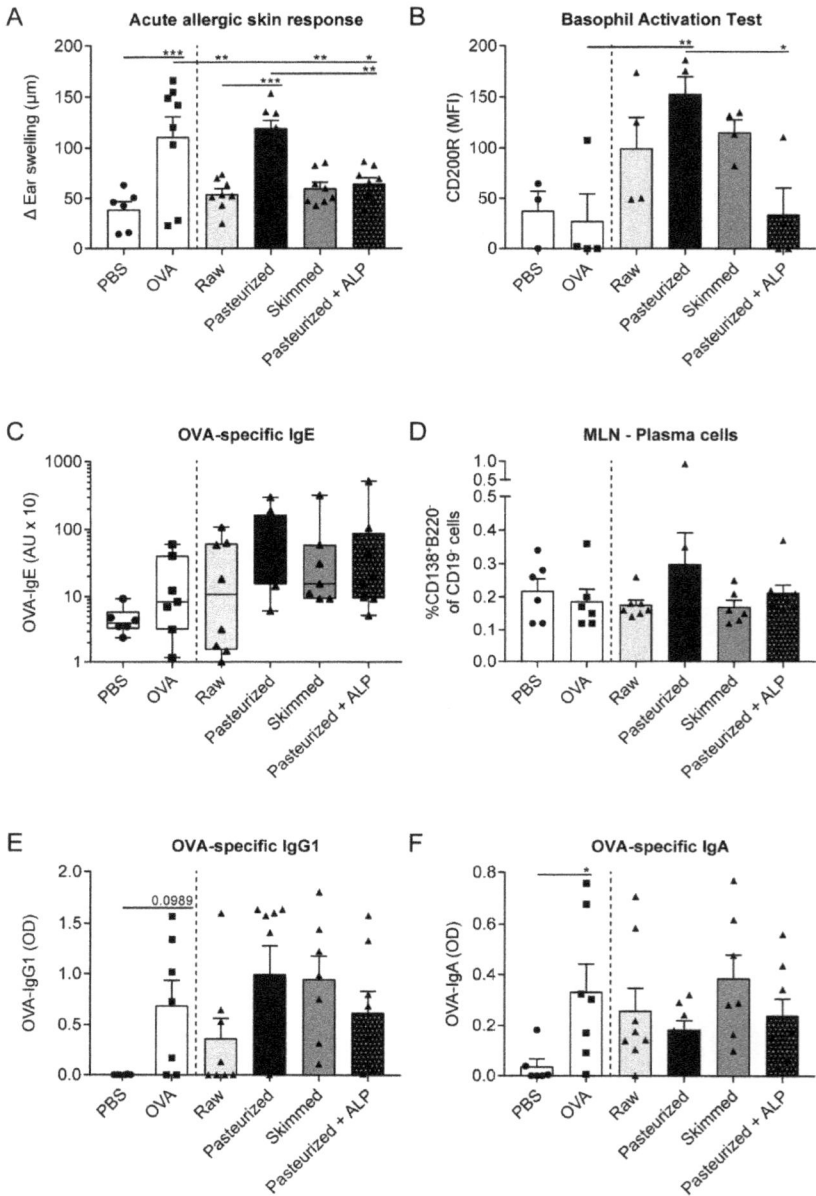

Figure 2. The protective effect of raw milk on the allergic effector response is retained by skimming but abolished by pasteurization of the milk. (**A**) The acute allergic skin response, expressed as Δ ear swelling, measured after intradermal challenge in the ear pinnae of both ears with OVA. (**B**) Basophil activation determined at Day 27 by surface expression of CD200R upon stimulation of whole blood with OVA (after subtracting baseline basophil activation). (**C**) Serum OVA-specific IgE levels measured 16 h after oral challenge. (**D**) Plasma cell (CD138$^+$B220$^-$ of CD19$^-$ cells) frequency assessed in the MLN. (**E**) Serum OVA-specific IgG1 and (**F**) IgA levels measured 16 h after oral challenge. Data are presented

as mean ± SEM or as box-and-whisker Tukey plot when data were not normally distributed. In addition, individual data points are displayed, $n = 6$ in PBS group and $n = 6$–8 in all other groups. For the basophil activation test (**B**), blood samples from two mice were pooled, $n = 3$ in the PBS group and $n = 4$ in all other groups. * $P < 0.05$, ** $P < 0.01$, *** $P < 0.001$ as analyzed with one-way ANOVA followed by Bonferroni's multiple comparisons test for pre-selected groups (**A,B,D,F**) or Kruskal-Wallis test for non-parametric data followed by Dunn's multiple comparisons test for pre-selected groups (**C**). PBS, phosphate-buffered saline; OVA, ovalbumin; raw, raw cow's milk; pasteurized, pasteurized cow's milk; skimmed, skimmed raw cow's milk; pasteurized + ALP, pasteurized milk spiked with alkaline phosphatase; MFI, median fluorescence intensity; AU, arbitrary units; MLN, mesenteric lymph nodes; OD, optical density.

3.2. Low Th2-Related Cytokine Production by Splenocytes from Raw Milk- and Skimmed Milk-Treated Mice after ex vivo Stimulation with OVA

To investigate whether different milk types affect T helper cell phenotype, spleen and MLN cells were isolated and analyzed by flow cytometry. Percentages of Th1 and Th2 cells were not affected in OVA-sensitized allergic mice compared to PBS-sensitized control mice (Figure 3A–D). However, in the spleen of mice treated with pasteurized milk, Th1 cells tended to decrease compared to allergic mice treated with PBS (Figure 3A). Th2 cell frequency in the spleen did not differ between milk groups (Figure 3B). In the MLN, the percentage of Th1 cells did not differ between milk groups, whereas Th2 cell frequency was increased in mice treated with pasteurized milk + ALP compared to allergic mice treated with PBS (Figure 3D). To determine the functional response of splenocytes and MLN cells upon exposure to OVA, cytokine production was determined. Th1-related IFNγ production by splenocytes was not affected by the different milk types (Figure 3E). For the Th2-related cytokine IL-13, low concentrations were observed in the raw milk, skimmed milk and pasteurized milk + ALP group (Figure 3F), which coincided with the effects observed on the acute allergic skin response (Figure 2A). Pasteurized milk treatment tended to increase the IL-13 production compared to raw milk treatment, whereas adding ALP to pasteurized milk tended to restore the low IL-13 levels (Figure 3F). Compared to raw milk, pasteurized milk also increased the production of IL-10 (Figure 3G), which was previously shown to act as a Th2 cytokine in this OVA-induced food allergy model [19]. Ex vivo stimulation of MLN cells with OVA did not induce detectable cytokine production (data not shown).

3.3. Raw Milk and Skimmed Milk Induce Tolerance-Associated Cell Types in the MLN

To assess whether the prevention of OVA-induced food allergic symptoms by raw milk, skimmed milk and pasteurized milk + ALP was associated with the induction of tolerance-associated cell types, changes in different dendritic cell (DC) and regulatory T cell (Treg) subsets were determined in the MLN. DC (CD11c$^+$MHCII$^+$) numbers tended to increase in raw milk-treated mice and increased in skimmed milk-treated mice compared to PBS-treated allergic mice (Figure 4A). More specific assessment of the DC subsets affected, revealed that both milk types mainly increased the tolerogenic CD103$^+$CD11b$^+$ DC subpopulation (Figure 4B). Although CD103$^+$ DCs are known for their capacity to induce FoxP3$^+$ Tregs in the MLN [20], no differences between groups were observed in the percentage of CD25$^+$FoxP3$^+$ Treg cells (Figure 4C). However, interestingly, the Treg subtype secreting TGF-β, also known as Th3 cells, tended to increase in the raw milk group compared to the pasteurized milk group (Figure 4D).

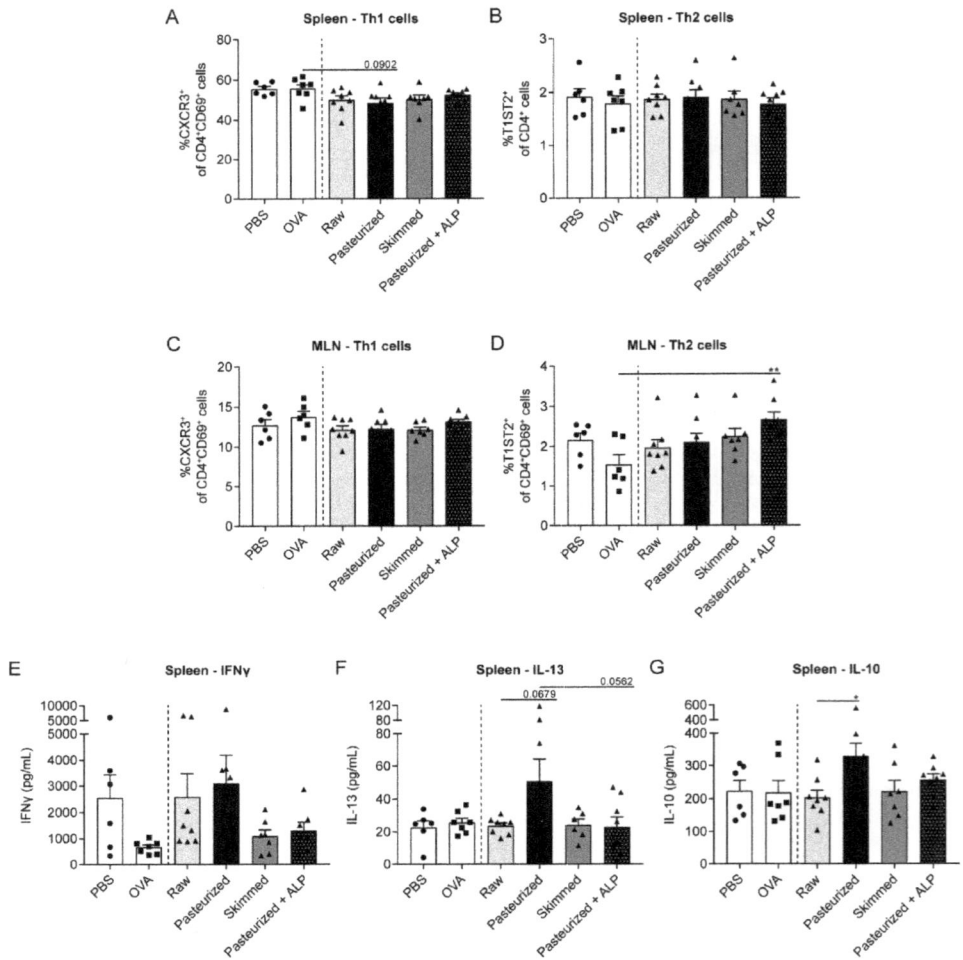

Figure 3. Th2-related cytokine concentrations produced by OVA-stimulated splenocytes were low in raw milk- and skimmed milk-treated mice. (**A**) The percentage of activated Th1 cells (CXCR3$^+$ of CD4$^+$CD69$^+$ cells) and (**B**) Th2 cells (T1ST2$^+$ of CD4$^+$ cells) in the spleen. (**C**) The percentage of activated Th1 cells (CXCR3$^+$ of CD4$^+$CD69$^+$ cells) and (**D**) activated Th2 cells (T1ST2$^+$ of CD4$^+$CD69$^+$) in the MLN. (**E**) IFNγ, (**F**) IL-13 and (**G**) IL-10 concentrations measured in supernatant of ex vivo stimulated splenocytes with OVA (stimulated for four days, 37 °C, 5% CO$_2$). Data are presented as mean ± SEM, including individual data points, n = 6 in PBS group and n = 6–8 in all other groups. * $P < 0.05$, ** $P < 0.01$ as analyzed with one-way ANOVA followed by Bonferroni's multiple comparisons test for pre-selected groups. PBS, phosphate-buffered saline; OVA, ovalbumin; raw, raw cow's milk; pasteurized, pasteurized cow's milk; skimmed, skimmed raw cow's milk; pasteurized + ALP, pasteurized milk spiked with alkaline phosphatase; MLN, mesenteric lymph nodes.

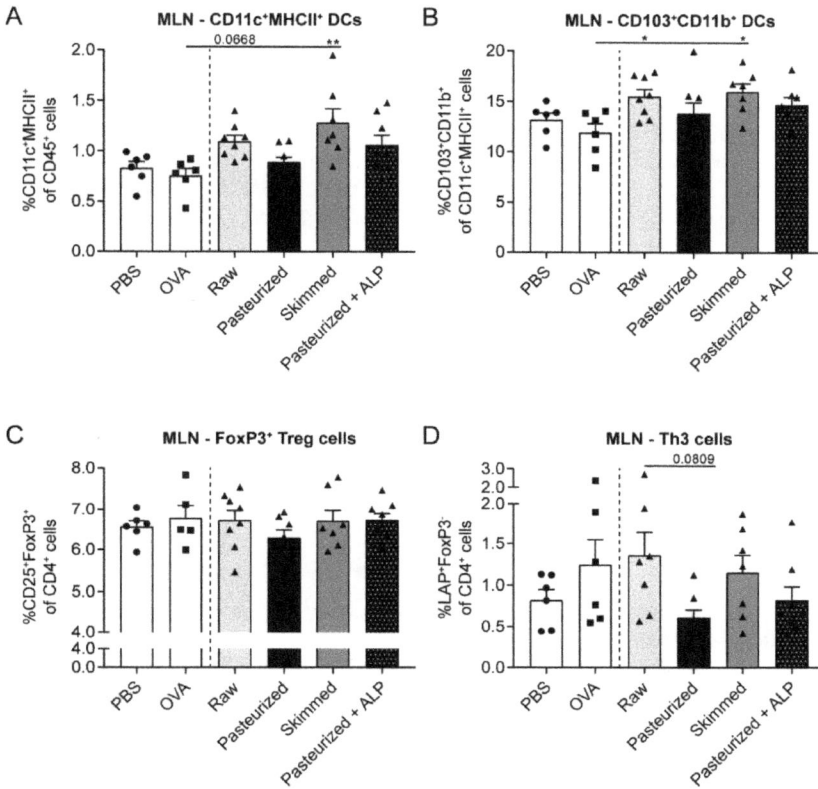

Figure 4. Induction of tolerance-associated cell types in the MLN of raw milk- and skimmed milk-treated mice. (**A**) Percentage of CD11c$^+$MHCII$^+$ DCs (CD11c$^+$MHCII$^+$ of CD45$^+$ cells), (**B**) CD103$^+$CD11b$^+$ DCs (CD103$^+$CD11b$^+$ of CD11c$^+$MHCII$^+$ cells), (**C**) FoxP3$^+$ Treg cells (CD25$^+$FoxP3$^+$ of CD4$^+$ cells) and (**D**) Th3 cells (LAP$^+$FoxP3$^-$ of CD4$^+$ cells) in the MLN. Data are presented as mean ± SEM, including individual data points, n = 6 in PBS group and n = 5–8 in all other groups. * $P < 0.05$, ** $P < 0.01$ as analyzed with one-way ANOVA followed by Bonferroni's multiple comparisons test for pre-selected groups. MLN, mesenteric lymph nodes; PBS, phosphate-buffered saline; OVA, ovalbumin; raw, raw cow's milk; pasteurized, pasteurized cow's milk; skimmed, skimmed raw cow's milk; pasteurized + ALP, pasteurized milk spiked with alkaline phosphatase.

3.4. Increased Percentage of Tolerogenic DCs in MLN of Raw Milk- and Skimmed Milk-Treated Mice is Not Associated with Increased Treg Cell Frequency in the LP

Besides promoting the differentiation of naïve T cells into Treg cells, CD103$^+$ DCs also induce the expression of gut-homing receptors on the surface of Treg cells [20]. To investigate whether the increased tolerogenic CD103$^+$CD11b$^+$ DC subpopulation in the MLN of raw milk- and skimmed milk-treated mice was associated with increased Treg cell trafficking to the gut, lamina propria cells were isolated and analyzed by flow cytometry. However, CD25$^+$FoxP3$^+$ Treg frequency did not differ between milk groups (Figure 5C) and also CD11c$^+$MHCII$^+$ DCs and the CD103$^+$CD11b$^+$ subset showed no differences in the LP (Figure 5A,B).

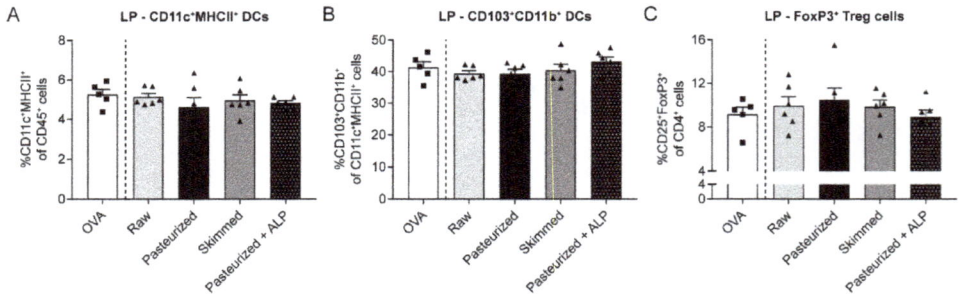

Figure 5. Different milk types did not affect tolerogenic DC and Treg cell frequency in the LP. (**A**) Percentage of CD11c$^+$MHCII$^+$ DCs (CD11c$^+$MHCII$^+$ of CD45$^+$ cells), (**B**) CD103$^+$CD11b$^+$ DCs (CD103$^+$CD11b$^+$ of CD11c$^+$MHCII$^+$ cells) and (**C**) FoxP3$^+$ Treg cells (CD25$^+$FoxP3$^+$ of CD4$^+$ cells) in the LP. Data are presented as mean ± SEM, including individual data points, *n* = 5–6/group. No significant differences were observed. LP, lamina propria; PBS, phosphate-buffered saline; OVA, ovalbumin; raw, raw cow's milk; pasteurized, pasteurized cow's milk; skimmed, skimmed raw cow's milk; pasteurized + ALP, pasteurized milk spiked with alkaline phosphatase.

3.5. Different Milk Types Did Not Affect SCFA Concentrations

Since modulation of the gut microbiome might be a way in which raw milk, skimmed milk and pasteurized milk + ALP induced tolerance to OVA, metabolic activity of the gut microbiome was assessed by determining SCFA concentrations in the caecum of the mice. Total SCFA concentrations were not significantly different between groups, but skimmed milk- and pasteurized milk + ALP-treated mice showed the highest levels (Figure 6A). Regarding individual SCFA, a similar pattern was observed for butyric acid and acetic acid concentrations, although again differences did not reach significance (Figure 6B,C). For propionic acid, concentrations were comparable in each milk group (Figure 6D).

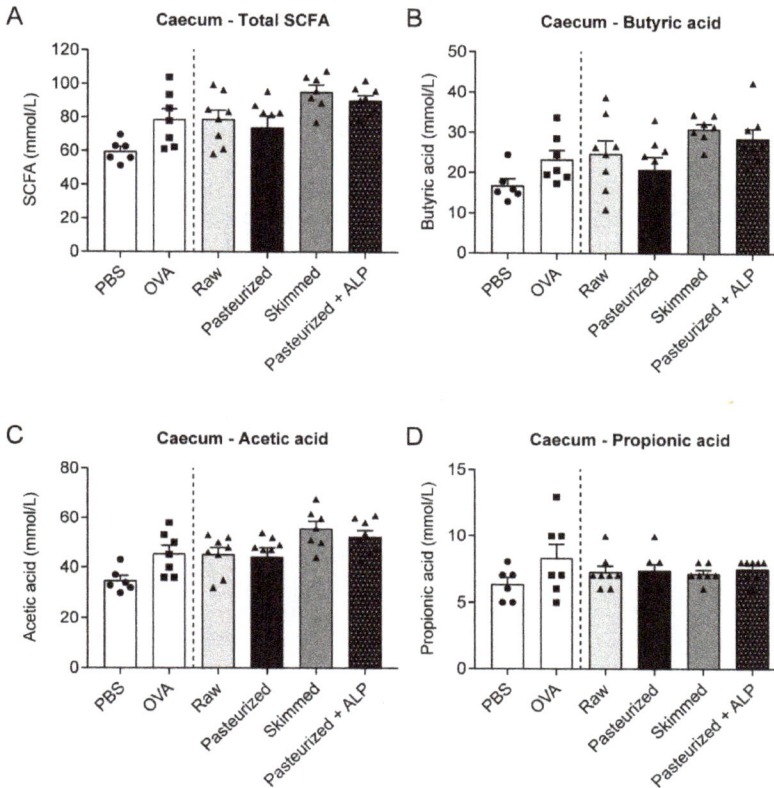

Figure 6. No differences in SCFA concentrations between milk groups. (**A**) Total SCFA concentrations and individual concentrations of (**B**) butyric acid, (**C**) acetic acid and (**D**) propionic acid measured in caecal content. Data are presented as mean ± SEM, including individual data points, $n = 6$ in PBS group and $n = 7$–8 in all other groups. No significant differences were observed. SCFA, short-chain fatty acids; PBS, phosphate-buffered saline; OVA, ovalbumin; raw, raw cow's milk; pasteurized, pasteurized cow's milk; skimmed, skimmed raw cow's milk; pasteurized + ALP, pasteurized milk spiked with alkaline phosphatase.

4. Discussion

We previously showed that raw, unprocessed cow's milk induces tolerance to OVA, an unrelated, non-milk, food allergen, in a murine food allergy model [11]. In the present study, we demonstrated that this protective effect is retained after skimming but abolished after pasteurization of the milk. Similar to raw cow's milk, skimmed raw milk reduced the acute allergic skin response after intradermal challenge with OVA. This coincided with low levels of OVA-specific IgE and Th2-related cytokines. An increase in CD103⁺CD11b⁺ DCs and TGF-β-producing Treg cells in the MLN, both associated with tolerance induction, might underlie the allergy-protective effects of raw and skimmed raw cow's milk. In addition, this study provides a first indication that adding ALP to heat-treated milk might be an interesting preventive strategy since spiking pasteurized milk with ALP restored the allergy-protective effects.

Although several epidemiological studies have shown the potency of raw cow's milk to reduce/prevent allergic diseases [5–9], its consumption is limited due to the potential presence of pathogens. The risks of diseases outbreaks by pathogens such as *Mycobacterium tuberculosis*, *Listeria*,

Salmonella, *Campylobacter*, Enterohemorrhagic *Escherichia coli* and Shigatoxigenic *Escherichia coli* are the reason for governmental agencies to prohibit the sale of raw cow's milk [12]. To prevent these potential risks, milk used for commercial purposes is processed. This means that, upon collection, raw milk undergoes various processing steps such as milk fat standardization, homogenization and heat treatment. These processing steps have profound effects on the milk structure and are shown to be detrimental to the allergy-protective effects [5,10,14].

Milk processing predominantly affects the fat content of the milk and the heat-sensitive milk components [21,22]. Since milk processing also abolishes the allergy-protective effects of raw cow's milk, this suggests that these constituents contribute to the observed protection. Indeed, both Wijga et al. (2003) and Waser et al. (2006) showed that frequent consumption of products containing milk fat was associated with a reduced asthma risk [8,23]. In addition, Brick et al. (2016) concluded that part of the asthma-protective effect of raw cow's milk was explained by a higher fat content and, particularly, higher *n*-3 polyunsaturated fatty acids levels compared to shop milk [14]. However, at the same time, there are also studies where the total fat content was not significantly related to asthma [5]. Epidemiological evidence also exists for a potential contribution of heat-sensitive raw milk components. Loss et al. (2011) demonstrated that raw farm milk, but not boiled farm milk, was inversely associated with asthma, hay fever and atopy. The heat-sensitive whey protein fraction of raw milk was implied to underlie these effects [5]. However, since these are all associations, proof of causality is needed to confirm the protective effects of these different raw milk constituents.

In the present study, we therefore investigated the effect of skimmed raw milk and pasteurized milk in a murine OVA-induced food allergy model. Skimmed milk was as allergy-protective as raw milk, suggesting that the fat content of the milk does not contribute to a large extent to the allergy-protective effects of raw cow's milk. Our results are in contrast with most of the epidemiological findings [8,14,23], emphasizing the importance of demonstrating a cause–effect relationship. On the other hand, the discrepancy could also be caused by the fact that these epidemiological studies mainly focused on asthma, whereas our study focused on food allergy. Different disease pathogenesis might underlie the different outcomes.

In contrast to skimming, pasteurization abolished the allergy-protective effects of raw cow's milk in the murine food allergy model used. This is in accordance with epidemiological evidence and with our previous results in a murine asthma model, both showing a loss of protection after heat treatment [5,10]. By comparing milk from the same origin, differing in only one processing step we can conclude with certainty that pasteurization is harmful to the allergy-protective capacity of raw cow's milk. The importance of heat-sensitive milk components, such as proteins, microRNAs and microbes, is thereby emphasized. Particularly, the heat-sensitive whey protein fraction of raw milk is often mentioned as source of the allergy-protective components. The major whey proteins, namely α-lactalbumin, β-lactoglobulin and bovine serum albumin, do not have immunomodulatory functionalities that can directly be linked to the allergy-protective effects of raw cow's milk, but several less abundant whey proteins such as immunoglobulins, lactoferrin, TGF-β and IL-10 theoretically do [24–26].

A first step towards identifying the potential allergy-protective whey proteins was made by Brick et al. (2017), who investigated the effect of processing intensity on immunologically active milk proteins [27]. As expected, a decrease in the number and abundance of detectable native whey proteins was observed with increasing heat load. Interestingly, the subsequent proteomic analysis categorized the milk samples into two clusters; high (boiled, ultra-high temperature and extended shelf life) and no/low heat treatment (raw, skimmed, pasteurized) [27]. Although pasteurized milk clustered together with raw and skimmed milk, indicating similar native protein patterns, it did not confer protection in our study. One could therefore argue that the overall native protein pattern looks similar but that minor differences still have major consequences for the allergy-protective capacity of the milk. One could also argue that even though pasteurization does not lead to denaturation or chemical modifications of whey proteins, it might lead to loss of functionality.

Although the effect of processing intensity on immunologically active whey proteins is very relevant, it does not provide a direct link to allergic diseases. To provide this link, specific whey proteins can be added to heat-treated milk to see whether they could restore the allergy-protective effect. As a first proof-of-concept, we spiked pasteurized milk with ALP and we assessed the effects on OVA-induced food allergic symptoms. ALP is probably best-known for its function in dairy industry as indicator of successful pasteurization. Upon pasteurization, ALP becomes inactivated and loses its activity, making it an ideal indicator of product safety [28]. Since ALP is one of the first bioactive raw milk components losing activity upon heat treatment, it is also a likely allergy-protective candidate. Oral administration of ALP was already shown to be effective in reducing inflammatory diseases [29–32], but whether it can also affect allergic diseases has, to our knowledge, never been studied.

Surprisingly, ALP was able to fully restore the allergy-protective effect in the food allergy model used. On practically every parameter assessed, ALP added to pasteurized milk showed similar protective effects as raw milk and skimmed raw milk. As this was a first proof-of-concept, 10 times higher ALP concentrations than present in raw cow's milk were added to pasteurized milk. We can therefore not yet conclude that ALP is the component underlying the allergy-protective effects of raw cow's milk, but it seems to be a promising candidate to be used as supplement to heat-treated milk.

In addition to the components involved, this study also provides some indication of the underlying mechanisms. The fact that mice orally treated for eight days with raw cow's milk were protected against OVA-induced allergic symptoms indicates that they developed oral tolerance to OVA. This oral tolerance was induced in the absence of the allergen, demonstrating that raw cow's milk has the capacity to induce tolerance via generic immunomodulation. The many immunomodulatory components present in raw cow's milk are likely to create a tolerogenic environment favoring unresponsiveness upon allergen exposure. Raw cow's milk is hypothesized to promote Treg cell development, to modulate the gut microbiome and to enhance intestinal barrier function [24–26]. However, none of these effects have actually been demonstrated after drinking raw milk.

The present study therefore tried to get more insight into some of these proposed mechanisms. Treg cells are identified as key players in inducing and maintaining oral tolerance [33]. However, in our study, the percentage of CD25$^+$FoxP3$^+$ Treg cells in the MLN was not affected by raw milk treatment. Interestingly, raw milk did increase the Treg subtype secreting TGF-β compared to pasteurized milk. The importance of these Th3 cells is demonstrated in a study showing reduced numbers in the intestine of food allergic children [34].

Induction of FoxP3$^+$ Treg cells occurs in the MLN by CD103$^+$ DCs under the influence of retinoic acid and TGF-β [35,36]. CD103$^+$ DCs originate in the LP and migrate to the MLNs in a CCR7-dependent manner after acquiring antigen [37]. This DC trafficking from the intestinal mucosa to the MLNs is crucial for oral tolerance induction [38,39]. Interestingly, while raw milk exposure did not significantly affect CD25$^+$FoxP3$^+$ Treg cells in the MLN, it did increase the tolerogenic CD103$^+$CD11b$^+$ DC subpopulation. Besides promoting the development of FoxP3$^+$ Treg cells, these DCs also induce the expression of gut-homing receptors on the cell surface of FoxP3$^+$ Treg cells [35], indicating that the Treg cells might have migrated to the gut. However, also in the LP, FoxP3$^+$ Treg cell frequency was not increased by raw milk. Examining effects on Treg cell populations directly after raw milk exposure, instead of at the end of the study, might be of importance, since farm milk exposure was previously shown to be associated with increased FoxP3$^+$ Treg cell numbers in children [40].

Regarding the potential immune modulation via the gut microbiome, results were not convincing for raw milk. Caecal SCFA concentrations, as indicator of metabolic activity of the gut microbiome, were not altered compared to other milk groups. However, effects on the gut microbiota itself were not assessed and the timing of measuring SCFA levels might also be crucial in this case. Highest SCFA concentrations, particularly butyric acid and acetic acid, were observed after exposure to skimmed raw milk and ALP. Since oral administration of ALP was previously shown to preserve normal gut microbiome homeostasis [41–43], it is tempting to speculate that this feature contributes to its allergy-protective effect.

A limitation of the current study is the lack of a significant IgE response in OVA-sensitized allergic mice compared to PBS-sensitized control mice. However, although significance was not reached, most of the animals in the OVA group did show higher OVA-IgE levels than animals in the PBS group. We would like to emphasize that serum IgE levels do not always correlate with the severity of the allergic response and that allergic symptoms are not solely induced by IgE [44–46]. The acute allergic skin response is the primary parameter of food allergic symptoms in the validated mouse model used. This response is acknowledged as a true acute allergic response and recognized as a translatable readout [47,48].

In summary, we demonstrated that the suppression of food allergic symptoms by raw cow's milk is retained after skimming but abolished after pasteurization of the milk. The data presented therefore indicate that not the fat content, but the heat-sensitive milk components are underlying the allergy-protective effects of raw cow's milk. The protection by raw and skimmed raw cow's milk was accompanied by an induction of tolerance-associated cell types in the MLN. In addition, we showed that ALP has the capacity to restore the allergy-protective effects abolished by heat treatment. This study thereby provides, for the first time, a direct link between one of the immunologically active whey proteins present in raw cow's milk and the suppression of allergic symptoms. Although its potency and mechanism of action still need to be determined, ALP is a promising raw milk component to be added to heat-treated milk. Hence, this research represents an attractive preventive strategy for allergic diseases as alternative to raw milk consumption.

Author Contributions: Conceptualization, S.A., J.T.R., G.H., J.G. and B.C.A.M.v.E.; Funding acquisition, G.H., J.G. and B.C.A.M.v.E.; Investigation, S.A., J.T.R. and M.A.P.D.; Methodology, S.A. and B.C.A.M.v.E.; Project administration, S.A. and B.C.A.M.v.E.; Supervision, S.A., J.G. and B.C.A.M.v.E.; Visualization, S.A.; Writing—original draft, S.A.; and Writing—review and editing, J.T.R., G.H., J.G. and B.C.A.M.v.E.

Funding: This research was financially supported by Danone Nutricia Research.

Acknowledgments: The authors would like to thank W. Seinen (Utrecht University, Utrecht, The Netherlands) for kindly providing bovine intestinal alkaline phosphatase and B.R.J. Blokhuis (Utrecht University) for his technical assistance.

Conflicts of Interest: J.T.R. and G.H. are employed at Danone Nutricia Research. J.G. and B.C.A.M.v.E. are partly employed at Danone Nutricia Research. All other authors report no potential conflicts of interest.

References

1. West, C.E.; D'Vaz, N.; Prescott, S.L. Dietary immunomodulatory factors in the development of immune tolerance. *Curr. Allergy Asthma Rep.* **2011**, *11*, 325–333. [CrossRef] [PubMed]

2. Thijs, C.; Muller, A.; Rist, L.; Kummeling, I.; Snijders, B.E.; Huber, M.; van Ree, R.; Simoes-Wust, A.P.; Dagnelie, P.C.; Van Den Brandt, P.A. Fatty acids in breast milk and development of atopic eczema and allergic sensitisation in infancy. *Allergy* **2011**, *66*, 58–67. [CrossRef] [PubMed]

3. Verhasselt, V.; Milcent, V.; Cazareth, J.; Kanda, A.; Fleury, S.; Dombrowicz, D.; Glaichenhaus, N.; Julia, V. Breast milk-mediated transfer of an antigen induces tolerance and protection from allergic asthma. *Nat. Med.* **2008**, *14*, 170–175. [CrossRef] [PubMed]

4. Moro, G.; Arslanoglu, S.; Stahl, B.; Jelinek, J.; Wahn, U.; Boehm, G. A mixture of prebiotic oligosaccharides reduces the incidence of atopic dermatitis during the first six months of age. *Arch. Dis. Child.* **2006**, *91*, 814–819. [CrossRef] [PubMed]

5. Loss, G.; Apprich, S.; Waser, M.; Kneifel, W.; Genuneit, J.; Buchele, G.; Weber, J.; Sozanska, B.; Danielewicz, H.; Horak, E.; et al. The protective effect of farm milk consumption on childhood asthma and atopy: The GABRIELA study. *J. Allergy Clin. Immunol.* **2011**, *128*. [CrossRef] [PubMed]

6. Ege, M.J.; Frei, R.; Bieli, C.; Schram-Bijkerk, D.; Waser, M.; Benz, M.R.; Weiss, G.; Nyberg, F.; van Hage, M.; Pershagen, G.; et al. Not all farming environments protect against the development of asthma and wheeze in children. *J. Allergy Clin. Immunol.* **2007**, *119*, 1140–1147. [CrossRef] [PubMed]

7. Riedler, J.; Braun-Fahrlander, C.; Eder, W.; Schreuer, M.; Waser, M.; Maisch, S.; Carr, D.; Schierl, R.; Nowak, D.; von Mutius, E.; et al. Exposure to farming in early life and development of asthma and allergy: A cross-sectional survey. *Lancet* **2001**, *358*, 1129–1133. [CrossRef]

8. Waser, M.; Michels, K.B.; Bieli, C.; Floistrup, H.; Pershagen, G.; von Mutius, E.; Ege, M.; Riedler, J.; Schram-Bijkerk, D.; Brunekreef, B.; et al. Inverse association of farm milk consumption with asthma and allergy in rural and suburban populations across Europe. *Clin. Exp. Allergy* **2007**, *37*, 661–670. [CrossRef]

9. Perkin, M.R.; Strachan, D.P. Which aspects of the farming lifestyle explain the inverse association with childhood allergy? *J. Allergy Clin. Immunol.* **2006**, *117*, 1374–1381. [CrossRef]

10. Abbring, S.; Verheijden, K.A.T.; Diks, M.A.P.; Leusink-Muis, A.; Hols, G.; Baars, T.; Garssen, J.; van Esch, B.C.A.M. Raw Cow's Milk Prevents the Development of Airway Inflammation in a Murine House Dust Mite-Induced Asthma Model. *Front. Immunol.* **2017**, *8*, 1045. [CrossRef]

11. Abbring, S.; Wolf, J.; Ayechu Muruzabal, V.; Diks, M.A.P.; Alashkar Alhamwe, B.; Alhamdan, F.; Harb, H.; Renz, H.; Garn, H.; Garssen, J.; et al. Raw cow's milk suppresses allergic symptoms in a murine model for food allergy—A potential role for epigenetic modifications. *Nutrients* (under review).

12. Committee on Infectious Diseases & Committee on Nutrition & American Academy of Pediatrics. Consumption of Raw or Unpasteurized Milk and Milk Products by Pregnant Women and Children. *Pediatrics* **2014**, *133*, 175–179. [CrossRef] [PubMed]

13. Verordnung über die Güteprüfung und Bezahlung der Anlieferungsmilch (Milch-Güteverordnung). Available online: http://www.gesetze-im-internet.de/milchg_v/index.html (accessed on 26 March 2019).

14. Brick, T.; Schober, Y.; Bocking, C.; Pekkanen, J.; Genuneit, J.; Loss, G.; Dalphin, J.C.; Riedler, J.; Lauener, R.; Nockher, W.A.; et al. Omega-3 fatty acids contribute to the asthma-protective effect of unprocessed cow's milk. *J. Allergy Clin. Immunol.* **2016**, *137*, 1699–1706.e13. [CrossRef] [PubMed]

15. Torrero, M.N.; Larson, D.; Hubner, M.P.; Mitre, E. CD200R surface expression as a marker of murine basophil activation. *Clin. Exp. Allergy* **2009**, *39*, 361–369. [CrossRef] [PubMed]

16. Deurloo, D.T.; van Esch, B.C.; Hofstra, C.L.; Nijkamp, F.P.; van Oosterhout, A.J. CTLA4-IgG reverses asthma manifestations in a mild but not in a more "severe" ongoing murine model. *Am. J. Respir. Cell Mol. Biol.* **2001**, *25*, 751–760. [CrossRef] [PubMed]

17. Abbring, S.; Kusche, D.; Roos, T.C.; Diks, M.A.P.; Hols, G.; Garssen, J.; Baars, T.; van Esch, B.C.A.M. Milk processing increases the allergenicity of cow's milk-preclinical evidence supported by a human proof-of-concept provocation pilot. *Clin. Exp. Allergy* **2019**. [CrossRef] [PubMed]

18. Bakker-Zierikzee, A.M.; Alles, M.S.; Knol, J.; Kok, F.J.; Tolboom, J.J.; Bindels, J.G. Effects of infant formula containing a mixture of galacto- and fructo-oligosaccharides or viable Bifidobacterium animalis on the intestinal microflora during the first 4 months of life. *Br. J. Nutr.* **2005**, *94*, 783–790. [CrossRef]

19. Polukort, S.H.; Rovatti, J.; Carlson, L.; Thompson, C.; Ser-Dolansky, J.; Kinney, S.R.; Schneider, S.S.; Mathias, C.B. IL-10 Enhances IgE-Mediated Mast Cell Responses and Is Essential for the Development of Experimental Food Allergy in IL-10-Deficient Mice. *J. Immunol.* **2016**, *196*, 4865–4876. [CrossRef]

20. Coombes, J.L.; Siddiqui, K.R.; Arancibia-Carcamo, C.V.; Hall, J.; Sun, C.M.; Belkaid, Y.; Powrie, F. A functionally specialized population of mucosal CD103+ DCs induces Foxp3+ regulatory T cells via a TGF-beta and retinoic acid-dependent mechanism. *J. Exp. Med.* **2007**, *204*, 1757–1764. [CrossRef]

21. Michalski, M.C.; Januel, C. Does homogenization affect the human health properties of cow's milk? *Trends Food Sci. Technol.* **2006**, *17*, 423–437. [CrossRef]

22. Niero, G.; Penasa, M.; Berard, J.; Kreuzer, M.; Cassandro, M.; De Marchi, M. Technical note: Development and validation of an HPLC method for the quantification of tocopherols in different types of commercial cow milk. *J. Dairy Sci.* **2018**, *101*, 6866–6871. [CrossRef] [PubMed]

23. Wijga, A.H.; Smit, H.A.; Kerkhof, M.; de Jongste, J.C.; Gerritsen, J.; Neijens, H.J.; Boshuizen, H.C.; Brunekreef, B.; PIAMA. Association of consumption of products containing milk fat with reduced asthma risk in pre-school children: The PIAMA birth cohort study. *Thorax* **2003**, *58*, 567–572. [CrossRef] [PubMed]

24. Van Neerven, R.J.; Knol, E.F.; Heck, J.M.; Savelkoul, H.F. Which factors in raw cow's milk contribute to protection against allergies? *J. Allergy Clin. Immunol.* **2012**, *130*, 853–858. [CrossRef] [PubMed]

25. Abbring, S.; Hols, G.; Garssen, J.; van Esch, B.C.A.M. Raw cow's milk consumption and allergic diseases—The potential role of bioactive whey proteins. *Eur. J. Pharmacol.* **2019**, *843*, 55–65. [CrossRef] [PubMed]

26. Perdijk, O.; van Splunter, M.; Savelkoul, H.F.J.; Brugman, S.; van Neerven, R.J.J. Cow's Milk and Immune Function in the Respiratory Tract: Potential Mechanisms. *Front. Immunol.* **2018**, *9*, 143. [CrossRef]

27. Brick, T.; Ege, M.; Boeren, S.; Bock, A.; von Mutius, E.; Vervoort, J.; Hettinga, K. Effect of Processing Intensity on Immunologically Active Bovine Milk Serum Proteins. *Nutrients* **2017**, *9*, 963. [CrossRef]

28. Rankin, S.A.; Christiansen, A.; Lee, W.; Banavara, D.S.; Lopez-Hernandez, A. Invited review: The application of alkaline phosphatase assays for the validation of milk product pasteurization. *J. Dairy Sci.* **2010**, *93*, 5538–5551. [CrossRef] [PubMed]

29. Beumer, C.; Wulferink, M.; Raaben, W.; Fiechter, D.; Brands, R.; Seinen, W. Calf intestinal alkaline phosphatase, a novel therapeutic drug for lipopolysaccharide (LPS)-mediated diseases, attenuates LPS toxicity in mice and piglets. *J. Pharmacol. Exp. Ther.* **2003**, *307*, 737–744. [CrossRef]

30. Tuin, A.; Poelstra, K.; de Jager-Krikken, A.; Bok, L.; Raaben, W.; Velders, M.P.; Dijkstra, G. Role of alkaline phosphatase in colitis in man and rats. *Gut* **2009**, *58*, 379–387. [CrossRef]

31. Whitehouse, J.S.; Riggle, K.M.; Purpi, D.P.; Mayer, A.N.; Pritchard, K.A., Jr.; Oldham, K.T.; Gourlay, D.M. The protective role of intestinal alkaline phosphatase in necrotizing enterocolitis. *J. Surg. Res.* **2010**, *163*, 79–85. [CrossRef]

32. Lalles, J.P. Intestinal alkaline phosphatase: Multiple biological roles in maintenance of intestinal homeostasis and modulation by diet. *Nutr. Rev.* **2010**, *68*, 323–332. [CrossRef] [PubMed]

33. Akdis, M.; Blaser, K.; Akdis, C.A. T regulatory cells in allergy: Novel concepts in the pathogenesis, prevention, and treatment of allergic diseases. *J. Allergy Clin. Immunol.* **2005**, *116*, 961–968. [CrossRef] [PubMed]

34. Perez-Machado, M.A.; Ashwood, P.; Thomson, M.A.; Latcham, F.; Sim, R.; Walker-Smith, J.A.; Murch, S.H. Reduced transforming growth factor-beta1-producing T cells in the duodenal mucosa of children with food allergy. *Eur. J. Immunol.* **2003**, *33*, 2307–2315. [CrossRef]

35. Scott, C.L.; Aumeunier, A.M.; Mowat, A.M. Intestinal CD103+ dendritic cells: Master regulators of tolerance? *Trends Immunol.* **2011**, *32*, 412–419. [CrossRef] [PubMed]

36. Mucida, D.; Pino-Lagos, K.; Kim, G.; Nowak, E.; Benson, M.J.; Kronenberg, M.; Noelle, R.J.; Cheroutre, H. Retinoic acid can directly promote TGF-beta-mediated Foxp3(+) Treg cell conversion of naive T cells. *Immunity* **2009**, *30*, 471–472. [CrossRef] [PubMed]

37. Jang, M.H.; Sougawa, N.; Tanaka, T.; Hirata, T.; Hiroi, T.; Tohya, K.; Guo, Z.; Umemoto, E.; Ebisuno, Y.; Yang, B.G.; et al. CCR7 is critically important for migration of dendritic cells in intestinal lamina propria to mesenteric lymph nodes. *J. Immunol.* **2006**, *176*, 803–810. [CrossRef]

38. Worbs, T.; Forster, R. A key role for CCR7 in establishing central and peripheral tolerance. *Trends Immunol.* **2007**, *28*, 274–280. [CrossRef] [PubMed]

39. Spahn, T.W.; Weiner, H.L.; Rennert, P.D.; Lugering, N.; Fontana, A.; Domschke, W.; Kucharzik, T. Mesenteric lymph nodes are critical for the induction of high-dose oral tolerance in the absence of Peyer's patches. *Eur. J. Immunol.* **2002**, *32*, 1109–1113. [CrossRef]

40. Lluis, A.; Depner, M.; Gaugler, B.; Saas, P.; Casaca, V.I.; Raedler, D.; Michel, S.; Tost, J.; Liu, J.; Genuneit, J.; et al. Increased regulatory T-cell numbers are associated with farm milk exposure and lower atopic sensitization and asthma in childhood. *J. Allergy Clin. Immunol.* **2014**, *133*, 551–559. [CrossRef]

41. Bates, J.M.; Akerlund, J.; Mittge, E.; Guillemin, K. Intestinal alkaline phosphatase detoxifies lipopolysaccharide and prevents inflammation in zebrafish in response to the gut microbiota. *Cell Host Microbe* **2007**, *2*, 371–382. [CrossRef]

42. Malo, M.S.; Alam, S.N.; Mostafa, G.; Zeller, S.J.; Johnson, P.V.; Mohammad, N.; Chen, K.T.; Moss, A.K.; Ramasamy, S.; Faruqui, A.; et al. Intestinal alkaline phosphatase preserves the normal homeostasis of gut microbiota. *Gut* **2010**, *59*, 1476–1484. [CrossRef] [PubMed]

43. Lalles, J.P. Intestinal alkaline phosphatase: Novel functions and protective effects. *Nutr. Rev.* **2014**, *72*, 82–94. [CrossRef]

44. Schouten, B.; van Esch, B.C.; van Thuijl, A.O.; Blokhuis, B.R.; Groot Kormelink, T.; Hofman, G.A.; Moro, G.E.; Boehm, G.; Arslanoglu, S.; Sprikkelman, A.B.; et al. Contribution of IgE and immunoglobulin free light chain in the allergic reaction to cow's milk proteins. *J Allergy Clin Immunol.* **2010**, *125*, 1308–1314. [CrossRef]

45. Sampson, H.A.; Aceves, S.; Bock, S.A.; James, J.; Jones, S.; Lang, D.; Nadeau, K.; Nowak-Wegrzyn, A.; Oppenheimer, J.; Perry, T.T.; et al. Food allergy: A practice parameter update-2014. *J. Allergy Clin. Immunol.* **2014**, *134*. [CrossRef]

46. Schouten, B.; van Esch, B.C.; Hofman, G.A.; van den Elsen, L.W.; Willemsen, L.E.; Garssen, J. Acute allergic skin reactions and intestinal contractility changes in mice orally sensitized against casein or whey. *Int. Arch. Allergy Immunol.* **2008**, *147*, 125–134. [CrossRef]

47. Van Esch, B.C.; Schouten, B.; Hofman, G.A.; van Baalen, T.; Nijkamp, F.P.; Knippels, L.M.; Willemsen, L.E.; Garssen, J. Acute allergic skin response as a new tool to evaluate the allergenicity of whey hydrolysates in a mouse model of orally induced cow's milk allergy. *Pediatr. Allergy Immunol.* **2010**, *21*, e780–e786. [CrossRef] [PubMed]

48. Van Esch, B.C.; van Bilsen, J.H.; Jeurink, P.V.; Garssen, J.; Penninks, A.H.; Smit, J.J.; Pieters, R.H.; Knippels, L.M. Interlaboratory evaluation of a cow's milk allergy mouse model to assess the allergenicity of hydrolysed cow's milk based infant formulas. *Toxicol. Lett.* **2013**, *220*, 95–102. [CrossRef] [PubMed]

nutrients

MDPI

Article

Modulation of Milk Allergenicity by Baking Milk in Foods: A Proteomic Investigation

Simona L. Bavaro [1], Elisabetta De Angelis [1], Simona Barni [2], Rosa Pilolli [1], Francesca Mori [2], Elio. M. Novembre [2] and Linda Monaci [1,*]

[1] Institute of Sciences of Food Production, Italian National Research Council (ISPA-CNR), Via Amendola 122/O, 70126 Bari, Italy

[2] Allergy Unit, Department of Pediatrics, Anna Meyer Children's University Hospital, University of Florence, 50139 Florence, Italy

* Correspondence: linda.monaci@ispa.cnr.it; Tel.: +39-080-592-9343

Received: 31 May 2019; Accepted: 1 July 2019; Published: 6 July 2019

Abstract: Cow's milk is considered the best wholesome supplement for children since it is highly enriched with micro and macro nutrients. Although the protein fraction is composed of more than 25 proteins, only a few of them are capable of triggering allergic reactions in sensitive consumers. The balance in protein composition plays an important role in the sensitization capacity of cow's milk, and its modification can increase the immunological response in allergic patients. In particular, the heating treatments in the presence of a food matrix have demonstrated a decrease in the milk allergenicity and this has also proved to play a pivotal role in developing tolerance towards milk. In this paper we investigated the effect of thermal treatment like baking of cow's milk proteins that were employed as ingredients in the preparation of muffins. A proteomic workflow was applied to the analysis of the protein bands highlighted along the SDS gel followed by western blot analyses with sera of milk allergic children in order to have deeper information on the impact of the heating on the epitopes and consequent IgE recognition. Our results show that incorporating milk in muffins might promote the formation of complex milk–food components and induce a modulation of the immunoreactivity towards milk allergens compared to milk baked in the oven at 180 °C for ten minutes. The interactions between milk proteins and food components during heating proved to play a role in the potential reduction of allergenicity as assessed by in vitro tests. This would help, in perspective, in designing strategies for improving milk tolerance in young patients affected from severe milk allergies.

Keywords: milk allergen; baked milk; cow's milk; allergenicity modulation; proteomics

---

1. Introduction

The introduction of cow's milk (CM) in the human diet has been a very long tradition, for approximately 9000 years. Since then, the incidence of adverse reactions to CM is constantly increasing, becoming one of the first and most common causes of food allergies in early childhood in Europe [1]. The reported prevalence of cow's milk allergy (CMA) varies considerably between studies probably due to the different methods used for diagnosis or the differences in the ages of the studied populations [2]. It has been reported that nowadays 0.6% to 3% of children under the age of 6 years, 0.3% of older children and teens, and less than 0.5% of adults suffer from CMA [3]. Although 15% of affected children remain allergic throughout adulthood, the majority of milk allergic infants seems to be able to consume milk and its by-products with a total resolution of CMA in 19% of the children by 4 years of age, in 42% by 8 years of age, in 64% by 12 years of age, and in 79% by 16 years of age. Despite these encouraging data, the mechanisms underpinning the development of clinical tolerance are not fully understood [4,5].

Cow's milk contains approximately 30–35 g of proteins per liter encompassing more than 25 different proteins, although only some of them are capable of triggering allergic reactions. Proteins composing milk typically belong to two different categories: Caseins (αS1-casein, αS2-casein, β-casein and k-casein) and whey proteins (β-lactoglobulin [β-LG], α-lactalbumin [ALA], bovine lactoferrin [LF], bovine serum albumin [BSA] and bovine immunoglobulins [Ig]), accounting for respectively the 80% and 20% of the total cow milk protein content [6,7]. According to the World Health Organization and International Union of Immunological Societies (WHO/IUIS) official list of allergens, milk allergen proteins are classified with the following designation: Bos d 5 (β-LG), Bos d 4 (ALA), Bos d 6 (BSA), Bos d 7 (Ig), Bos d 9 (αS1-casein), Bos d 10 (αS2-casein), Bos d 11 (β-casein), Bos d 12 (κ-casein). From a biochemical point of view, caseins are phosphoproteins that exist as colloidal aggregates known as casein micelles, and whose function mainly consists in binding essential minerals, such as calcium phosphate, that would otherwise precipitate resulting difficult in being ingested [8]. In cow's milk, four main casein phosphoproteins have been identified namely αS1-casein, αS2-casein, β-casein, and κ-casein, approximately in proportions respectively of 4:1:4:1 by weight. Their molecular weights range between 19 and 25 kDa, with an average isoelectric point (pI) comprised of between 4.6 and 4.8. Moreover, all caseins are amphiphilic and have well-defined structures, with a little primary structure homology while sharing biophysical features such as heat resistance [9]. On the contrary, whey proteins (WPs) consist mostly of β-LG (about 44.70%) and ALA (about 14.22%), along with Ig and BSA (about 1.5%) [10], with molecular weight between 14 and 80 kDa. Their structure is mainly composed of nine anti-parallel β-sheets and one α-helix, with two intra-molecular disulphide bonds and one free potentially reactive sulfhydryl group [11] that confers high stability against proteases and acidic hydrolysis [12].

The balance between these two different groups of protein plays an important role in the sensitization capacity of cow's milk, and its modification can increase the immunological response in allergic patients [13]. However, the control of the daily intake of caseins/WPs ratio in allergic consumers is not easy, and the only useful action to protect them from developing adverse reactions due to CM consumption remains the strict avoidance of milk and dairy products. Nonetheless several strategies have been developed to promote safe consumption of milk and its derived products in allergic patients and efforts have been also directed to design the best procedures for inducing cow's milk oral tolerance [14] or to calculate the safest dose to start from for oral desensitization studies [15]. Among these it has been demonstrated that the consumption of baked milk as such or as an ingredient included into the food matrix might induce milk tolerance in 50–70% of CM allergic children enrolled in the study [16,17]. This is also confirmed by other more recent works reporting that the consumption of baked goods containing egg or cow's milk may hasten the development of tolerance to these foods in an unheated form [18–20]. This paves the way for setting up oral food challenge studies using milk-including baked goods to be administered to milk allergic patients.

The heating treatment and the matrix where the allergen is contained, play a pivotal role in developing tolerance towards milk. High temperature and prolonged cooking time, as well as intrinsic characteristics and physicochemical conditions of the cooking environment, can induce significant changes in proteins structure, such as destruction of conformational epitopes, alteration of allergens tridimensional structure, with a consequent decrease of the IgE-binding. It has been observed that heat treatments commonly applied during industrial processing could deeply affect milk protein stability. For instance, whey proteins tend to aggregate due to the interaction of a free–SH group with the S–S bond of cysteine-containing proteins, such as β-LG, κ-casein, ALA, and BSA via –SH/S–S interchange reactions [21]. In addition, an extensive interaction between matrix components and milk proteins could occur after heating, originating the so-called "matrix effect", with consequent alteration of the final allergenicity. It has been hypothesized that the interaction between proteins and other components of the food matrix (fat or sugar) can alter protein structure and hide IgE binding sites. Schulten et al. demonstrated in 2011 that complex food matrices such as hazelnuts and peanuts can significantly reduce the gastrointestinal digestibility and the epithelial transport of cow's milk and apple allergens, thereby reducing their final allergenicity [22].

In other investigations carried out in our group we demonstrated that application of thermal treatments can induce relevant changes in the protein structure hiding or destroying specific epitopes with promising results on the reduction of the allergenic potential [23,24].

According to the data obtained in this work, it can be speculated that all interactions leading to an irreversible aggregation of proteins into complexes of various molecular size as a consequence of heating and/or protein composition, can influence the allergic response. In support of this, several authors, studying the IgE- and IgG-binding affinity and stability of different allergic subjects, observed that the types and severity of reactions displayed in the same individual depended on the different physical and chemical modifications that proteins underwent upon food processing [17,25].

In this context, the aim of the current study is to widen, from an allergenic point of view, the knowledge about the effect of thermal treatment on cow's milk proteins that were employed as ingredients in the preparation of muffins for infants. Any change in the protein profile was investigated by means of electrophoresis technique, and any possible protein–protein aggregation, polymerization or co-migration with the food matrix was highlighted. Furthermore, western blot analyses with sera of milk allergic children were performed in order to obtain deeper information on the impact of the applied heating on epitopes and consequent IgE recognition. This would help in better understanding the phenomena occurring along the protein structure and/or amino acid modification and their role in improving milk tolerance in young patients affected from severe milk allergies.

2. Materials and Methods

2.1. Chemicals

Trizma-base, sodium chloride, Tween-20, Triton X-100, ammonium bicarbonate (AMBIC), iodoacetamide (IAA), bovine serum albumin (BSA), along with other chemicals for electrophoresis namely dithiothreitol (DTT), sodium dodecyl sulfate (SDS), glycine, glycerol, Coomassie brilliant blue-G 250 were obtained from Sigma-Aldrich (Milan, Italy). Bromophenol blue was provided by Carlo Erba Reagents (Cornaredo, Italy) while phosphate buffer saline (PBS) was purchased from VWR International s.r.l. (Milan, Italy). Syringe filters in cellulose acetate (CA) from 1.2 μm were obtained from Labochem Science S.r.l. (Catania, Italy) whilst 0.45 μm syringe filters in polytetrafluoroethylene (PTFE) were purchased from Sartorius (Göttingem, Germania). Acetonitrile (Gold HPLC ultragradient), and trifluoroacetic acid (TFA) were purchased from Carlo Erba Reagents (Cornaredo, Milan, Italy) and ultrapure water was produced by a Millipore Milli-Q system (Millipore, Bedford, MA, USA). Formic acid (MS grade) was provided by Fluka (Milan, Italy) while trypsin (proteomic grade) for in gel protein digestion, from Promega (Milan, Italy).

2.2. Sera of Milk Allergic Patients

Sera were obtained from a total of 6 milk allergic children with levels of total IgE ranging from 203 to 5000 KU/L with an age comprised of between 5 and 16 years, according to ethical requirements. Tests were conducted in accordance with the Declaration of Helsinki and all procedures of the study were approved by the local Ethics Committee (code 2018/128). Permission to participate in the study of all children was obtained and the written informed consent was signed by the parents. The allergy symptoms in general ranged from urticaria to angioedema and anaphylaxis. The clinical features of the allergic individuals enrolled in this study are reported in Table 1. Diagnosis of IgE-mediated allergy to CM was confirmed by skin prick test (SPT) and serum-specific IgE (ImmunoCAP, Phadia, Uppsala, Sweden) to CM and CM proteins (s-IgE to CM, β-LG, ALA, caseins, total serum), allowing for a reliable diagnosis of IgE-mediated CMA. All of the serum sera samples were stored at −80 °C before their use.

Table 1. The clinical features of the allergic individuals enrolled in this study.

Serum	Age (Years)	IgE Total (KU/L)	IgE to Cow's Milk (KU/L)	IgE to Casein (KU/L)	Allergic Reaction Displayed
1	8	5000	62	44	anaphylaxis
2	5	203	100	100	anaphylaxis
3	11	433	54	56	anaphylaxis
4	16	370	87	80	anaphylaxis
5	6	4786	56	34	urthicaria
6	5	4662	100	100	vomit

2.3. Samples Preparation

Commercial fresh whole cow's milk (submitted to High Temperature Short Time-HTST) used in the present study was purchased from a local store shortly after delivery. Muffins baked with cow's milk, were prepared according to the following recipe: 60 g of wheat flour, 100 g of sugar, 1 sachet of vanillin, 8 g of baking powder, and 100 mL of fresh cow's milk (approximately 0.85 g of milk proteins for each muffin). The muffin was baked in an oven for 30 min at 180 °C. Blank muffin samples were also produced by replacing milk with water (a total of 100 mL). In addition, in order to have additional information on the effects of heating on milk proteins stability/structural–chemical modifications, the same amount of milk used for muffin preparation (100 mL) was baked at 180 °C for 10 min in an oven in absence of the food matrix.

2.4. Protein Extraction and Quantification

Blank muffins and CM incurred muffins were coarsely ground by hand and submitted to protein extraction procedure along with pasteurized and baked liquid milk. Briefly, 1 part of ground muffins or milk was mixed with 2 parts of extraction buffer (PBS, pH 7.4 containing 1% of Tween 20 (v/v) and 0.4% of Triton X-100 (v/v)), homogenized for 35 s (5 cycles of 7 s each) in a blender (Sterilmixer 12 model 6805-50; PBI International) and then shaken overnight at room temperature in an orbital shaker (KS 4000 i-control shaker, IKA Works GmbH & Co. KG, Staufen, Germany). Afterwards, samples were centrifuged for 20 min at 12,000 g at 4 °C, the upper phase was discarded, and the supernatant was carefully collected and filtered through 1.2 μm CA syringe filters. Protein concentration of samples was calculated as mg/albumin equivalent by Bradford assay (Quick Start™ Bradford Protein Assay). Samples were stored at −20 °C until use and filtered through 0.45 μm PTFE filters just before electrophoretic analysis.

2.5. SDS-PAGE Analysis

Fifteen microgram of protein extracts from muffin and milk samples, were separated, under reducing conditions, by means of sodium dodecyl sulfate-polyacrylamide gel electrophoresis (SDS-PAGE) on an 8–16% polyacrylamide pre-cast gels (8.6 cm × 6.7 cm × 1 mm) using a Mini-Protean Tetra Cell equipment (Bio-rad Laboratories, Segrate, Milano, Italy). Samples were dissolved in a Laemmli buffer (62.5 mM TrisHCl, pH 6.8, 25% glycerol, 2% SDS, 0.01% Bromophenol Blue, 100 mM DTT) (1:1 ratio) and denatured for 5 min at 95 °C. As running buffer, a TGS (25 mM Tris, 192 mM Glycine, 0.1% SDS) solution was employed while electrophoretic separation was performed at 100 V. Gels were stained by using a Coomassie Brilliant Blue G-250 solution and the bands were detected on a ChemiDOC™ MP Imaging system (Bio-Rad Laboratories, Segrate, Milano, Italy) and analyzed by using the software ImageLab 4.1. Precision Plus Protein™ all blue standards (10–250 kDa, Bio-Rad Laboratories, Hercules, CA, USA) was used as protein molecular weight referencing.

2.6. In-Gel Protein Digestion

Selected protein bands were cut from the polyacrylamide gel and submitted to in gel-digestion procedure according to the protocol reported in our previous work [26]. Finally, each sample was resuspended in 70 μL of H$_2$O/ACN, 90/10 + 0.1% formic acid (*v/v*) and 3 μL were further injected into LC/MS apparatus.

2.7. Protein Identification by Untargeted HR MS/MS Analysis

Protein bands were analyzed by using a Q-Exactive™ Plus Hybrid Quadrupole-Orbitrap™ Mass Spectrometer coupled to a Ultra-High-Performance Liquid Chromatography (UHPLC)pump systems (Thermo Fisher Scientific, San Josè, CA, USA). Peptides mixture was separated on an Acclaim™ PepMap analytical column (1 mm × 15 cm × 3 μm, 100 Å porosity, Thermo Fisher Scientific) at a flow rate of 60 μL/min, using a binary gradient composed of H$_2$O + 0.1% formic acid (solvent A) and CH$_3$CN/H$_2$O 80:20 + 0.1% formic acid (solvent B). The gradient elution program was as follows: 0–60 min linear from 10% to 60% B; quick increase to 80% B and isocratic for 10 min; then returning to 10% B and isocratic for 20 min for column re-conditioning. MS spectra were acquired in positive ion mode. The Heated Electrospray Ionization(HESI) ion source settings are reported here: Spray voltage at 3.4 kV, capillary temperature at 320 °C, sheath gas flow rate at 25 arbitrary units and S-lens at 55. The other MS settings are the same of what was reported in Bavaro et al. [24]. Raw data were processed via the commercial software Proteome Discoverer™ version 2.1 (Thermo-Fisher-Scientific, San Josè, CA, USA) and protein identification was achieved by Sequest HT search against a milk customized database extracted by Uniprot DB basing on the taxonomy code of *Bos Taurus* (ID: 9913) and containing about 44,000 sequences. The identification of tryptic peptides produced by in gel digestion with trypsin was accomplished by setting at 5 ppm and 0.05 Da the mass tolerance on the precursor and fragment ions, respectively. Only trustful peptide–spectrum matches were accepted and in particular a minimum of three peptides or higher were the minimum criteria for protein identification by selecting a high confidence (FDR < 1%).

2.8. In Silico Analysis to Assess the Immunoreactivity of Milk Proteins after Baking Process

Peptide sequences identified from the excised and digested protein bands were finally screened by interrogating in the Immune Epitope Database (IEDB) database (https://www.iedb.org/) in order to detect epitope linear sequences involved in IgE immunoreactivity. The following filters were applied for IEDB screening: Linear sequence for epitope structure, exact match for Basic Local Alignment Search Tool (BLAST) option and human as host.

2.9. Immunoblot for IgE-Binding Assay

Six μg of proteins extracted from allergen free and allergen incurred muffins and pasteurized/baked liquid milk were separated by electrophoresis under reducing conditions as already described in Section 2.5, and subsequently electroblotted on an immuno-blot low-fluorescence polyvinylidene fluoride (PVDF) membranes in 7 min (1.3 A, 25 V) using the Trans-Blot Turbo Transfer System (Bio-Rad Laboratories, Segrate, Milano, Italy).

Membranes were washed for 30 min (3 cycles of 10 min each) in TBS buffer containing 0.1% of Tween-20 (TBS-T) and then blocked for 2 h at room temperature with 3% BSA solution (prepared in TBS–T buffer). The membranes were incubated with pooled sera of a total of 7 young allergic patients previously diluted in TBS-T at 1/25 ratio and kept shaking overnight at 4 °C. After washing with TBS-T (3 cycles of 10 min each), membranes were incubated with monoclonal peroxidase-conjugated mouse anti-human IgE antibody (Sigma Aldrich, Milan, Italy) diluted in blocking solution (1/5000) and left shaking for 2 h at room temperature. Successively, membranes were washed with TBS-T (3 cycles of 10 min each) and then with TBS (30 min) before being incubated with Clarity chemiluminescence substrate (Bio-Rad Laboratories, Segrate, MI, Italy), 5 min prior to UV exposition. Images were acquired on a ChemiDoc™ MP Imaging System.

3. Results and Discussion

3.1. Effects of Baking and Matrix on Milk Protein Profiles

At first, the stability of milk proteins submitted to baking treatment either as matrix-free liquid or included into a food matrix, such as a muffin, was investigated by SDS-PAGE analysis. After a deep inspection of the electrophoretic protein profiles, the most relevant bands were digested in-gel by trypsin and the resulting peptide pool subjected to discovery analysis by LC-HR-MS/MS platform. The MS spectra obtained for each individual band were processed via commercial software and the respective proteins identified by interrogating a refined *Bos Taurus* database available on-line from the UniProt portal. In Figure 1 the SDS-PAGE protein profiles of cow's milk pasteurized (Lane 1), baked at 180 °C for 10 min (Lane 2), and blank muffin (allergen-free Lane 3), and CM incurred muffin, (Lane 4) were illustrated. Protein bands selected for identification were labeled from *a–h* (Lane 1) and *i-o* (Lane 4) (Figure 1). The results retrieved by the software for each spot analyzed were summarized in Table 2.

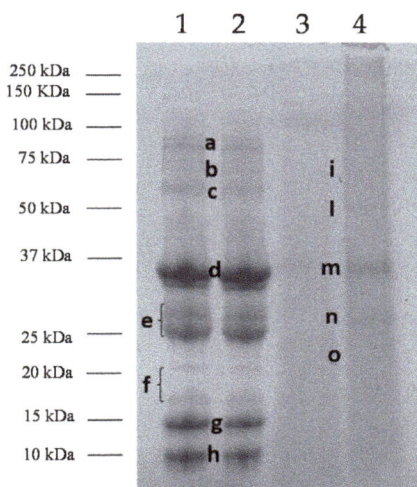

Figure 1. SDS-PAGE of cow's milk (CM) submitted to the different treatments: Pasteurized CM (Lane 1), baked CM at 180 °C for 10 min (Lane 2), blank muffin (Lane 3) and CM incurred muffin (Lane 4).

As displayed in Figure 1, no significant difference in the protein profile was observed between pasteurized and baked CM, suggesting that heating weakly affects the final stability of milk proteins in absence of the food matrix. In particular, protein bands with MW comprised between 75 and 50 kDa (Figure 1, Lanes 1 and 2, Band *c*), detectable in both extracts were mainly attributed to the whey protein BSA also named Bos d 6 that appeared as faint band in the first two lanes underlying the susceptible feature to the applied heating [27,28], while *a* and *b* did not lead to a univocal identification due to the low intensity of the protein bands.

Other bands uttermost intense displayed in pasteurized and baked milk were detected in the range of 37 and 25 kDa, namely *d* and *e* (Lane 1), and were assigned to caseins. Specifically, band *d* was attributed to αS1- and αS2 casein, (Bos d 9 and Bos d 10, respectively) while band *e* was identified as a mix of αS2-casein, β-casein (Bos d 11), and κ-casein (Bos d 12). Although clearly formed by two individual signals (see Figure 1, Lane 1, Band *e*), band *e* was experimentally processed as one and this could be the explanation for the different proteins identified. Our results highlight a good stability of caseins to heat treatments, as already reported in literature [7]. The heat resistance of the caseins group seems ascribable to a well-defined disordered mobile structure (rheomorphic) and to the lack of co-operative transition of unfolding, or partial folding, during heating [29]. Indeed, caseins lack a rigid tertiary structure, which confers stability and develop a "random coil" conformation stabilized by hydrophobic interactions [30]. Consequently, caseins are very stable to heat treatments, showing only a partial reduction or no change in their allergenicity [31]. Bloom and others demonstrated that the heat treatment after 60 min at 95 °C, did not affect the immunoreactivity of caseins [32]. Besides the time of heat exposure, the temperature or the presence of the food matrix (for example, wheat) during the heat process, casein allergenicity can be influenced also by digestion processes. Indeed, Morisawa et al. showed that α-caseins submitted to thermal treatment did not affect the histamine released from basophils, on the contrary, the combination of heat treatment with enzymatic digestion proved to decrease histamine released, reducing the interaction between α-casein specific-IgE and its linear epitopes [33]. In another study, Chatchatee et al. identified six major and three minor IgE epitopes of β-casein in persistent CMA patients. Among those, epitope 83–92 was the most frequently recognized (found in 13 out of 15 patients) and was identified by Dupont et al., in a tract highly resistant to digestion [34,35].

Table 2. Identification of protein bands excised from the SDS gel and analyzed by LC-HR-MS/MS through detection of the proteotypic peptides.

Sample	Band	Accession Number	Allergenic Proteins	Allergen Code	Coverage	Filtered Peptides
	c	A0A140T897	Serum albumin	Bos d 6	38.22	26
	d	P02662	αS1-casein	Bos d 9	33.17	1
		P02663	αS2-casein	Bos d 10	40.54	12
		B5B3R8	αS1-casein	Bos d 9	33.17	1
Pasteurized Milk/Baked milk		A0A140T8A9	κ-casein	Bos d 12	23.15	4
	e	A0A1Y0KDJ6	β-casein	Bos d 11	25	4
		J9UHS4	β-casein	Bos d 11	28.57	1
		P02663	αS2-casein	Bos d 10	22.97	6
	f	B5B0D4	β-lactoglobulin	Bos d 5	30.89	6
		Q28049	α-lactalbumin	Bos d 4	23.,57	4
	g	B5B0D4	β-lactoglobulin	Bos d 5	71.91	12
	h	Q28049	α-lactalbumin	Bos d 4	44.71	5
	i	P02662	αS1-casein	Bos d 9	17.28	2
		A0A1Y0KDJ6	β-casein	Bos d 11	9.37	3
CM incurred muffins	m	B5B3R8	αS1-casein	Bos d 9	39.71	1
		P02662	αS1-casein	Bos d 9	41.58	2
	n	A0A1Y0KDJ6	β-casein	Bos d 11	23.,21	4
	o	P02662	αS1-casein	Bos d 9	8.87	1

Other bands approximately comprised of between 20 and 10 kDa (Lanes 1, Bands *f, g* and *h*) were displayed in both pasteurized and baked milk protein profile (Figure 1, Lanes 1 and 2). By bioinformatic search, Band *f* was assigned to a mix of ALA (Bos d 4) and β-LG (Bos d 5), while bands *g* and *h* were singly attributed to β-LG and ALA respectively. Similarly to what already observed for other classes of proteins, baking seems not to affect β-LG and ALA stability. As reported in literature whey proteins are thermolabile with a consequent change in their allergenicity [36]. It is well known that β-LG increased its antigenicity and allergenicity when subjected to temperatures ranging from 50 to 90 °C, on the contrary a decrease was observed after 90 °C [37]. The fluctuated phenomena of β-LG thermal denaturation are characterized by well-defined temperature thresholds and lead to reversible and irreversible modifications [38,39]. Regarding ALA, several authors observed that this protein is more heat-stable than β-LG showing a greater decrease in antigenicity only when subjected to high temperatures, likely due to the loss of conformational epitopes that are more IgE-reactive [37,40]. It has been reported that the presence of fat and lactose enhanced β-LG denaturation in cow's milk. Due to Maillard reaction, protein–lactose interactions occur during heating with consequent stabilization and increase of hydration of protein molecules and/or irreversible aggregation of the whey proteins with casein. Therefore, Maillard reaction may lead to a loss of β-LG linear epitopes and consequently reduces the antigenicity of the protein [32,37].

In Figure 1, protein profiles of blank muffin and milk incurred muffin extracts were also displayed (Lanes 3 and 4). At a glance, the electrophoretic pattern of milk proteins extracted from muffin (Lane 4) appears significantly different from those of pasteurized and baked milk. This different behavior is also due to the lower extraction yield of proteins processed and embedded into a complex food matrix. In particular, the bands markedly detectable in pasteurized and baked milk (Lanes 1 and 2, Bands *b*, *d*, and *e*) were visible in the milk containing muffin SDS-PAGE profile as weak signals (Lane 4, Bands *i*, *m*, and *n*), while bands *c*, *f*, *g*, and *h* of pasteurized milk (Lane 1) seems to disappear in the milk incurred muffin. Interestingly, a new band with MW comprised of between 20 and 25 kDa appeared in the milk muffin pattern (Band *o*, Lane 4). For identification purpose, bands *i-o* of the milk muffin profile were submitted to in-gel protein digestion, mass spectrometry analysis, and bioinformatic search. Specifically, band *i* was attributed to a mix of αS1- and β-casein, while bands *m* and *n* (Lane 4) were singly assigned to αS1- and β-casein, respectively. Additional band *o* (Lane 4) was identified as αS1-casein (see Table 2). As for blank muffin (Lane 3), no milk proteins were detected as expected. The differences observable between pasteurized/baked milk and CM incurred muffin protein profiles highlight well the importance of food matrix on protein stability to heating. In general, almost all classes of milk proteins (caseins and whey proteins) seems to be deeply affected by baking in the presence of the food matrix and this is confirmed by the reduction of signal intensity of the most intense bands in milk-incurred muffin or by the disappearance of specific bands. The only proteins detected in incurred muffin extract, although at a lower concentration, belong to casein (αS1- and ß-casein, bands *i*, *m*, *n*, *o*, Lane 4) which previously studies reported to be very stable to heat treatment. Whey proteins, such as BSA or ALA and ß-LG were not displayed in the gel. It should be hypothesized that the presence of the food matrix improved the heat dispersion within the environment where proteins are dispersed with a consequent increase of the temperature that induces protein degradation. On the other hand, it is well known that heating is likely to promote interaction between protein and other food components causing important structural and chemical changes in the proteins involved (denaturation, aggregation, and Maillard reaction) and altering the pH and the solubility and/or structure of allergenic proteins. In the light of this, it is reasonable to assume that the interaction of proteins with food components or alteration of their solubility might decrease the extraction efficiency preventing their detection in the final extract. In addition, the occurrence of some degradation phenomena during heating could explain the appearance of the new band at MW approximately below 22 kDa (Band *o*, Lane 4), putatively attributed to αS1- casein.. It was largely demonstrated that the interactions between milk proteins and matrix ingredients including proteins, fats, and sugars that are ingredients typically used in bakery products, can also decrease the bioavailability of allergic proteins to immune system and consequently reduce their allergenicity [41]. The interaction with matrix components has led some authors to recommend to young allergic patients to follow a diet including milk-containing bakery products like muffins. In these studies, about 70% of tested children were able to ingest a muffin containing baked milk without displaying any immediate clinical symptoms [17]. They suggest to add baked milk products into the daily diet in order to accelerate the rise of tolerance to unheated milk rather than to avoid strictly such allergenic food [20].

3.2. Immunoblot of Milk Products with Sera of Allergic Patients

In order to study the effect of baking in the final immunoreactivity of milk muffins, immunoblot analysis with sera of allergic young patients (mean age ± standard deviation: 8.5 ± 4.3 years) was performed. Specifically, pasteurized and baked milk along with muffins prepared with or without milk were separated on monodimensional electrophoresis, blotted on a PVDF membrane and detected by chemiluminescence reaction. A picture reporting the western blot analysis performed with pool sera of patients is shown in Figure 2. As appearing, pasteurized and baked milk showed a similar immunoreactivity profile (Lanes 1 and 2) suggesting that baking under these conditions did not alter the final allergenicity of that food. In particular, IgE immunoreactivity was detected in correspondence of bands comprised between 75 and 100 kDa, while a strong antibody reactivity was displayed for a band with MW of approximately 60 kDa (Figure 2, Lanes 1 and 2) experimentally assigned to BSA (Figure 1, Lane 1, Band *c*). The clinical relevance of BSA in milk is difficult to evaluate, as allergic children are generally sensitized to two or more milk allergen proteins [1], and BSA is a minor allergen [42]. Nevertheless, it was reported that 70% of patients with persistent milk allergy, have a greater risk to develop sensitization to bovine serum albumin (BSA). It was reported that heating treatment of milk, as boiling at 100 °C for 10 min, determines an increase of dimeric, trimeric and higher polymeric BSA forms, which maintain strong IgE-binding properties. Our results demonstrated that baking milk at 180 °C in the oven for 10 min did not significantly reduce BSA immunoreactivity, likely due to the ineffective spread of the temperature within milk during heating.

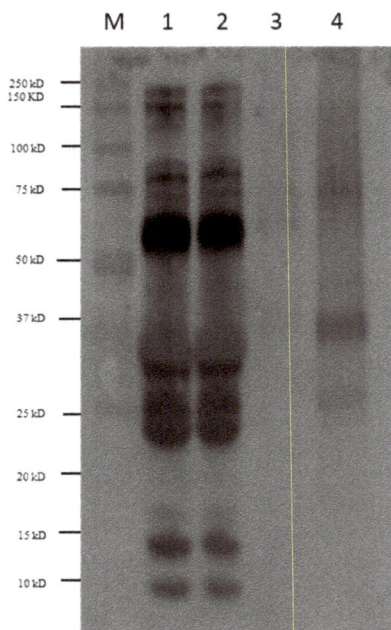

Figure 2. Immunoblot of CM sample extracts under reducing conditions referred to pasteurized CM (Lane 1), baked CM at 180 °C for 10 min (Lane 2), blank muffin (Lane 3), and CM incurred muffin (Lane 4). M: MW reference standard The immunoblot was carried out on a pool of sera of young patients (mean age ± standard deviation: 8.5 ± 4.3 years) with a clinical allergy to CM proteins.

An IgE reactivity lower than the BSA band was instead observed in pasteurized and baked milk corresponding to bands with MW in the range 37–22 kDa (Figure 2 Lanes 1 and 2) namely bands labeled as *d* and *e* in the respective SDS-PAGE pattern (Figure 1, Lane 1) and assigned to the casein group (αS1/αS2-casein, β-casein and κ-casein). Our findings are in accordance with what was reported in literature. Indeed, as known among cow's milk proteins, caseins showed a high heat stability with persistence of IgE-binding properties. Nevertheless, IgE binding depends on the clinical history of patients. Indeed, several authors observed that casein heated for a prolonged time produced consistent reactivity in some milk-reactive subjects, while reduced adverse reactions were observed in other milk-tolerant patients, even if comparable milk-specific IgE concentrations was revealed in both groups [32].

A weak immunoreactivity was also observed in the range of 10–15 kDa both for pasteurized and baked milk, corresponding to band *g* and *h* in the relative electrophoretic profile illustrated in Figure 1 and putatively attributed to whey proteins (ALA and β-LG). Heat treatments were demonstrated to affect the antigenicity of ALA and β-LG in whey protein isolate. Umesh Kumar Shandilya et al., found that the consumption of sterilized cow milk by WP-sensitized animals caused a significant reduction ($p \leq 0.01$) of total IgE levels by 43% compared with raw milk WP [43]. Moreover, the aggregation phenomena of β-LG and ALA during heat treatment, could also cause the inhibition of protein uptake by intestinal epithelial cells.

Different results in IgE reactivity were displayed when milk proteins were cooked within a food matrix, such as muffin (Figure 2, Lane 4). In this case, IgE reactivity of milk proteins appeared drastically reduced and only proteins banding between 37 kDa and 20 kDa (Figure 2, Lane 4, corresponding to bands *m*, *n*, and *o* in the respective SDS-PAGE profile) showed a very weak IgE response. As previously described, these bands belong to αS1 casein and β-casein. The scarce IgE reactivity observed for milk proteins baked within a muffin matrix compared to what displayed for baked milk, well highlights the importance of the food matrix in the modulation of the final immunoreactivity of an allergenic food. As already discussed, the occurrence of interaction between proteins and food components could lead to chemical/structural modifications on the protein moiety, with consequent masking of active epitopes and reduction of IgE binding sites. This behavior was already reported in literature. Indeed, in a prospective study, Nowak-Wegrzyn et al. found that heated milk–reactive subjects had significantly larger skin prick test wheals and higher milk-specific and casein-specific IgE levels than other groups. After 3 months of ingesting heated milk products, reactive subjects had significantly smaller skin prick test wheals compared to time 0 and higher casein-IgG4. In conclusion the paper demonstrates that the majority (75%) of children with milk allergy tolerate heated milk [17]. In another study, Kim et al. report that after sequential food challenges with baked cheese and unheated milk in a test population of children previously found tolerant to extensively heated (baked) milk products, approximately 28% and 60% of them were able to tolerate baked milk/baked cheese and unheated milk, respectively with no difference in milk-specific IgE levels between groups [20].

On the other hand, heating milk allergens contained into a food matrix does not completely eliminate the risk of an allergic reaction, because according to the works published, only the reactivity of a few allergens showed to be significantly reduced after the heating applied [20,44]. In this regard, Bloom et al. demonstrated that matrix has an effect on the decrease of IgE immunoreactivity. This was demonstrated by comparing immunoblot analysis of pooled and individual sera of subjects fed with wheat food matrix enriched with milk vs. milk heated under the same conditions. The authors speculated that low reactivity observed in milk containing food matrix was due to the formation of complexes between wheat and milk proteins. However, RBL assay and tests of stimulation of peripheral mononuclear cells obtained from allergic children, showed that there were no differences between heated and unheated proteins, with higher mediator release and higher T-cell stimulation index in tolerant subjects. They hypothesized that the formation of protein complexes during heating could enhance the allergenicity in in vitro systems [32].

3.3. In Silico Analyses to Assess IgE Binding Capacity of Milk Products

In the final section of our work we also investigated the immunoreactive epitopes spread along the protein moiety found positive to immunoblot with allergic sera (Figure 2). To this purpose, all peptides obtained from in gel tryptic digestion of protein bands of pasteurized/baked milk and milk-incurred muffin excised from the gel (see Figure 1, Lane 1, Bands *c–h*; Lane 4, Bands *m–o*) were taken into consideration. The IEDB database was screened to find a match with known milk linear epitopes recognized by *Homo sapiens* as host. The results are summarized in Table 3. In Table S1 of supplementary data, peptide sequences searched in IEDB database along with the epitopic sequences were shown. As for milk (pasteurized and baked) several peptide sequences were found to match with intact epitopes, specifically LGEYGFQNALIVR that was previously attributed to BSA protein

(Figure 1, Lane 1, Band *c*) and TPEVDDEALEK, VLVLDTTDYK, and VYVEELKPTPEGDLEILLQK all attributed to β-LG (Figure 1, Lane 1, Band g). Similarly, peptides FFVAPFPEVFGK, EGIHAQQK, and HIQKEDVPSER, obtained from the digestion of selected bands referred to CM incurred muffin sample and assigned to αS1-casein (Figure 1, Lane 4, Band *m*) were found to fully match with immunoreactive epitopes. On the contrary, most of the peptides retrieved by the software were found to overlap with a small portion of the epitope sequence and this was observed both in pasteurized milk and incurred muffin (Table 3, Lane 1 Bands *d, e, g, h* and Lane 4, Bands *m* and *n*). Finally, some peptides belonging to β-LG and identified in band *g* (Figure 1, Lane 1, pasteurized milk) together with those assigned to αS1 casein and β-casein by tryptic digestion of bands *m* and *n* (Figure 1, Lane 4, milk muffin) were found to include short epitopic sequences. Interestingly, two peptides namely YLGYLEQLLR and FFVAPFPEVFGK, identified in bands *m* and *o* (Figure 1, Lane 4) of milk-containing muffin and attributed to αS1 casein, were recognized as immunodominant epitopes since no differences in the epitope specificity between IgG and IgE were highlighted. This directly translates into a higher T cell stimulation capacity than other epitope regions of the αS1 casein [45,46]. In the light of these results, we could conclude that different epitope sequences spread along caseins and whey proteins show to survive to baking at 180 °C (10 min) in heated milk (Figure 2, Lanes 1 and 2). On the contrary, the final immunoreactivity of muffin incurred with milk appeared consistently reduced compared to the not baked food product and the only immunoreactivity observable was ascribable to resistant epitopes belonging to αS1, as demonstrated by proteomic analysis and IEDB search. It is worth to be noted that, even in the aggregated form (MW of band *m* seems slightly higher than the corresponding band *d* of pasteurized milk, see Figure 1), αS1 casein retains its allergenicity.

Table 3. List of potential immunogenic sequences recognized in the peptides identified in specific electrophoretic bands of the milk extracts along with the relevant cow's milk epitope ID reported in Immune Epitope Database (IEDB).

Protein Band	Peptide Sequence	Epitope ID
c	LGEYGFQNALIVR	235209
d	LHSMK	70444, 115236, 11860, 35531, 56749, 109484, 109828, 115343, 115476
	EDVPSER	663659, 28169, 109358, 30333, 30334, 31120, 31121, 115310, 48707, 78245, 115440, 190571, 68322
	ITVDDK	78138, 115315, 115477, 115479, 115532, 606543
	LNFLK	45706, 78144, 95351, 95560, 115226, 115512
e	EAMAPK	115216
	FFSDK	15893, 30141, 6173, 78257, 115305, 115404, 115449, 115465, 115733, 229682, 229689
	GPFPIIV	115251
g	TPEVDDEALEK	65565, 78111, 96146, 13583, 56146, 95306, 95369, 95922, 96628, 115172, 115519, 146504, 222188
	TKIPAVFK	33732, 95498, 96388, 33733, 46987, 78279, 96064, 96968, 115173, 115185, 115313, 222020
	VLVLDTDYKK	96517, 69827, 95545, 95579, 222193, 223163
	IPAVFK	33732, 96388, 31382, 33733, 46987, 78279, 96064, 96968, 98849, 115173, 115185, 115313, 222020
	VLVLDTDYK	96091, 96517, 69827, 95545, 95579, 222193, 223163
	LSFNPTQLEEQCHI	39349, 2820, 115382, 24090, 95389, 95574, 98777, 98893, 115427
	VYVEELKPTPEGDLEILLQK	72178, 32907, 96569, 97098, 32908, 72177, 95347, 96219, 98752, 98760, 99028, 99036, 224315
h	EQLTK	115234, 227758, 558421

Table 3. Cont.

Protein Band	Peptide Sequence	Epitope ID
m	FFVAPFPEVFGK	38207, 43705, 15930, 15931, 44794, 67707, 69660, 110049, 115396, 115467, 190478, 659427, 659428
	HQGLPQEVLNENLLR	115282, 31145, 50721, 50900, 109844, 115311, 675165
	EPMIGVNQELAYFYPELFR	12961, 13714, 13715, 13716, 23078, 45538, 45539
	YLGYLEQLLR	74689, 30334, 74687, 74688, 115482, 14100, 109358, 110060, 115060, 115122, 115213, 115440, 190580, 229693
	VNELSK	68473, 115253, 12896, 20548, 68472, 70058, 70059, 78158, 108948, 110049, 115068, 115531, 115544, 190572, 229694
	EGIHAQQK	24814, 12187, 30400, 41811, 115236, 24813, 109484, 109828, 115306, 115476, 190445, 606414
	HIQKEDVPSER	68322, 78245, 28169, 31120, 31121, 48707, 109358, 115310, 190571, 663657, 663658, 663659
	EDVPSER	68322, 30334, 78245, 28169, 30333, 31120, 31121, 48707, 109358, 115310, 115440, 190571, 663659
n	GPFPIIV	115251, 658276, 670213, 671639, 673180, 673307, 688313
	DMPIQAFLLYQEPVLGPVR	42283, 75481, 115675, 115796, 115847, 115866
	AVPYPQR	51169, 70443, 115430, 115439, 115495, 115694, 657013, 657014
	VLPVPQK	52358, 52359, 115495, 115835, 161678, 227654, 679766, 735655
o	FFVAPFPEVFGK	15930, 15931, 659427, 659428, 38207, 190478, 43705, 44794, 115396, 115467, 67707, 69660, 110049

4. Conclusions

In this study, the effects of thermal treatment on cow milk proteins included into a food like muffins, were investigated from a proteomic point of view and for their potential to reduce the allergenic response by immunoblot analysis. To this purpose, pasteurized and baked milk and the inclusion or not into a food was studied in terms of allergenicity retention. The analysis of the protein profile showed that the presence or absence of the food matrix might account for differences detected in the allergen pattern and the resulting immunoreactivity as assessed by SDS-PAGE and immunoblot analysis using a pool of sera of cow's milk allergic patients.

In the light of our results, milk baked within the muffin matrix might promote formation of complexes with food components inducing a modulation of the immunoreactivity towards milk allergens compared to milk baked in the oven at 180 °C for ten minutes. The interactions between milk proteins and some components of the food matrix during heating seemed to play a role in the possible reduction of allergenicity as assessed by in vitro tests. Further studies employing simulated in vitro gastrointestinal digestion systems will be necessary to better investigate the slight residual IgE reactivity displayed in the CM incurred muffin. Indeed, assessing the fate of allergenic proteins subjected to heat processing techniques in food matrices upon gastrointestinal digestion can help to understand the immunomodulatory effects and the total tolerance of these types of foods in cow milk allergic patients.

Supplementary Materials: The following are available online at http://www.mdpi.com/2072-6643/11/7/1536/s1.

Author Contributions: Conceptualization, L.M. and E.M.N.; methodology, S.L.B., E.D.A., software, S.L.B., E.D.A., R.P.; data curation, E.M.N., L.M.; writing—original draft preparation, S.L.B., E.D.A.; writing—review and editing, E.M.N., R.P., L.M., F.M., S.B.; visualization, E.D.A., S.B.L.; supervision, L.M.; funding acquisition, L.M.

Funding: The equipment used in this work was supported by the Biodiversità per la valorizzazione e sicurezza delle produzioni alimentari tipiche pugliesi, BioNet-PTP project (Cod. 73) funded by Programma Operativo Regionale Puglia FESR 2000–2006—Risorse liberate—Obiettivo Convergenza.

Conflicts of Interest: The authors declare no conflict of interest.

References

1. Hochwallner, H.; Schulmeister, U.; Swoboda, I.; Spitzauer, S.; Valenta, R. Cow's Milk Allergy: From Allergens to New Forms of Diagnosis, Therapy and Prevention. *Methods* **2014**, *66*, 22–33. [CrossRef] [PubMed]
2. Venter, C.; Arshad, S.H. Epidemiology of Food Allergy. *Pediatr. Clin. N. Am.* **2011**, *58*, 327–349. [CrossRef] [PubMed]
3. Nwaru, B.I.; Hickstein, L.; Panesar, S.S.; Roberts, G.; Muraro, A.; Sheikh, A. Prevalence of Common Food Allergies in Europe: A Systematic Review and Meta-Analysis. *Allergy Eur. J. Allergy Clin. Immunol.* **2014**, *69*, 992–1007. [CrossRef] [PubMed]
4. Fiocchi, A.; Brozek, J.; Schunemann, H.; Bahna, S.L.; Von Berg, A.; Beyer, K.; Bozzola, M.; Bradsher, J.B.; Compalati, E.; Ebisawa, M.; et al. World Allergy Organization (WAO) Diagnosis and Rationale for Action Against Cow's Milk Allergy (DRACMA) Guidelines. *World Allergy Organ. J.* **2010**, *3*, 57–161. [CrossRef] [PubMed]
5. Skripak, J.M.; Matsui, E.C.; Mudd, K.; Wood, R.A. The Natural History of IGE-mediated Cow's Milk Allergy. *J. Allergy Clin. Immunol.* **2007**, *120*, 1172–1177. [CrossRef] [PubMed]
6. Martorell, A.; Alonso, E.; Bone, J.; Echeverria, L.; Lopez, M.C.; Martin, F.; Nevot, S.; Plaza, A.M. Position Document: IGE-mediated Allergy to Egg Protein Food Allergy Committee of SEICAP. *Allergol Immunopathol. (Madr)* **2013**, *41*, 320–336. [CrossRef] [PubMed]
7. Villa, C.; Costa, J.; Oliveira, M.B.P.P.; Mafra, I. Bovine Milk Allergens: A Comprehensive Review. *Compr. Rev. Food Sci. Food Saf.* **2018**, *17*, 137–164. [CrossRef]
8. Marchesseau, S.; Mani, J.-C.; Martineau, P.; Roquet, F.; Cuq, J.-L.; Pugniere, M. Casein Interactions Studied by the Surface Plasmon Resonance Technique. *J. Dairy Sci.* **2002**, *85*, 2711–2721. [CrossRef]
9. Broyard, C.; Gaucheron, F. Modifications of Structures and Functions of Caseins: A Scientific and Technological Challenge. *Dairy Sci. Technol.* **2015**, *95*, 831–862. [CrossRef]
10. Foegeding, E.A.; Mleko, S.W. Whey Protein Products. U.S. Patent 6,383,551, 7 May 2002.
11. Kontopidis, G.; Holt, C.; Sawyer, L. Lactoglobulin: Binding Properties, Structure, and Function. *J. Dairy Sci.* **2004**, *87*, 785–796. [CrossRef]
12. Mulcahy, E.M.; Fargier-Lagrange, M.; Mulvihill, D.M.; O'Mahony, J.A. Characterisation of Heat-induced Protein Aggregation in Whey Protein Isolate and the Influence of Aggregation on the Availability of Amino Groups as Measured by the ORTHO-Phthaldialdehyde (OPA) and Trinitrobenzenesulfonic Acid (TNBS) methods. *Food Chem.* **2017**, *229*, 66–74. [CrossRef] [PubMed]
13. Vitaliti, G.; Cimino, C.; Coco, A.; Pratico, A.D.; Lionetti, E. The Immunopathogenesis of Cows Milk Protein Allergy (CMPA). *Ital. J. Pediatr.* **2012**, *38*, 35. [CrossRef] [PubMed]
14. Chen, M.; Land, M. Baked Milk and Baked Egg Oral Immunotherapy. *Immunotherapy* **2017**, *9*, 1201–1204. [CrossRef] [PubMed]
15. Mori, F.; Cianferoni, A.; Brambilla, A.; Barni, S.; Sarti, L.; Pucci, N.; De Martino, M.; Novembre, E. Side Effects and their Impact on the Success of Milk Oral Immunotherapy (OIT) in Children. *Int. J. Immunopathol. Pharmacol.* **2017**, *30*, 182–187. [CrossRef] [PubMed]
16. Caubet, J.C.; Nowak-Węgrzyn, A.; Moshier, E.; Godbold, J.; Wang, J.; Sampson, H.A. Utility of Casein-specific IGE Levels in Predicting Reactivity to Baked Milk. *J. Allergy Clin. Immunol.* **2013**, *131*, 222–224. [CrossRef] [PubMed]
17. Nowak-Wegrzyn, A.; Bloom, K.A.; Sicherer, S.H.; Shreffler, W.G.; Noone, S.; Wanich, N.; Sampson, H.A. Tolerance to Extensively Heated Milk in Children with Cow's Milk Allergy. *J. Allergy Clin. Immunol.* **2008**, *122*, 342–347. [CrossRef] [PubMed]
18. Robinson, M.L.; Lanser, B.J. The Role of Baked Egg and Milk in the Diets of Allergic Children. *Immunol. Allergy Clin.* **2018**, *38*, 65–76. [CrossRef]
19. Lambert, R.; Grimshaw, K.E.C.; Ellis, B.; Jaitly, J.; Roberts, G. Evidence that Eating Baked Egg or Milk Influences Egg or Milk Allergy Resolution: A Systematic Review. *Clin. Exp. Allergy* **2017**, *47*, 829–837. [CrossRef]
20. Kim, J.S.; Nowak-Węgrzyn, A.; Sicherer, S.H.; Noone, S.; Moshier, E.L.; Sampson, H.A. Dietary Baked Milk Accelerates the Resolution of Cow's Milk Allergy in Children. *J. Allergy Clin. Immunol.* **2011**, *128*, 125–131. [CrossRef]

21. Considine, T.; Patel, H.A.; Anema, S.G.; Singh, H.; Creamer, L.K. Interactions of Milk Proteins during Heat and High Hydrostatic Pressure Treatments-A Review. *Innov. Food Sci. Emerg. Technol.* **2007**, *8*, 1–23. [CrossRef]

22. Schulten, V.; Lauer, I.; Scheurer, S.; Thalhammer, T.; Bohle, B. A Food Matrix Reduces Digestion and Absorption of Food Allergens in Vivo. *Mol. Nutr. Food Res.* **2011**, *55*, 1484–1491. [CrossRef] [PubMed]

23. De Angelis, E.; Bavaro, S.L.; Forte, G.; Pilolli, R.; Monaci, L. Heat and Pressure Treatments on Almond Protein Stability and Chane in Immunoreactiveity after Simulated Human Digestion. *Nutrients* **2018**, *10*, 1679. [CrossRef] [PubMed]

24. Bavaro, S.L.; Di Stasio, L.; Mamone, G.; De Angelis, E.; Nocerino, R.; Canani Berni, R.; Logrieco, A.F.; Montemurro, N.; Monaci, L. Effect of Thermal/Pressure Processing and Simulated Human Digestion on the Immunoreactivity of Extractable Peanut Alllergens. *Food Res. Int.* **2018**, *109*, 126–137. [CrossRef] [PubMed]

25. Nowak-Węgrzyn, A.; Lawson, K.; Masilamani, M.; Kattan, J.; Bahnson, H.T.; Sampson, H.A. Increased Tolerance to Less Extensively Heat-Denatured (Baked) Milk Products in Milk-Allergic Children. *J. Allergy Clin. Immunol. Pract.* **2018**, *6*, 486–495. [CrossRef] [PubMed]

26. De Angelis, E.; Pilolli, R.; Bavaro, S.L.; Monaci, L. Insight Into the Gastro-Duodenal Digestion Resistance of Soybean Proteins and Potential Implications for Residual Immunogenicity. *Food Funct.* **2017**, *8*, 1599–1610. [CrossRef] [PubMed]

27. Kleber, N.; Hinrichs, J. Antigenic Response of Beta-lactoglobulin in Thermally Treated Bovine Skim Milk and Sweet Whey. *Milchwissenschaft* **2007**, *62*, 121–124.

28. Wijayanti, H.B.; Bansal, N.; Deeth, H.C. Stability of Whey Proteins during Thermal Processing: A Review. *Compr. Rev. Food Sci. Food Saf.* **2014**, *13*, 1235–1251. [CrossRef]

29. Perticaroli, S.; Nickels, J.D.; Ehlers, G.; Mamontov, E.; Sokolov, A.P. Dynamics and Rigidity in an Intrinsically Disordered Protein, β-casein. *J. Phys. Chem. B* **2014**, *118*, 7317–7326. [CrossRef]

30. Tsabouri, S.; Douros, K.; Priftis, K. Cow's Milk Allergenicity. *Endocr. Metab. Immune Disord. Targets* **2014**, *14*, 16–26. [CrossRef]

31. Bhat, M.Y.; Dar, T.A.; Singh, L.R. Casein Proteins: Structural and Functional Aspects. In *Milk Proteins-From Structure to Biological Properties and Health Aspects*; Gigli, I., Ed.; IntechOpen: London, UK, 2016; Chapter 1, pp. 4–18. [CrossRef]

32. Bloom, K.A.; Huang, F.R.; Bencharitiwong, R.; Bardina, L.; Ross, A.; Sampson, H.A.; Nowak-Wegrzyn, A. Effect of Heat Treatment on Milk and Egg Proteins Allergenicity. *Pediatr. Allergy Immunol.* **2014**, *25*, 740–746. [CrossRef]

33. Morisawa, Y.; Kitamura, A.; Ujihara, T.; Zushi, N.; Kuzume, K.; Shimanouchi, Y.; Tamura, S.; Wakiguchi, H.; Saito, H.; Matsumoto, K. Effect of Heat Treatment and Enzymatic Digestion on the B Cell Epitopes of Cow's Milk Proteins. *Clin. Exp. Allergy* **2009**, *39*, 918–925. [CrossRef] [PubMed]

34. Chatchatee, P.; Jarvinen, K.M.; Bardina, L.; Vila, L.; Beyer, K.; Sampson, H.A. Identification of IGE and IGG Binding Epitopes on β- and κ-casein in Cow's Milk Allergic Patients. *Clin. Exp. Allergy* **2001**, *31*, 1256–1262. [CrossRef] [PubMed]

35. Dupont, D.; Mandalari, G.; Molle, D.; Jardin, J.; Rolet-Repecaud, O.; Duboz, G.; Leonil, J.; Mills, C.E.N.; Mackie, A.R. Food Processing Increases Casein Resistance to Simulated Infant Digestion. *Mol. Nutr. Food Res.* **2010**, *54*, 1677–1689. [CrossRef] [PubMed]

36. Verhoeckx, K.C.; Vissers, Y.M.; Baumert, J.L.; Faludi, R.; Feys, M.; Flanagan, S.; Herouet-Guicheneyg, C.; Holzhauserh, T.; Shimojoi, R.; Van der Boltj, N.; et al. Food Processing and Allergenicity. *Food Chem. Toxicol.* **2015**, *80*, 223–240. [CrossRef] [PubMed]

37. Bu, G.; Luo, Y.; Zheng, Z.; Zheng, H. Effect of Heat Treatment on the Antigenicity of Bovine α-lactalbumin and β-lactoglobulin in Whey Protein Isolate. *Food Agric. Immunol.* **2009**, *20*, 195–206. [CrossRef]

38. Dumitraşcu, L.; Moschopoulou, E.; Aprodu, I.; Stanciu, S.; Rapeanu, G.; Stanciuc, N. Assessing the Heat Induced Changes in Major Cow and Non-cow Whey Proteins Conformation on Kinetic and Thermodynamic Basis. *Small Rumin. Res.* **2013**, *111*, 129–138. [CrossRef]

39. Kleber, N.; Krause, I.; Illgner, S.; Hinrichs, J. The Antigenic Response of β-lactoglobulin is Modulated by Thermally Induced Aggregation. *Eur. Food Res. Technol.* **2004**, *219*, 105–110. [CrossRef]

40. Jarvinen, K.M.; Chatchatee, P.; Bardina, L.; Beyer, K.; Sampson, H.A. IGE and IGG Binding Epitopes on α-lactalbumin and β-lactoglobulin in Cow's Milk Allergy. *Int. Arch. Allergy Immunol.* **2001**, *126*, 111–118. [CrossRef]

41. Rahaman, T.; Vasiljevic, T.; Ramchandran, L. Effect of Processing on Conformational Changes of Food Proteins Related to Allergenicity. *Trends Food Sci. Technol.* **2016**, *49*, 24–34. [CrossRef]

42. Hochwallner, H.; Schulmeister, U.; Swoboda, I.; Balic, N.; Geller, B.; Nystrand, M.; Harlin, A.; Thalhamer, J.; Scheiblhofer, S.; Niggemann, B.; et al. Microarray and Allergenic Activity Assessment of Milk Allergens. *Clin. Exp. Allergy* **2010**, *40*, 1809–1818. [CrossRef]

43. Shandilya, U.K.; Kapila, R.; Haq, R.M.; Kapila, S.; Kansal, V. Effect of Thermal Processing of Cow and Buffalo Milk on the Allergenic Response to Caseins and Whey Proteins in Mice. *J. Sci. Food Agric.* **2013**, *93*, 2287–2292. [CrossRef] [PubMed]

44. Mills, E.N.C.; Sancho, A.I.; Rigby, N.M.; Jenkins, J.A.; Mackie, A.R. Impact of Food Processing on the Structural and Allergenic Properties of Food Allergens. *Mol. Nutr. Food Res.* **2009**, *53*, 963–969. [CrossRef] [PubMed]

45. Elsayed, S.; Eriksen, J.; Øysæd, L.K.; Idsøe, R.; Hill, D.J. T Cell Recognition Pattern of Bovine Milk αS1-Casein and its Peptides. *Mol. Immunol.* **2004**, *41*, 1225–1234. [CrossRef] [PubMed]

46. Elsayed, S.; Hill, D.J.; Do, T.V. Evaluation of the Allergenicity and Antigenicity of Bovine-milk αs1-casein Using Extensively Purified Synthetic Peptides. *Scand. J. Immunol.* **2004**, *60*, 486–493. [CrossRef] [PubMed]

![nutrients logo] *nutrients*

MDPI

Article

Differential Effects of Dry vs. Wet Heating of β-Lactoglobulin on Formation of sRAGE Binding Ligands and sIgE Epitope Recognition

Hannah E. Zenker [1,*], Arifa Ewaz [2], Ying Deng [3,4], Huub F. J. Savelkoul [2], R.J. Joost van Neerven [2,5], Nicolette W. De Jong [6], Harry J. Wichers [3,4], Kasper A. Hettinga [1] and Malgorzata Teodorowicz [2]

[1] Food Quality & Design Group, Wageningen University & Research Centre, 6700 AA Wageningen, The Netherlands; kasper.hettinga@wur.nl
[2] Cell Biology & Immunology, Wageningen University & Research Centre, 6700 AA Wageningen, The Netherlands; arifa.ewaz@wur.nl (A.E.); huub.savelkoul@wur.nl (H.F.J.S.); joost.vanneerven@frieslandcampina.com (R.J.J.v.N.); gosia.teodorowicz@wur.nl (M.T.)
[3] Wageningen Food & Biobased Research, Wageningen University & Research Centre, 6700 AA Wageningen, The Netherlands; ying.deng@wur.nl (Y.D.); harry.wichers@wur.nl (H.J.W.)
[4] Laboratory of Food Chemistry, Wageningen University and Research, 6700 AA Wageningen, The Netherlands
[5] FrieslandCampina, 3800 BN Amersfoort, The Netherlands
[6] Erasmus University medical Centre Rotterdam, Dept. Internal Medicine, 3000 CA Rotterdam, The Netherlands; n.w.dejong@erasmusmc.nl
[*] Correspondence: hannah.zenker@wur.nl

Received: 17 May 2019; Accepted: 17 June 2019; Published: 25 June 2019

Abstract: The effect of glycation and aggregation of thermally processed β-lactoglobulin (BLG) on binding to sRAGE and specific immunoglobulin E (sIgE) from cow milk allergic (CMA) patients were investigated. BLG was heated under dry conditions (water activity < 0.7) and wet conditions (in phosphate buffer at pH 7.4) at low temperature (<73 °C) and high temperatures (>90 °C) in the presence or absence of the milk sugar lactose. Nε-(carboxymethyl)-l-lysine (CML) western blot and glycation staining were used to directly identify glycation structures on the protein fractions on SDS-PAGE. Western blot was used to specify sRAGE and sIgE binding fractions. sRAGE binding was highest under wet-heated BLG independent of the presence of the milk sugar lactose. Under wet heating, high-molecular-weight aggregates were most potent and did not require the presence of CML to generate sRAGE binding ligands. In the dry system, sRAGE binding was observed only in the presence of lactose. sIgE binding affinity showed large individual differences and revealed four binding profiles. Dependent on the individual, sIgE binding decreased or increased by wet heating independent of the presence of lactose. Dry heating required the presence of lactose to show increased binding to aggregates in most individuals. This study highlights an important role of heating condition-dependent protein aggregation and glycation in changing the immunogenicity and antigenicity of cow's milk BLG.

Keywords: aggregation; allergenicity; β-lactoglobulin; CML; glycation; sRAGE; IgE binding

1. Introduction

The manufacturing of many dairy products implies heating at moderate (<73 °C) or high temperatures (>90 °C) to ensure product safety or to produce powdered products such as infant formula. Maillard reaction (MR)-induced glycation, as well as protein aggregation, are the most abundant protein modifications during milk processing. The formation of aggregates and the degree of glycation strongly depends on the applied processing conditions and may strongly differ between wet and dry heating conditions as they are applied in industrial processing of milk or dried dairy products. Wet heating of β-lactoglobulin (BLG) results in relatively more aggregation and the formation of polymers with a molecular weight (MW) > 10 kDa compared to dry heating [1]. At the same time, dry heating induces more glycation as low moisture content (water activity between 0.2 and 0.8) enhances protein glycation via the MR [2]. These protein modifications are known to change reactivity of food allergens to innate cell surface receptors and specific immunoglobulin E (sIgE) binding affinity by either destruction and masking of epitopes or formation of neo-epitopes via exposure of interior structures and amino acid side chain modifications [3]. The applied processing conditions, such as pH, humidity, and temperature are crucial determinants for epitope destruction/formation. For example, pasteurization of cow's milk BLG enhances its allergenicity by redirecting its epithelial uptake towards Peyer's patches [4]. At the same time, clinical studies report that "baked milk products" can be tolerated by ~70% of children with IgE-mediated cow's milk allergy and that consumption of these products potentially facilitates the development of oral tolerance towards raw milk [5–7]. Both glycation and aggregation result in decreased binding of sIgE from cow's milk allergic (CMA) children if heated above 90 °C in a wet system [8,9]. At the same time, cellular signaling can be promoted by the formation of glycation structures on the food allergen [10].

The receptor for advanced glycation end products (RAGE) is one of the most studied cell surface receptors in relation to protein glycation. It is known to bind to advanced glycation end products (AGEs) that are formed via the MR. RAGE is expressed by vascular endothelial cells and cells of the innate and adaptive immune system [10]. RAGE-ligand interaction activates an intracellular signaling cascade, resulting in the release of pro-inflammatory cytokines via NF-κB activation. Its soluble isoform (sRAGE) binds to the same ligands; however it is considered to be a decoy for RAGE in the peripheral system [11]. Besides AGEs, sRAGE binds to several other ligands, such as lipopolysaccharide, amyloid-β, and S100 protein. Those ligands have the common property to act as oligomers [12]. Liu et al. [13] indicated that not only the level of glycation but also the aggregation occurring during dry heating of BLG can promote sRAGE binding affinity. Perkins et al. [14] recently reported the involvement of RAGE in type 2 cytokine signal transduction in the lungs of mice. Type 2 cytokines, e.g., interleukin (IL)-4 are involved in B-cell class switch and sIgE production [15]. These findings indicate the direct involvement of RAGE with the clinical manifestation of the allergic reaction towards the allergen. MR not only leads to the formation of AGEs, as commonly described RAGE ligands, but also promotes protein aggregation [16]. By using controlled processing conditions, protein modification can be directed towards either the formation of AGEs or the formation of aggregates enabling a distinction between the functional effects of these two distinct protein modifications. This study aimed to investigate the effect of protein aggregation and glycation on sRAGE binding affinity and the potential allergenic impact. Therefore, BLG was heated under controlled wet and dry conditions to investigate whether AGE formation or rather aggregation (glycation or non-glycation-induced) contributes to changes in sRAGE and sIgE binding affinity to BLG.

2. Materials and Methods

2.1. Chemicals

Acetonitril uHPLC-MS grade was purchased from VWR chemicals (Radnor, PA, USA). Nε-(Carboxymethyl)-l-lysine (CML) and Nε-(Carboxy [2H2]methyl)-l-lysine (d2-CML) were purchased from Polypeptide laboratories (Strasbourg, France) NuPAGE® LDS sample buffer (4× conc.), and NuPAGE™ MOPS SDS running buffer (20×), NuPAGE™ 12% Bis-Tris protein gel, 1.0 mm. gels. Soluble AGE Product-Specific Receptor Human *E. coli* (RD172116100) was obtained from Biovendor (Brno, Czech Republic). Anti-RAGE antibody (monoclonal mouse IgG$_2$B clone, MAB11451) purchased from R&D systems (Minneapolis, MN, USA). HRP conjugated anti-mouse polyclonal goat (P0447) was purchased from Dako (Glostrup, Denmark). TMB substrate (3,3′,5,5′-tetramethylbenzidine) for high sensitivity ELISA was purchased from sdt-reagents (Baesweiler, Germany). WesternBright™ ECL western blotting detection kit was obtained from Advansta (San Jose, CA, USA). Ovalbumin (OVA) was purchased from InvivoGen (San Diego, CA, USA). Bovine Serum Albumin Fraction V (BSA) was obtained from Roche (Basel, Switzerland). Amyloid-β (1-42) ultrapure HFIP was purchased from Westburg (Leusden, The Netherlands). (N-Epsilon)-Carboxymethyl-Lysine primary antibody was purchased from Nordic-MUbio (Susteren, The Netherlands). Pro-Q® Emerald 300 Glycoprotein Gel and Blot Stain Kit (Thermo Fisher Scientific, Waltham, MA, USA).

BlueRay prestained protein marker was obtained from Jena Bioscience GmbH (Jena, Germany). Coomassie brilliant blue R-250 was purchased from Bio-Rad (Hercules, CA, USA). Ultrapure water was prepared by an Purelab® Ultra water system from ELGA LabWaters (Celle, Germany). Three Plasma were purchased from PlasmaLab International (Everett, WA, USA). Two sera were provided by the Queen Beatrix Hospital (Winters Wijk, The Netherlands), all other sera were provided by the archival serumbank of the Erasmus Medical Centre (Rotterdam, The Netherlands). All other chemicals were purchased from Sigma Aldrich (St Louise, MO, USA) unless mentioned otherwise.

2.2. BLG Isolation and Purification

Raw bulk milk obtained from the Department of Animal Sciences, Wageningen University & Research (Wageningen, The Netherlands). BLG was purified and isolated as described by De Jongh et al. [17] using anion exchange chromatography DEAE Sepharose C-6B (GE healthcare, Chicago, IL, USA). Isolated BLG was lyophilised and a purity >94% was measured as described by Deng et al. [1].

2.3. Heat Treatment of BLG

BLG was heated in a wet system above the denaturation temperature by heating it in phosphate buffer (PBS) at pH 7.4 applying 100 °C for 90 min in the presence of lactose (W-HT-La) and in the absence of lactose (W-HT). Three additional treatments were conducted as described by Deng et al. [1]. Briefly, a low-temperature wet heating was conducted by heating BLG below its denaturation temperature in 10 mM PBS (pH 7.4) at 60 °C for 72 h in the presence of lactose (W-LT-La) and in the absence of lactose (W-LT). High temperature dry heating of BLG was conducted at 130 °C for 10 min in the presence of lactose (D-HT-La) and in the absence of lactose (D-HT). Prior to heat treatment, the BLG solution was lyophilised and a_w-level was adjusted to 0.53 over saturated sodium bromide solution. For low-temperature dry heating, BLG was heated at 50 °C for 9 h at aw 0.65 in the absence of lactose (D-LT) and the presence of lactose (D-LT-La). Humidity was monitored with a humidity control chamber (HCP108, Memmert, Schwabach, Germany). After heat processing, dry-heated samples were dissolved in water to starting protein concentration. All samples were centrifuged at $2900 \times g$ for 30 min to remove insoluble material and unreacted lactose was removed by dialysis. Protein concentration of samples showing the formation of insoluble material was determined with DUMAS as described by Deng et al. [1].

2.4. Quantification of CML Using uHPLC-MS/MS

CML was quantified using uHPLC-ESI-MS/MS according to a method described by Troise et al. [18]. Samples were diluted to 2.5 mg/mL in ultrapure water and mixed with hydrochloric acid to a final ratio of 0.63 mg protein/1 mL 6 M hydrochloric acid. Solutions were saturated with nitrogen and heated for 22 h at 110 °C. Hydrolysates were centrifuged ($4500 \times g$, 10 min, 20 °C) using a Heraeus multifuge X3R (Thermo Fisher Scientific, Waltham, MA, USA) and filtered through a 0.2 μm Polytetrafluoroethylene (PTFE) syringe filter (Phenomenex, Torrance, CA, USA). An aliquot was dried under nitrogen and dissolved to the same volume in ultrapure water. Samples were centrifuge ($10,000 \times g$, 20 min, 20 °C) using an Eppendorf multifuge 5430R (Eppendorf, Hamburg, Germany). Subsequently, they were diluted with acetonitrile to reach 50% acetonitrile and spiked with internal standard CML-d2.

Standard solutions were prepared in a concentration range between 25 ng/mL and 750 ng/mL and spiked with CML-d2. Final concentration of CML-d2 in all sample and standard solutions was 250 ng/mL. CML was separated on a Kinetex 2.6 μ HILCI 100A, 100 × 2.1 mm (Phenomenex, Torrance, CA, USA) at 35 °C column temperature. Eluent A was ultrapure water with 0.1% formic acid, eluent B was acetonitrile with 0.1%, and eluent C was 50 mM ammonium formate. Flow rate was set to 0.4 mL/min using the following gradient (time [min]/eluent B [%]/eluent C [%]): (0/80/10), (0.8/80/10), (3.5/40/10), (6.5/80/10), (8.0/80/10), (11/80/10). Electron ionization was conducted in positive mode. Spray voltage was set to 3500 °C, vaporizing temperature was 250 °C, and sheath gas pressure was 60 psig. Capillary temperature was set to 290 °C. Parent mass $[M+H]^+$ 205.3 m/z were selected in Q1 and the characteristic product ions 130.0 m/z (CE: 12 V; tube lens: 78) and 84.0 m/z (CE: 22 V; tube lens: 78) were recorded in Q3.

2.5. SDS-PAGE Gelelectorphoresis

Gel electrophoresis under non-reducing conditions was performed to monitor protein aggregates formed during heat treatment of BLG. Samples, NuPAGE® LDS sample buffer (4× conc.) and ultrapure water were mixed in a ratio 5/5/10 (*v/v/v*), centrifuged (1 min, 500× *g*, 20 °C) on a Eppendorf multifuge 5430R (Eppendorf, Hamburg, Germany) and incubated for 10 min at 70 °C. BlueRay prestained protein marker was used as MW marker. Six microgram protein of each sample were loaded on NuPAGE™ 12% Bis-Tris protein gel, 1.0 mm. Gels were run at 120 V for 1.5 h using NuPAGE™ 1× MOPS SDS running buffer and stained with Coomassie Brilliant Blue R-250 or using Pro-Q® Emerald 300 Glycoprotein staining kit according to the manufacturer's instructions. Images of the stained gels were obtained using a Universal Hood III (Bio-Rad, Hercules, CA, USA) and Image Lab 4.1 software (Bio-Rad, Hercules, CA, USA).

2.6. Thioflavin-T Assay

Thioflavin-T (ThT) assay was conducted to monitor the formation of fibril structures during heating. Protein concentration was adjusted to 0.25 mg/mL and mixed with 3.9 mM aqueous ThT solution in a ratio 5.8/1.0 (*v/v*). All samples were prepared in duplicate. The solution was transferred in a 96-well black greiner polystyrene plate (Greiner CELLSTAR®, Kremsmünster, Austria). After incubated for 10 min in the dark, fluorescence emission was measured ($\lambda_{exitation}$ = 450 nm, $\lambda_{emission}$ = 485 nm, gain 100) using Infinite® 200 PRO NanoQuant with i-control software (Tecan, Männedorf, Switzerland). The fluorescence intensity [a.U.] of heated and glycated BLG was corrected for the blank (PBS at pH 7.5).

2.7. Inhibition sRAGE ELISA

Inhibition sRAGE ELISA was conducted to determine sRAGE binding affinity as described by Liu et al. [13] with some modifications. Briefly, soy protein extract glycated with glucose (90 min, 100 °C, wet conditions) was used as coating material. Transparent high binding ELISA plate (Greiner Bio-One, Kremsmuenster, Austria) were coated with G90 for 12 h at 4 °C. Sample protein concentration was adjusted to 25 μg/mL with 1.5% BSA (*v/w*) in 0.025% tween in 10 mM PBS (PBST). The optimal protein concentration was chosen based on a dilution curve of BLG-NT obtained from a pre-experiment. The samples were pre-incubated with 1 μg/mL sRAGE in a ratio 1:1 (*v/v*) for 45 min at 37 °C on a NuncTM polystyrene plate (Thermo Fisher Scientific, MA, USA) before addition to the ELISA plate. The coated ELISA plate was blocked with PBS with 3% BSA (*v/w*) for 1 h at room temperature and washed with PBST. The washing step was repeated after each step of ELISA. After blocking, the pre-incubated sRAGE/sample mixture was transferred into the ELISA plate and incubation was continued for 1 h at 37 °C. After washing, anti-sRAGE antibody was added at a concentration of 0.25 μg/mL and the plate was incubated under shaking for 30 min at room temperature. After washing, anti-mouse polyclonal goat HRP conjugated antibody at a concentration 0.25 μg/mL was added and the incubation was continued for 30 min at room temperature. The signal was detected with TMB. The color reaction was measured at 450 nm vs. 620–650 nm reference using a Filter Max F5 multi-mode microplate reader (Molecular Devices, San Jose, CA, USA). Each sample was measured in triplicate. Amyloid-β was used as a positive control, while ovalbumin was used as a negative control.

Inhibition was calculated using the following formula:

$$\text{Inhibition [\%]} = (\text{Abs}_{\text{max}} - (\text{Abs}_{\text{sample}} - \text{Abs}_{\text{Min}}))/\text{Abs}_{\text{Max}} \times 100$$

where Abs_{Max} is the absorbance obtained from sRAGE without competition agent and Abs_{Min} is the absorbance obtained from blank sample (PBS) without sRAGE, $\text{Abs}_{\text{sample}}$ is the absorbance obtained from the mixture of sRAGE and each sample. High inhibition indicates high sRAGE binding affinity.

2.8. sRAGE Western Blot

SDS-PAGE was conducted as described before. However, protein concentration increased to 20 µg protein for wet-heated BLG at high temperature, 25 µg protein for wet-heated BLG at low temperature, 10 µg protein for dry-heated BLG in the presence of lactose, and 4 µg protein for BLG-NT and BLG heated in a dry system in the absence of lactose, to achieve similar band densities on the gel. Gels were blotted on Amersham™ Protran™ 0.45 µm nitrocellulose membrane (GE Healthcare Life science, Marlborough, MA, USA). Gels were blotted with semi-dry western blot blotting buffer at 15 V for 35 min. Membranes were washed for 2 × 5 min with 1x Tris buffered saline (TBS) with 0.2% tween (TBST) and incubated for 1 h at room temperature with 3% BSA in TBST. Subsequently, membranes were washed 2 × 10 min with TBST and incubated at 4 °C for 12 h with sRAGE diluted to 1 µg/mL with 1.5% BSA in TBS (*w/v*). Membranes were washed 4 × 7 min with 1 × TBST/Triton and 2 × 5 min with TBST. Anti-RAGE antibody was diluted to 0.25 µg/mL with 1.5% BSA in TBS and added to the membrane for 1 h at room temperature. Subsequently, the membrane was washed as described before and incubated for 1 h with anti-mouse polyclonal goat HRP conjugated antibody diluted to 0.25 µg/mL with 1.5% BSA in TBST. After incubation, membranes were washed 4 × 7 min with 1 × TBST/Triton and 2 × 5 min with TBS. ECL western blot detection reagent was added for 30 s. Chemiluminescence was visualized in ChemHighsensitivity mode using an Universal Hood III (Bio-Rad, Hercules, CA, USA) and Image Lab 4.1 software (Bio-Rad, Hercules, CA, USA)

2.9. CML Western Blot

SDS-PAGE western blotting was performed as described before. Blotted membranes were washed 2 × 5 min with TBST and incubated 1 h in TBST with 3% BSA. After washing 2 × 10 min with TBST, membranes were incubated with 0.25 µg/mL anti-CML-antibody with 1% BSA and 0.5% raw whey protein in TBS at 4 °C for 12 h. Membranes were washed 4 × 7 min with TBST/Triton and 2 × 5 min with TBST. Anti-mouse polyclonal goat HRP conjugated antibody was diluted to 0.25 µg/mL with 1% BSA and 1% raw whey protein in TBST. The membrane was incubated for 30 min with the antibody. Subsequently, it was washed 4 × 7 min with TBST/Triton and 2 × 5 min with TBS. ECL chemiluminescence detection was conducted as described before.

2.10. sIgE Binding Dot Blot

sIgE Dot blot was conducted to screen the available sera and plasma for their binding affinity to BLG heated under dry conditions at high temperature and BLG heated under wet conditions at high temperature, each time in the presence of lactose or absence of lactose.

Each time, 5 µg protein was spotted on the membranes. The membranes were washed 5 min with TBST and incubated for with 3% BSA in TBST for 1 h at room temperature. Membranes were washed 2 × 5 min in TBST and incubated with sera/plasma diluted in 1% BSA/TBST for 12 h at room temperature. Sera dilutions were prepared in the ratios as shown in Table 1. Membranes were washed 4 × 7 min with TBST/Triton and 2 × 5 min with TBST. Mouse anti-human sIgE antibody was diluted 0.5 µg/mL with 0.5% non-fat dry milk (NFDM) in TBST and added to the membranes for 1 h at room temperature. Subsequently, the membranes were washed 4 × 7 min with TBST/Triton and 2 × 5 min with TBST. Anti-mouse polyclonal goat HRP conjugated antibody was diluted to 0.25 μg/mL with 0.5% NFDM in TBST and added to the membranes for 30 min at room temperature. Subsequently, they were washed 4 × 7 min with TBST/Triton and 2 × 5 min with TBS. Chemiluminescence detection was conducted as described before.

Table 1. sIgE levels of sera from cow's milk allergic patients. sIgE levels as measured by ImmunoCap.

Patient #	sIgE-Level Cow's Milk Proteins [kU/L]	Specimen Type	Dilution
1	52.2	Serum	1:5
2	0.73	Serum	1:3
3	0.96	Serum	1:3
4	0.53	Serum	1:3
5	0.96	Serum	1:3
6	1.69	Serum	1:3
7	1.55	Serum	1:3
8	>100	Plasma	1:5
9	91.0	Plasma	1:5
10	94.8	Plasma	1:5
11	28.4	Serum	1:5
12	6.6	Serum	1:5

2.11. sIgE Binding Western Blot

sIgE western blot was conducted to directly identify the bands on the SDS-PAGE that show sIgE binding. The specimen was selected depending their activity observed in the sIgE Dot blot and pooled for similar sIgE binding according to the scheme in Table 2.

Table 2. Pooling of sera and plasma for IgE-western blot and used dilution. Membrane number indicates the membrane that was incubated with the specific pooled specimen.

Pool/Serum	Patient Serum/Plasma	Dilution
Pool 1	7, 11, and 12	1:5
Serum 6	6	1:7
Pool 2	8 and 9	1:7
Pool 3	1 and 10	1:5

SDS-PAGE was prepared and blotted as described for sRAGE western blot. All other steps were conducted as described for sIgE Dot blot.

2.12. Statistical Analysis

Statistical analysis was conducted using SPSS version 23. For multiple sample comparison, one-way analysis of variance (ANOVA) and Tukey post hoc comparison test was used. Results were considered statistical different at $p < 0.05$ if not mentioned otherwise.

3. Results

To investigate the effect of different heat treatments of BLG on aggregation, formation of MR products, binding to sRAGE, and binding of sIgE, BLG was heated in solution (wet conditions) and in powdered form with controlled water activity (dry conditions). For each humidity condition, BLG was heated in the presence or absence of the milk sugar lactose to distinguish between the effect of glycation and heating. Additionally, each treatment was conducted below the denaturation temperature (low-temperature heating) and above the denaturation temperature (high temperature heating) of BLG.

3.1. Solubility of BLG after Heat Treatments

Solubility of BLG was impaired after heating in a dry system, but only in the presence of lactose at 130 °C. This was observed by the formation of insoluble material after dissolving BLG in PBS. Therefore, the insoluble aggregates were removed by centrifugation and the soluble fraction of was used for further characterization. Analysis of the total nitrogen content with DUMAS showed a recovery of soluble protein of 76.5% of the total protein. BLG heated in a wet system and at low temperature in a dry system, did not show the formation of insoluble material, independent of the presence of lactose.

3.2. Quantification of Nε-carboxymethyl-L-lysine

CML was quantified in all samples to determine the level of the advanced stage of the MR. Table 3 compares CML quantities of BLG heated under different conditions in the presence or absence of lactose. No CML was found in the BLG samples that were heated without lactose, because CML formation requires the presence of either a reducing sugar moiety or polyunsaturated fatty acids. CML quantities were on average ~30% lower when BLG was heated in a wet system than when it was heated in a dry system. The quantities of CML were positively correlated with the temperature of heating in both systems. Additionally, CML levels were relatively higher after dry heating of BLG compared to wet heating.

Table 3. CML quantities determined by uHPLC-ESI-MS/MS in the soluble fraction of BLG heated in the presence of lactose (La) using wet (W) or dry (D) heat treatment at either low or high temperature (LT/HT), BLG-NT: unheated BLG, ND: quantities below the detection limit.

Sample	BLG-NT	W-HT-La	W-HT	W-LT-La	W-LT	D-HT-La	D-HT	D-LT-La	D-LT
CML mg/100 g protein	ND	129 ± 10 a	ND	77 ± 4 b	ND	182 ± 7 c	ND	158 ± 9 d	ND

Data are expressed in means of technical duplicates ± SD (*n* = 2). Letters indicate significant differences between groups (*p* < 0.05) as analyzed by one-way ANOVA and Tukey post hoc comparison test (SPSS).

The levels of CML, as a commonly used marker for the "advanced stage" of the MR, indicate that the level of glycation was higher when BLG was heated under dry conditions than under wet conditions. The formation of CML was further facilitated at higher temperature heating within the same humidity conditions.

3.3. SDS-PAGE

Protein aggregation was monitored using SDS-PAGE under non-reducing conditions, as shown in Figure 1.

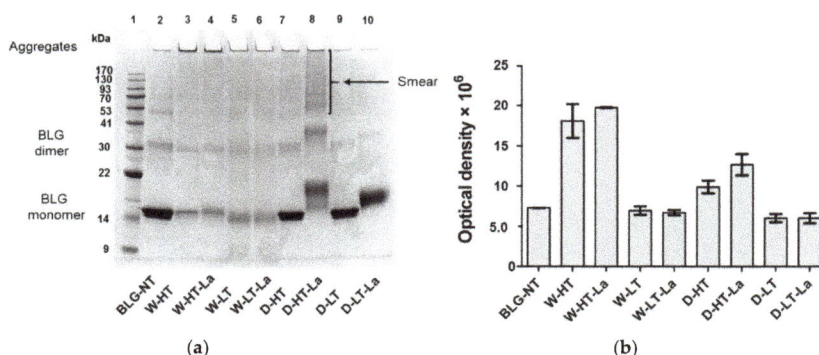

Figure 1. (**a**) Non-reducing SDS-PAGE image of soluble BLG heated and glycated in a wet system (W) and in a dry system (D), respectively in the absence of lactose or in the presence of lactose (La) and at either high temperatures (HT) or low temperatures (LT). (**b**) optical density of the aggregates visible in the top of each lane (>170 kDa). Error bars represent standard deviation of the optical density of two SDS-PAGE gels prepared independently from each other.

When heated in a wet system, BLG formed high-MW aggregates that did not migrate into the gel. Aggregate formation occurred below (60 °C, Lanes 5–6) as well as above the denaturation temperature of BLG (100 °C, Lanes 3–4); however, they were more abundant when BLG was heated above its denaturation temperature. The relative aggregate quantities, based on the optical density, showed this effect of heating temperature, but not of lactose presence for the wet-heated samples (Lanes 3–4 and 5–6 of Figure 1a, corresponding bars of Figure 1b). Dry-heated BLG aggregates showed 23% higher optical density in the presence of lactose compared to the heating in the absence of lactose, however only when BLG was heated at high temperature.

Heating of BLG in a dry system resulted in protein aggregation showing clear differences between the samples with and without lactose. After heating of BLG with lactose at high temperature (Figure 1a, Lane 8), the MW of the BLG monomer increased by approx. 4 kDa while the BLG dimer showed an increasing MW by approx. 6 kDa. The relative intensity of the BLG dimer increased by 55% compared to the unheated control, indicating a shift of the monomer-dimer equilibrium of BLG. Additionally, D-HT-La showed a smear starting from an apparent MW > 53 kDa until the top of the gel (Lane 8). D-LT (Lane 9) showed monomeric BLG at an apparent MW of 14 kDa, which is similar to the non-treated BLG (BLG-NT). However, a faint smear, similar to the one of D-HT-La but with much lower intensity, was also observed for this sample. Dry heating at low temperatures did not induce the formation of aggregates, neither in the presence or absence of lactose, in the soluble fraction. However, a shift of apparent MW of the BLG monomer by 2 kDa was observed in D-LT-La.

These data indicate that whereas wet heating favors formation of aggregates, dry heating favors MR.

3.4. Detection of Glycation Strucutres on SDS-PAGE Visible Proteins

To directly identify which of the bands visible on the SDS-PAGE are modified by MR, the Pro-Q™ Emerald 300 glycoprotein staining kit was used. The staining reacts with structures carrying a reducing carbonyl moiety and can therefore indicate the presence of Maillard reaction products (MRPs) from all three stages of the MR. Stained glycated protein will appear as white bands on the SDS-PAGE image (Figure 2).

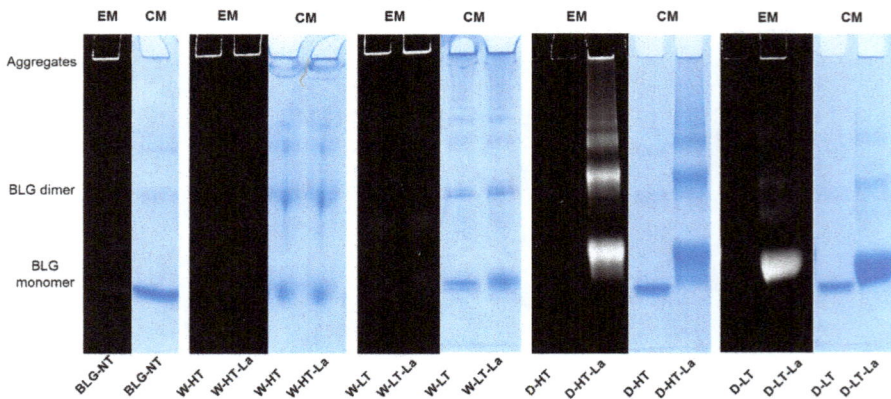

Figure 2. Non-reducing SDS-PAGE image of soluble BLG heated and glycated in a wet system (W) and in a dry system (D), respectively in the absence of lactose or in the presence of lactose (La) and at either high temperatures (HT) or low temperatures (LT). Gels were stained with Q™ Emerald 300 staining for glycoproteins (EM) and Coomassie staining (CM) for total protein staining.

Wet-heated BLG showed a fluorescence signal of the aggregates in the top of the gel, independent of the presence of lactose and the temperature. Based on CML western blot (Figure 3) it can be concluded that under wet heating conditions, the glycation structures are mainly present in the high-MW aggregates; however it was not clearly visualized by glycation staining. It may be explained by the fluorescent specking at the edges of the gel which is an intrinsic property of the dye and/or the fluorescence signal from tryptophan residues of native BLG. Both effects can contribute to the positive signal for unheated BLG in the top of the gel. It may also partly contribute to the positive signal of the high-MW aggregates of wet-heated BLG making difficult to quantitatively interpret the presence of reducing carbonyl groups in the wet-heated samples.

Dry heating showed high fluorescence when heated in the presence of lactose at high and low temperatures, and not when heated in the absence of lactose. The signal for glycation structures was observed for all proteins that were also visible with Coomassie staining. A positive signal with glycation staining was also observed for the BLG monomer of D-HT and D-HT-La and the high-MW aggregates. The same is also true for BLG-NT. However, the relative glycation staining intensity of the aggregates from heated and glycated BLG is higher than the high-MW band visible in BLG-NT.

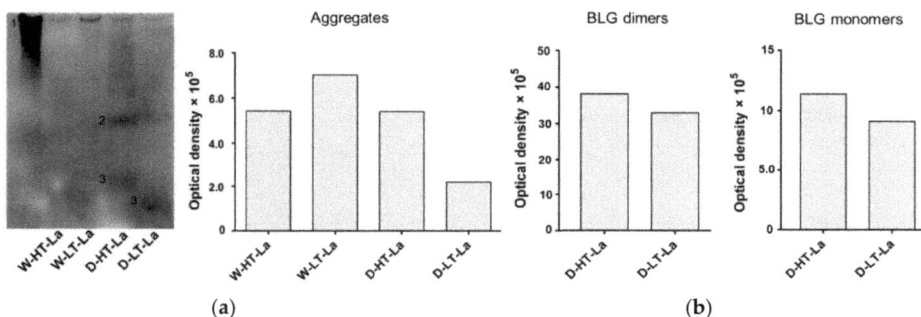

Figure 3. (**a**) Membrane image of CML western blot measured of BLG heated in the presence of lactose (La) in a wet system (W) or a dry system (D) at low temperatures (LT) or high temperatures (HT). G90: soy protein extract glycated with glucose for 90 min (positive control). Numbers indicate aggregates (1), BLG dimers (2), and BLG monomers (3). (**b**) Optical density of bands visible on the CML western blot categorized in high-MW aggregates, BLG dimers, and BLG monomers as indicated on the membrane.

The band of wet-heated BLG at low temperature showed higher intensity than wet-heated BLG at high temperature, which is in contradiction to the total CML quantities. This can be explained by the lower accessibility of the CML antibody to CML structures if they are buried inside of the aggregate. This effect might have more impact in high temperature treated samples because of the more compact structure of the aggregates. These data show that under wet heating conditions glycation structures and CML are mainly detected in high-MW aggregates, while under dry heating conditions they are observed in all protein factions.

3.5. Formation of Fibril Structures

The ThT-assay was conducted to monitor the formation of fibril structures and exposure of β-sheet structures upon heating and glycation of BLG. Results are shown in Figure 4.

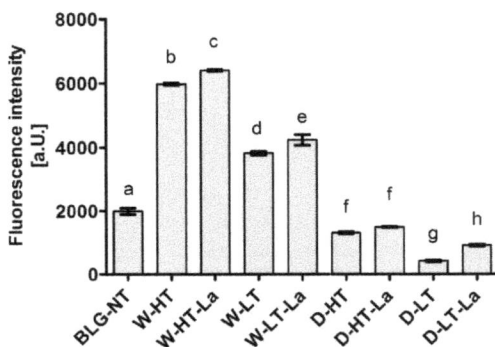

Figure 4. ThT fluorescence intensity after incubation with BLG heated in the absence of lactose or in the presence of lactose (La) in a wet system (W) or a dry system (D) under low temperatures (LT) or high temperatures (HT). BLG-NT: unheated BLG. Error bars represent standard deviation of technical duplicates. Letters indicate significant differences between the BLG groups. p-values < 0.05 are considered statistically significant, as analyzed with one-way ANOVA and Tukey post hoc comparison test (SPSS).

Highest fluorescence intensity was observed for wet-heated BLG and increased with higher heating temperature. When heated under dry conditions, the fluorescence intensity did not increase above the level of BLG-NT, indicating that no additional β-sheet structures were exposed. These results indicate that β-sheet structures are exposed and/or formed when BLG is heated under wet conditions independent of the presence of lactose. At the same time, this effect is not observed under dry heating conditions. This indicates that the exposure of β-sheet structures is facilitated under wet heating conditions while the presence of lactose plays a minor role independent of the humidity conditions.

3.6. Binding of sRAGE to Heated and Glycated BLG

Binding of sRAGE to heated and glycated BLG was monitored using sRAGE inhibition ELISA where high inhibition indicates high sRAGE binding affinity (Figure 5).

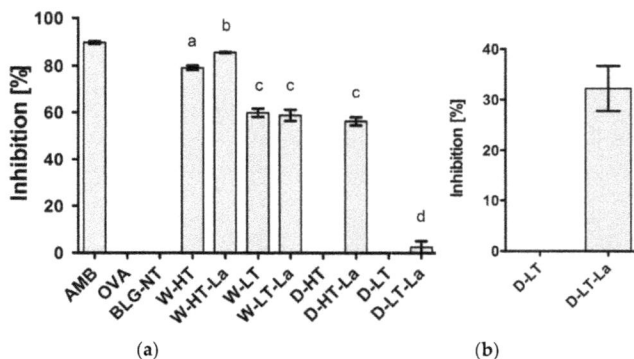

(a) (b)

Figure 5. (**a**) Inhibition of sRAGE binding in competition ELISA by BLG heated in wet system (W) and dry system (D) in the absence of lactose or presence of lactose (La) at high or low temperature (HT and LT) for all treatments at a protein concentration 25 µg/mL. AMB: amyloid-β (positive control), OVA: ovalbumin (negative control). (**b**) Results of inhibition sRAGE ELISA for D-BLG-LT and D-LT-LA at a protein concentration 100 µg/mL. Error bars represent standard deviation of technical triplicates. Letters indicate significant differences between the BLG groups. p-values < 0.05 are considered statistically significant, as analyzed with one-way ANOVA with Tukey post hoc comparison test (SPSS).

BLG-NT did not show any binding affinity to sRAGE while BLG heated in wet systems, both with and without lactose, showed increased sRAGE binding affinity compared to BLG-NT. Wet heating at higher temperature resulted in 28% higher sRAGE binding affinity than wet heating at lower temperature. At high temperature, sRAGE showed 8% higher binding to W-HT-La than W-HT, while at low temperatures no lactose dependent difference was observed. In contrast to the wet system, BLG heated without lactose in a dry system (at either high or low temperature) did not show any binding affinity for sRAGE. D-HT showed a similar sRAGE binding affinity as the low-temperature wet-heated samples. BLG heated in dry system with lactose at low temperature (D-LT-La) showed the lowest sRAGE binding compared to the other samples; however the binding increased with higher protein concentration (Figure 6b), indicating that sRAGE does show binding affinity but lower than for the other samples.

Western blot analysis was used to directly identify which fractions on the SDS-gel are responsible for the sRAGE binding determined for heated and glycated BLG in the sRAGE inhibition ELISA. These results are shown in Figure 6.

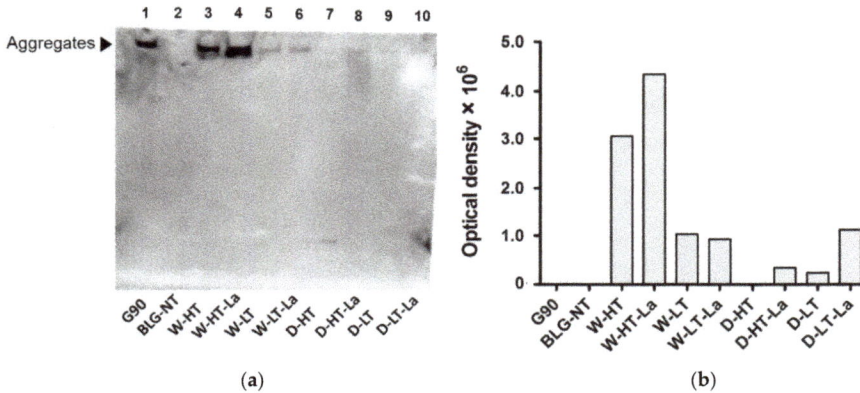

Figure 6. (**a**) Membrane of sRAGE western blot for BLG heated in the absence of lactose or presence of lactose (La) under wet conditions (W) or dry conditions (D) at high temperatures (HT) or low temperatures (LT). (**b**) Optical density of the bands showing sRAGE binding affinity. G90: soy protein extract glycated with glucose for 90 min, 100 °C, wet conditions (positive control).

No binding of sRAGE to the monomeric form of BLG was observed in any of the heat-treated BLG samples except for dry heating at low temperature (Figure 6a, Lane 9). The faint band in the lower part of Lane 7 is most likely an artefact, as no protein was seen at this position on the SDS-PAGE. The fraction showing highest affinity to sRAGE were high-MW aggregates of BLG observed in wet-heated samples, especially those heated at high temperature (Lanes 3–4). Aggregates of wet-heated samples with lactose showed ~30% higher intensity than samples heated without lactose, suggesting that both aggregation but also glycation play a role in sRAGE binding. The same tendency of slightly increasing sRAGE binding in the presence of lactose under wet conditions, was also observed in the ELISA. At low-temperature wet heating, no increased signal intensity was observed when lactose was present. In the dry system at high temperature, sRAGE binding to the smear formed at MW > 53 kDa was observed for lactose-containing sample. At low-temperature dry heating a faint band was observed for aggregates of lactose-free heated BLG. The signal intensity of BLG heated in the presence of lactose under dry conditions, based on the optical density (Figure 6b), increased by ~75%. The relative binding intensity of BLG heated under different conditions are in line with the ELISA results (Figure 5). These results indicate that aggregation promotes sRAGE binding affinity to a greater extent than glycation.

3.7. sIgE Binding to Heated and Glycated BLG

sIgE binding dot blot was conducted to screen the available plasma and sera for their sIgE binding affinity to untreated BLG as well as heated and glycated BLG under wet or dry conditions at high temperatures. The results of the 12 different sera/plasma samples from Table 1 are shown in Figure 7. Based on the dot blot results, the sera were pooled (see Table 2) to perform sIgE western blots.

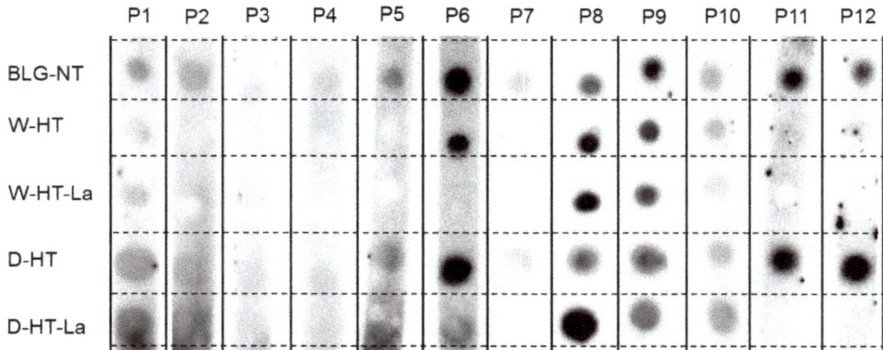

Figure 7. sIgE dot blot membranes of sera from 12 different patients (P) tested for their binding affinity to unheated BLG (BLG-NT) and BLG heated at high temperatures (HT) either under wet (W) or dry (D) conditions, both in the absence of lactose (BLG) or in the presence of lactose (Lac).

No binding to BLG was observed for the sera of Patient 3 and only weak binding to unheated BLG for the sera of Patient 4. All other sera showed a positive binding to BLG-NT.

Four different profiles for sIgE binding to heated and glycated BLG were observed. The sera described as sIgE binding profiles 1-4 were pooled (see Table 2 for details) to perform western blots (Figure 8). Sera 7, 11, and 12 showed only binding to unheated BLG and BLG heated in the absence of lactose under dry conditions. Serum 2 and Serum 5 also showed a similar profile of binding to BLG to Sera 7, 11, and 12; however, because of the high background, it could not be excluded that they also show binding to BLG heated in the presence of lactose under dry conditions. Therefore, they were not included in Pool 1.

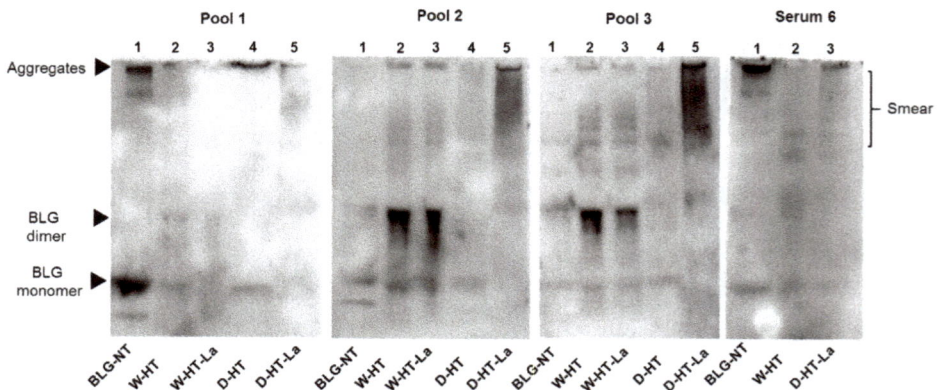

Figure 8. sIgE western blot membranes of unheated BLG (BLG-NT) as well as BLG heated at high temperatures (HT) either under wet (W) or dry (D) conditions, both in the absence of lactose or in the presence of lactose (La). Sera of cow's milk allergic patients were pooled. Pool 1: Sera 7, 11, 12. Pool 2: Serum 8 and Serum 9. Pool 3: Serum 1 and Serum 10.

Western blots confirmed the results observed in dot blot and indicated that the binding of sIgE occurs to both BLG monomer and the aggregates, with strongest binding for BLG-NT and D-HT (Pool 1, Line 4 and 5). A band below the BLG monomer band was also observed in unheated BLG although this band was not seen on the SDS-PAGE (Figure 1a). While the binding intensity for BLG-NT was equal between the monomer and the aggregates, in D-HT the binding intensity to the aggregates was ~30% higher than to the monomer (Pool 1, Lane 4). Weak binding was also observed for the BLG monomer in samples heated in the absence of lactose under wet conditions and to the smear of BLG heated in the presence of lactose under dry conditions (Pool 1, Lane 2 and 5). Pool 2 showed binding to all samples, with higher binding to wet-heated BLG independent of the presence of lactose and to dry-heated BLG in the presence of lactose. Western blots showed sIgE binding to the BLG monomer and the BLG dimer in unheated BLG. When heated under wet conditions (Pool 2, Lane 2 and Lane 3), binding was observed to the BLG monomer, BLG dimer, and the aggregates. The dimer showed the highest sIgE binding intensity with ~50% and ~60% of the total optical density of W-HT and W-HT-La, respectively. BLG heated under dry conditions showed binding to the BLG monomer and to a smear > 53 kDa, with higher binding when heated in the presence of lactose (Pool 2, Lane 5) compared to heating in the absence of lactose (Pool 2, Lane 4). Additionally, in the presence of lactose, the aggregates also showed high sIgE binding. Pool 3 showed equal binding to unheated and all heated BLG. This sample showed the same sIgE binding pattern as Pool 2, except for the smear of D-HT-La (Pool 3, Lane 5) that showed higher intensity than in Pool 2. Serum 6 showed a unique binding pattern with strong binding to unheated BLG, and BLG heated in the absence of lactose for dry and wet heating conditions. A weak binding was also observed for BLG heated under dry conditions in the presence of lactose but not to the equivalent sample heated under wet conditions. Interestingly, the binding intensity was higher to the aggregates than to the monomer, for both BLG-NT (Serum 6, Lane 1) and D-HT (Serum 6, Lane 3) with ~30% and ~55%, respectively. For BLG heated under wet conditions in the absence of lactose, the binding was also observed to the monomer and dimer but less to the aggregates.

To summarize, the results of sIgE dot blot indicate high individual variations in the sIgE binding profiles to heated and glycated BLG and allowed the grouping into four binding affinity profiles. Wet-heated BLG showed reduced sIgE binding when compared to non-treated BLG for 5 out of 10 sera giving the positive signal, independent of the presence of lactose. Dry heating of BLG without lactose did not affect sIgE binding compared to non-treated BLG. On the other hand, BLG dry heated in the presence of lactose increased sIgE binding compared to both untreated BLG and dry-heated BLG without lactose, revealing an effect of glycation on the sIgE binding capacity. Results showed which BLG fractions were responsible for sIgE binding observed in the dot blot analysis, indicating a shift of sIgE binding towards higher MW fractions (BLG dimers and aggregates) formed upon heating.

4. Discussion

Food processing promotes the changes of the immunogenic and antigenic properties of food allergens [3], among others via MR. In this study, we showed that during processing of BLG both the temperature and water activity (wet vs. dry heating) are important factors affecting its immunogenicity. Wet heating of BLG, independent of the presence of lactose in the sample, promotes the formation of aggregates recognized by sRAGE and sIgE from CMA patients. In contrast to wet heating the aggregates formed upon dry heating showed differences in an immunogenic profiles of the samples heated in the presence or absence of lactose, therefore revealing contribution of glycation to the formation of both sRAGE and sIgE binding epitopes.

Both glycation and aggregation of BLG depends on the applied heating conditions. Glycation of milk proteins is promoted at low water activity of 0.31–0.98 [19]. At the same time, heating at high humidity conditions facilitates the aggregation of BLG via disulfide linkages [20,21] but leads to a relative lower levels of glycation structures compared to dry heating [22,23]. This effect was also observed in this study where the formation of CML and detection of glycation structures on BLG was significantly higher in dry-heated BLG than in wet-heated BLG. This was confirmed by

CML western blot (Figure 3) showing highest levels of CML in D-HT-La detected in monomers, dimers, and aggregates of BLG while in the wet-heated samples the CML appeared only in the aggregates. Next to that, more glycation structures formed during heating of BLG under dry conditions in the presence of lactose were detected by the glycation staining (Figure 2). On the other hand, the formation of high-MW aggregates was independent of the presence of lactose in wet-heated samples compared to dry-heated samples. However, aggregation was also observed in the sample heated under dry conditions with lactose. It has already been described that glycation-induced modifications of lysine and arginine can increase the flexibility of the protein structure and leads to the formation of covalent cross linking between different amino acid residues. Both facilitates protein denaturation and aggregation [16,24] and influences the morphology of the protein aggregates [25,26]. Additionally, glycation strongly affects the size of these aggregates when heated under dry conditions but may also contributes to structural differences of the formed aggregates under both humidity conditions [1]. However, reactant mobility during the wet heating also affects the aggregation rate of proteins, where higher inter-molecular mobility results in higher protein aggregation [27]. The observed differences in aggregation of wet-heated BLG vs. dry-heated BLG are in line with the findings of Deng et al. [1] who showed increased formation of polymers and oligomers in wet-heated BLG compared to dry-heated BLG. The authors also showed that the dry heating of BLG at 130 °C increased its oligomerization and polymerization and was mainly affected by the presence of lactose. On the other hand, the wet heating led to more exposure of β-sheet structures in the soluble aggregates than dry heating. These findings stay in line with our results showing a higher content of surface-exposed β-sheets structures in the wet-heated BLG independent of the presence of lactose in the sample. The formation of aggregates in both wet and dry heating conditions was shown to have a great impact of the formation of sRAGE binding ligands. Three different patterns of sRAGE binding affinity (Figure 6) were observed, depending on the heating temperature and the water activity of the system. First, when heated in the wet system high sRAGE binding was observed to the formed aggregates, regardless of the presence of lactose. Secondly, sRAGE binding occurred independent of the presence of CML but increased with higher exposure of β-sheet structures after wet heating (Figures 3 and 4). Finally, in the dry-heated system, sRAGE binding was observed only to BLG heated in the presence of lactose (high-MW fractions), revealing an impact of glycation under dry heating conditions (Figure 6). Being a promiscuous receptor, sRAGE binds not only to AGEs but also to amyloid-β, S100 protein, and HGMB1 [11]. Those ligands share a common property to act as oligomers, highlighting the potential of sRAGE to recognize high-MW structures [12]. Next to the aggregation itself, the glycation-induced aggregation was already pointed out to be responsible for the formation of sRAGE ligands under dry heating conditions; however this was not directly proven [13]. With the western blot analysis of sRAGE binding we confirmed these findings, proving that under the dry heating conditions of BLG glycation without aggregation does not lead to increased sRAGE binding. Consequently, the changes in sRAGE binding were accompanied by the changes in sIgE binding affinity by (treatment-dependent) formation of neo-epitopes or destruction of existing epitopes. Several studies indicated that both glycation and aggregation, contribute to changes in cellular recognition and allergic sensitization [1,4,13,28,29]. The results of dot blot with the sera of CMA patients showed high individual variations in the sIgE binding to heated and glycated BLG. Four different profiles of sIgE binding affinity were observed which showed, depending on the used conditions, either increased or diminished sIgE binding to heat-treated BLG. The dot blot analysis revealed that wet heating reduces sIgE binding independent of the presence of lactose which was observed in 5 out of 10 sera giving a positive signal with non-treated BLG (Figure 7). Therefore, glycation appears to play a minor role in changing sIgE binding when BLG is heated under wet conditions. For dry-heated samples, sIgE binding was only decreased when lactose was present during heating (Figure 7), indicating the masking of epitopes by the attached sugar moiety. These differences indicate that humidity applied during the heat processing as well as glycation are important factors influencing sIgE binding affinity. The western blot results (Figure 8) obtained for four pools of sera to heated BLG allowed to detect the fractions and bands responsible for sIgE reactivity observed in dot blot. Three different fractions present in non-treated BLG were recognized by sIgE:BLG monomer, BLG dimer, and oligomers in the top of the gel. Binding of sIgE to the oligomers present in non-treated BLG was always accompanied with the positive binding to the oligomers present in dry-heated sample without lactose suggesting the existence of similar epitopes in these two samples (Figure 8, Pool 1 and Serum 6). Interestingly, the aggregates formed during the wet heating showed sIgE reactivity only in case of half of the tested sera indicating a lactose-independent decrease of sIgE binding affinity. It has been shown in the past that the free cysteine (C121) is strongly involved in the aggregation of BLG when heated in a wet system and this results in the formation of covalently linked aggregates [21]. Due to the higher translational motion in the wet system compared to the dry system the formation of these covalent linkages is favored in the wet system leading to higher extend of aggregation. This might also result in the embedding or exposure of different sIgE epitopes when comparing the two humidity conditions. The observation of sIgE binding for half of the patients stay in line with a previous finding that reported decreased sIgE binding to BLG when heated above 90 °C under wet conditions [9,30,31]. Additionally, Kleber and Hinrichs [31] postulated that protein aggregation may decreases sIgE binding affinity. However, our study points out that there are different MW fractions formed during wet and dry heating of BLG that can be recognized by sIgE of some patients. This was lactose-independent after wet heating of BLG, and was mostly found to higher MW fractions > BLG dimers. The same patients also show higher sIgE binding to dry-heated BLG in the presence of lactose compared to the lactose-free control. The most immuno-reactive fraction was a high-MW smear of dry-heated BLG in the presence of lactose which also showed the presence of glycation structures and CML (Figure 8). This was also true for the BLG monomer and dimer in this sample but sIgE binding was not observed to these fractions in the same pool of sera. This indicates that MR may has a double effect in sensitization. First, it can lead to masking of epitopes in the BLG monomer and dimer as described previously also for other food allergens [8,32,33]. Secondly, the formation of glycation-induced aggregates can enhance sIgE binding via neo-epitope formation. The study of Vissers et al. [34] showed reduced allergenicity of MR-modified peanut allergen Ara h 1 in sIgE binding test while enhanced β-hexominidase release from basophils upon incubation with the same MR-modified allergen. The authors suggested that MR-induced aggregates of Ara h 1 could be responsible for the observed increased capacity of antigen to cross-link the sIgE and initiate the mediator release from RBL-2H3 cells. Similar observations were also made for other protein families such as peanut proteins and scallop tropomyosin [8,35–37]. Thus, the influence of MR on sIgE binding seems to depend on physicochemical properties of proteins (hydrophobicity, size, amino acid composition, charge) as well as on the conditions of the MR (type of sugar, time, water activity, pH, temperature, presence of salts) [36,38].

Different sIgE binding profiles observed in this study indicate the high individual differences between CMA patients and demonstrates the importance of involving better characterized CMA patients' sera in the sIgE binding study. That would allow a prediction to be made towards the influence of aggregation and glycation of BLG on its allergenicity. This would help to optimize processing conditions towards less immunoreactive products which is especially relevant in the production of infant formula. Furthermore, it should be taken into account that both aggregation and glycation affect protein digestibility. Antigens with resistance to gastrointestinal digestion are commonly understood to be the most potent allergens in terms of sIgE binding [39]. It is known that glycation-induced aggregates are more resistant to in vitro gastrointestinal digestion [26]. Therefore, aggregates might have higher physiological relevance than aggregates formed in the absence of sugar even though they were as potent in stimulating sRAGE binding and sIgE binding as aggregates formed in the absence of sugar.

5. Conclusions

sRAGE binding is highest when BLG is heated under wet conditions and is mostly determined by aggregation rather than formation of MRPs. Under dry heating conditions sRAGE binding is mostly affected by the presence of lactose but a strict distinction between the effect of the formation of MRPs and glycation-induced aggregation could not be made. Also sIgE binding seems to be affected by aggregation and glycation; however high inter-individual differences do not allow clear differentiation between the effects of the used heating conditions on allergic sensitization. Therefore, more individuals need to be screened to get more insight in the role of aggregation and glycation on sRAGE binding and sIgE binding affinity of thermally processed BLG. However, this data suggests that different heating conditions of BLG result in the formation of different sRAGE ligands and sIgE epitopes and that the water activity during the processing of milk compounds contributes to the pattern of sensitization.

Author Contributions: H.E.Z.; methodology, investigation, validation, visualization, formal analysis, writing–original draft, A.E. methodology, formal analysis, Y.D.; methodology, resource, H.F.J.S.; resource, writing–review and editing, R.J.J.v.N.; writing–review and editing, N.W.D.J.; Resource, funding acquisition, writing–review and editing, H.J.W.: funding acquisition, writing–review and editing, K.A.H.; funding acquisition, writing–review and editing, supervision, M.T.; writing–review and editing, supervision, project administration.

Funding: This work is part of the research program iAGE/TTW with project number 14536, which is (partly) financed by the Netherlands Organisation for Scientific Research (NWO).

Acknowledgments: The authors thank C. W. Weykamp for providing us with the sera from 2014-08 and 2017-07. These sera were selected for this study by Shanna Bastiaan-Net. The authors also thank the Erasmus Medical Centre (Rotterdam, The Netherlands) for providing 7 sera and the iAGE/TTW for their support in this project.

Conflicts of Interest: R.J.J.v.N. is an employee of FrieslandCampina; all authors declare no conflict of interest.

References

1. Deng, Y.; Govers, C.; Bastiaan-Net, S.; van der Hulst, N.; Hettinga, K.; Wichers, H.J. Hydrophobicity and aggregation, but not glycation, are key determinants for uptake of thermally processed β-lactoglobulin by thp-1 macrophages. *Food Res. Int.* **2019**, *120*, 102–113. [CrossRef] [PubMed]

2. Ames, J.M. Control of the maillard reaction in food systems. *Trends Food Sci. Technol.* **1990**, *1*, 150–154. [CrossRef]

3. Rahaman, T.; Vasiljevic, T.; Ramchandran, L. Effect of processing on conformational changes of food proteins related to allergenicity. *Trends Food Sci. Technol.* **2016**, *49*, 24–34. [CrossRef]

4. Roth-Walter, F.; Berin, M.C.; Arnaboldi, P.; Escalante, C.R.; Dahan, S.; Rauch, J.; Jensen-Jarolim, E.; Mayer, L. Pasteurization of milk proteins promotes allergic sensitization by enhancing uptake through peyer's patches. *Eur. J. Allergy Clin. Immunol.* **2008**, *63*, 882–890. [CrossRef] [PubMed]

5. Kim, J.S.; Nowak-Wgrzyn, A.; Sicherer, S.H.; Noone, S.; Moshier, E.L.; Sampson, H.A. Dietary baked milk accelerates the resolution of cow's milk allergy in children. *J. Allergy Clin. Immunol.* **2011**, *128*, 125–131. [CrossRef]

6. Nowak-Wegrzyn, A.; Bloom, K.A.; Sicherer, S.H.; Shreffler, W.G.; Noone, S.; Wanich, N.; Sampson, H.A. Tolerance to extensively heated milk in children with cow's milk allergy. *J. Allergy Clin. Immunol.* **2008**, *122*, 342–347. [CrossRef]

7. Esmaeilzadeh, H.; Alyasin, S.; Haghighat, M.; Nabavizadeh, H.; Esmaeilzadeh, E.; Mosavat, F. The effect of baked milk on accelerating unheated cow's milk tolerance: A control randomized clinical trial. *Pediatr. Allergy Immunol.* **2018**, *29*, 747–753. [CrossRef]

8. Taheri-Kafrani, A.; Gaudin, J.C.; Rabesona, H.; Nioi, C.; Agarwal, D.; Drouet, M.; Chobert, J.M.; Bordbar, A.K.; Haertle, T. Effects of heating and glycation of β-lactoglobulin on its recognition by IgE of sera from cow milk allergy patients. *J. Agric. Food Chem.* **2009**, *57*, 4974–4982. [CrossRef]

9. Ehn, B.M.; Ekstrand, B.; Bengtsson, U.; Ahlstedt, S. Modification of IgE binding during heat processing of the cow's milk allergen β-lactoglobulin. *J. Agric. Food Chem.* **2004**, *52*, 1398–1403. [CrossRef]

10. Ott, C.; Jacobs, K.; Haucke, E.; Santos, A.N.; Grune, T.; Simm, A. Role of advanced glycation end products in cellular signaling. *Redox Biol.* **2014**, *2*, 411–429. [CrossRef]

11. Fritz, G. Rage: A single receptor fits multiple ligands. *Trends Biochem. Sci.* **2011**, *36*, 625–632. [CrossRef] [PubMed]

12. Kierdorf, K.; Fritz, G. Rage regulation and signaling in inflammation and beyond. *J. Leukoc. Biol.* **2013**, *94*, 55–68. [CrossRef] [PubMed]

13. Liu, F.; Teodorowicz, M.; Wichers, H.J.; Van Boekel, M.A.J.S.; Hettinga, K.A. Generation of soluble advanced glycation end products receptor (srage)-binding ligands during extensive heat treatment of whey protein/lactose mixtures is dependent on glycation and aggregation. *J. Agric. Food Chem.* **2016**, *64*, 6477–6486. [CrossRef] [PubMed]

14. Perkins, T.N.; Oczypok, E.A.; Dutz, R.E.; Donnell, M.L.; Myerburg, M.M.; Oury, T.D. The receptor for advanced glycation endproducts is a critical mediator of type 2 cytokine signaling in the lungs. *J. Allergy Clin. Immunol.* **2019**. [CrossRef] [PubMed]

15. Ngoc, L.P.; Gold, D.R.; Tzianabos, A.O.; Weiss, S.T.; Celedón, J.C. Cytokines, allergy, and asthma. *Curr. Opin. Allergy Clin. Immunol.* **2005**, *5*, 161–166. [CrossRef] [PubMed]

16. Cardoso, H.B.; Wierenga, P.A.; Gruppen, H.; Schols, H.A. Maillard induced aggregation of individual milk proteins and interactions involved. *Food Chem.* **2019**, *276*, 652–661. [CrossRef] [PubMed]

17. De Jongh, H.H.J.; Gröneveld, T.; De Groot, J. Mild isolation procedure discloses new protein structural properties of β-lactoglobulin. *J. Dairy Sci.* **2001**, *84*, 562–571. [CrossRef]

18. Troise, A.D.; Fiore, A.; Roviello, G.; Monti, S.M.; Fogliano, V. Simultaneous quantification of amino acids and amadori products in foods through ion-pairing liquid chromatography-high-resolution mass spectrometry. *Amino Acids* **2015**, *47*, 111–124. [CrossRef] [PubMed]

19. Gonzales, A.S.P.; Naranjo, G.B.; Leiva, G.E.; Malec, L.S. Maillard reaction kinetics in milk powder: Effect of water activity at mild temperatures. *Int. Dairy J.* **2010**, *20*, 40–45. [CrossRef]

20. Hoffmann, M.A.M.; Van Mil, P.J.J.M. Heat-induced aggregation of β-lactoglobulin: Role of the free thiol group and disulfide bonds. *J. Agric. Food Chem.* **1997**, *45*, 2942–2948. [CrossRef]

21. Roefs, S.P.F.M.; De Kruif, K.G. A model for the denaturation and aggregation of β-lactoglobulin. *Eur. J. Biochem.* **1994**, *226*, 883–889. [CrossRef] [PubMed]

22. Fenaille, F.; Morgan, F.; Parisod, V.; Tabet, J.C.; Guy, P.A. Solid-state glycation of β-lactoglobulin by lactose and galactose: Localization of the modified amino acids using mass spectrometric techniques. *J. Mass Spectrom.* **2004**, *39*, 16–28. [CrossRef] [PubMed]

23. Akıllıoglu, H.G.; Çelikbıçak, O.; Salih, B.; Gokmen, V. Monitoring protein glycation by electrospray ionization (ESI) quadrupole time-of-flight (Q-TOF) mass spectrometer. *Food Chem.* **2017**, *217*, 65–73. [CrossRef] [PubMed]

24. Iannuzzi, C.; Irace, G.; Sirangelo, I. Differential effects of glycation on protein aggregation and amyloid formation. *Front. Mol. Biosci.* **2014**, *1*, 9. [CrossRef] [PubMed]

25. Bouma, B.; Kroon-Batenburg, L.M.J.; Wu, Y.P.; Brünjes, B.; Posthuma, G.; Kranenburg, O.; De Groot, P.G.; Voest, E.E.; Gebbink, M.F.B.G. Glycation induces formation of amyloid cross-β structure in albumin. *J. Biol. Chem.* **2003**, *278*, 41810–41819. [CrossRef] [PubMed]

26. Pinto, M.S.; Léonil, J.; Henry, G.; Cauty, C.; Carvalho, A.F.; Bouhallab, S. Heating and glycation of β-lactoglobulin and β-casein: Aggregation and in vitro digestion. *Food Res. Int.* **2014**, *55*, 70–76. [CrossRef]

27. Gulzar, M.; Bouhallab, S.; Jardin, J.; Briard-Bion, V.; Croguennec, T. Structural consequences of dry heating on alpha-lactalbumin and beta-lactoglobulin at pH 6.5. *Food Res. Int.* **2013**, *51*, 899–906. [CrossRef]

28. Liu, F.; Teodorowicz, M.; Van Boekel, M.A.J.S.; Wichers, H.J.; Hettinga, K.A. The decrease in the igg-binding capacity of intensively dry heated whey proteins is associated with intense maillard reaction, structural changes of the proteins and formation of rage-ligands. *Food Funct.* **2016**, *7*, 239–249. [CrossRef]

29. Perusko, M.; van Roest, M.; Stanic-Vucinic, D.; Simons, P.J.; Pieters, R.H.H.; Cirkovic Velickovic, T.; Smit, J.J. Glycation of the major milk allergen β-lactoglobulin changes its allergenicity by alterations in cellular uptake and degradation. *Mol. Nutr. Food Res.* **2018**, *62*, 1800341. [CrossRef]

30. Bu, G.; Luo, Y.; Zheng, Z.; Zheng, H. Effect of heat treatment on the antigenicity of bovine α-lactalbumin and β-lactoglobulin in whey protein isolate. *Food Agric. Immunol.* **2009**, *20*, 195–206. [CrossRef]

31. Kleber, N.; Maier, S.; Hinrichs, J. Antigenic response of bovine β-lactoglobulin influenced by ultra-high pressure treatment and temperature. *Innov. Food Sci. Emerg. Technol.* **2007**, *8*, 39–45. [CrossRef]

32. Cellmer, T.; Bratko, D.; Prausnitz, J.M.; Blanch, H.W. Protein aggregation in silico. *Trends Biotechnol.* **2007**, *25*, 254–261. [CrossRef] [PubMed]

33. Scheurer, S.; Lauer, I.; Foetisch, K.; Moncin, M.S.M.; Retzek, M.; Hartz, C.; Enrique, E.; Lidholm, J.; Cistero-Bahima, A.; Vieths, S. Strong allergenicity of pru av 3, the lipid transfer protein from cherry, is related to high stability against thermal processing and digestion. *J. Allergy Clin. Immunol.* **2004**, *114*, 900–907. [CrossRef] [PubMed]

Nutrients **2019**, *11*, 1432

34. Vissers, Y.M.; Iwan, M.; Adel-Patient, K.; Skov, P.S.; Rigby, N.M.; Johnson, P.E.; Müller, P.M.; Przybylski-Nicaise, L.; Schaap, M.; Ruinemans-Koerts, J.; et al. Effect of roasting on the allergenicity of major peanut allergens ara h 1 and ara h 2/6: The necessity of degranulation assays. *Clin. Exp. Allergy* **2011**, *41*, 1631–1642. [CrossRef] [PubMed]

35. Gruber, P.; Vieths, S.; Wangorsch, A.; Nerkamp, J.; Hofmann, T. Maillard reaction and enzymatic browning affect the allergenicity of pru av 1, the major allergen from cherry (*Prunus avium*). *J. Agric. Food Chem.* **2004**, *52*, 4002–4007. [CrossRef] [PubMed]

36. Blanc, F.; Vissers, Y.M.; Adel-Patient, K.; Rigby, N.M.; Mackie, A.R.; Gunning, A.P.; Wellner, N.K.; Skov, P.S.; Przybylski-Nicaise, L.; Ballmer-Weber, B.; et al. Boiling peanut ara h 1 results in the formation of aggregates with reduced allergenicity. *Mol. Nutr. Food Res.* **2011**, *55*, 1887–1894. [CrossRef] [PubMed]

37. Rahaman, T.; Vasiljevic, T.; Ramchandran, L. Conformational changes of β-lactoglobulin induced by shear, heat, and pH-effects on antigenicity. *J. Dairy Sci.* **2015**, *98*, 4255–4265. [CrossRef]

38. Iwan, M.; Vissers, Y.M.; Fiedorowicz, E.; Kostyra, H.; Kostyra, E.; Savelkoul, H.F.J.; Wichers, H.J. Impact of maillard reaction on immunoreactivity and allergenicity of the hazelnut allergen cor a 11. *J. Agric. Food Chem.* **2011**, *59*, 7163–7171. [CrossRef]

39. Taylor, S.L.; Lemanske, R.F.; Bush, R.K.; Busse, W.W. Food allergens: Structure and immunologic properties. *Ann. Allergy* **1987**, *59*, 93–99.

nutrients

MDPI

Article

Higher Polygenetic Predisposition for Asthma in Cow's Milk Allergic Children

Philip R. Jansen [1,2], Nicole C. M. Petrus [3], Andrea Venema [4], Danielle Posthuma [1,4],
Marcel M. A. M. Mannens [4], Aline B. Sprikkelman [3,5] and Peter Henneman [4,*]

[1]   Department of Complex Trait Genetics, Center for Neuroscience and Cognitive Research, Amsterdam
     Neuroscience, VU University, 1081 HV Amsterdam, The Netherlands; p.r.jansen@vu.nl (P.R.J.);
     danielle.posthuma@vu.nl (D.P.)
[2]   Department of Child and Adolescent Psychiatry, Erasmus MC, 3015 GD Rotterdam, The Netherlands
[3]   Department of Pediatric Respiratory Medicine and Allergy, Emma Children's Hospital, AUMC,
     1105 AZ Amsterdam, The Netherlands; n.c.petrus@amc.nl (N.C.M.P.); a.b.sprikkelman@umcg.nl (A.B.S.)
[4]   Department Clinical Genetics, Genome Diagnostics Laboratory, AUMC, 1105 AZ Amsterdam, The Netherlands;
     a.venema@amc.uva.nl (A.V.); m.a.mannens@amc.nl (M.M.A.M.M.)
[5]   Department of Pediatric Pulmonology and Allergology, Beatrix Children's Hospital, UMCG,
     9713 GZ Groningen, The Netherlands
[*]   Correspondence: p.henneman@amc.uva.nl; Tel.: +31 566-8833; Fax: +31-565-669-389

Received: 6 September 2018; Accepted: 23 October 2018; Published: 27 October 2018

Abstract: Cow's milk allergy (CMA) is an early-onset allergy of which the underlying genetic factors remain largely undiscovered. CMA has been found to co-occur with other allergies and immunological hypersensitivity disorders, suggesting a shared genetic etiology. We aimed to (1) investigate and (2) validate whether CMA children carry a higher genetic susceptibility for other immunological hypersensitivity disorders using polygenic risk score analysis (PRS) and prospective phenotypic data. Twenty-two CMA patients of the Dutch EuroPrevall birth cohort study and 307 reference subjects were genotyped using single nucleotide polymorphism (SNP) array. Differentially genetic susceptibility was estimated using PRS, based on multiple P-value thresholds for SNP inclusion of previously reported genome-wide association studies (GWAS) on asthma, autism spectrum disorder, atopic dermatitis, inflammatory bowel disease and rheumatoid arthritis. These associations were validated with prospective data outcomes during a six-year follow-up in 19 patients. We observed robust and significantly higher PRSs of asthma in CMA children compared to the reference set. Association analyses using the prospective data indicated significant higher PRSs in former CMA patients suffering from asthma and related traits. Our results suggest a shared genetic etiology between CMA and asthma and a considerable predictive sensitivity potential for subsequent onset of asthma which indicates a potential use for early clinical asthma intervention programs.

Keywords: allergic march; cow's milk allergy; genome-wide association; polygenic risk score

---

1. Background

Cow's milk allergy (CMA) is among the most frequent food allergies in young children [1]. An exact incidence, however, is difficult to establish, since it has been shown that large discrepancies exist between self-reported and proper clinically diagnosed CMA [2–4]. As with most other allergies, CMA has a complex and heterogeneous clinical presentation. The heritability of CMA is estimated at 15%, which, when compared to other (food) allergies, represents a moderate genetic component [5,6]. Although most young children develop tolerance for cow's milk proteins within a few years, these children seem to have an increased risk for developing other diseases involving a hypersensitive immune system, including asthma and inflammatory bowel disease (IBD) [6–10], which may suggest common genetic pathways between these diseases. While several genome-wide surveys on asthma, allergic rhinitis (AR) and atopic dermatitis (AD) have been reported, genome-wide association studies (GWAS) of food allergy (FA) are currently still limited [11–14]. Recently, we reported a candidate-gene association study in CMA-children and matched controls [15]. Although the results of the latter study favoured the "Allergic March" hypothesis, the number of studied variants was limited and the specific direction of the allergic march was narrowly defined [11,16]. Asthma, AD, and rheumatoid arthritis (RA) have previously been suggested to be related to CMA [17,18]. In addition to these immunological hypersensitivity related disorders, human studies also revealed a clear link between autism spectrum disorder (ASD) and asthma, and related immunological sensitivities which strongly suggest that allergic diseases are common in patients with ASD [19,20]. In 2013 Theoharides suggested that activation of brain mast cells and immune factors are associated with behavioural and language development, which may lead to focal immune reactions in the brain and subsequent focal encephalitis [21].

Large-scale GWAS studies have identified large numbers of single nucleotide polymorphisms (SNPs) related to behavioural and disease-related traits [22]. Moreover, studies have shown that the vast majority of these traits are highly polygenic, showing evidence of a complex genetic architecture composed of many genetic variants of low individual effect. However, the variability that can be explained by results from current GWAS studies is much lower than the actual heritability of these traits, commonly referred to as the 'missing heritability' [23]. Despite a large gap between variance explained by GWAS and the total variance of the trait, it has been shown that marginally significant SNPs that do not reach genome-wide significance ($P < 5 \times 10^{-8}$) contribute to the explained variance in the trait [24]. The genetic signals of these sub-threshold markers can be collectively captured by a polygenic risk score (PRS) that includes significant and non-significant markers, quantifying the genome-wide genetic predisposition of an individual for a specific trait [25,26]. In addition, such a genome-wide score analysis can also be used to study a shared genetic susceptibility across traits and diseases.

In this study, we explore the contribution of common genetic variants associated with asthma, ASD, AD, IBD, and AR, captured by PRS, to patients that have suffered from CMA by comparing the PRSs of CMA children with the PRS of healthy controls. Secondly, using six-year follow-up data from these children we aim to link the PRS for these traits to phenotypes related to asthma, AR and AD. Put together, the results of this study may provide new insight into a possible shared genetic architecture between CMA and these traits.

2. Methods

Sample collection: In this study, we included 22 children, suffering from CMA at intake and participating in the Dutch EuroPrevall Birth Cohort study. All 22 children became cow's milk tolerant within a time frame of two years after diagnosis. The EuroPrevall study is described in detail in Supplementary File S1 and previously by others [4,27,28]. Follow-up data of the Dutch EuroPrevall Birth Cohort study were acquired between 2015 and 2016, when the children reached the age of approximately six years old. The follow-up study involved questionnaires regarding (food) allergy related symptoms and, when indicated, further medical examination. Prospective data were obtained for symptoms indicating (1) asthma: wheezing, dyspnoea, (nightly) coughing, clinically diagnosed asthma and asthma medication use over the last 12 months, (2) allergic rhinitis (AR): permanent irritated nasal mucosa, burning sensation of the eyes, clinically diagnosed AR and use of AR medication over the last 12 months, (3) atopic dermatitis (AD): eczema and the (cutaneous) use of topical steroids, and (4) any food allergy. The present study only focused on the questionnaire outcomes related to asthma, AR and AD and their clinical diagnosis and/or the use of anti-allergy medication. Complete follow-up data was available for 19 former CMA patients. The selection of this subset of CMA cases was done irrespective of health status, i.e., no information of the above described health status was available at the time of genotyping.

All children described in our manuscript are participating in the Dutch EuroPrevall birth cohort study. The Medical Ethics Committee of the AUMC approved the Dutch EuroPrevall Birth Cohort Study (METC 2006/005 and METC prospective data 2014/056). Written informed consent, for both the study and genetic sampling, was obtained from both parents of each child, unless only one of them had parental rights. Within the context of a reference group, we selected 307 anonymous famine unexposed controls from the "Dutch Famine cohort". The subjects of this reference set were, like our CMA cases, born in the vicinity of Amsterdam, The Netherlands, and born in the period between 1946 and 1947, sampled over the last two decades. This reference set is assumed to represent the general Dutch population rather than screen negatives for any studied trait in this study. According previous reports on the Dutch Famine cohort, we assumed for allergies and related phenotypes, that this reference set reflects the general prevalence's in the Dutch population [29] For the reference set, the Medical Ethics Committee of the AUMC approved the study (METC 2001/215) and a written informed consent was obtained for each participant [29,30].

Genotyping, quality control, and genotype imputation and polygenic scoring: amplified genomic (see Supplementary File S1) and genomic DNA of patients and reference set, respectively, were submitted for microarray-based genotyping at GenomeScan B.V. (Leiden, The Netherlands), using the Affymetrix Axiom UKB WCSG-96 array (Santa Clara, CA, USA). Imputation of the cleaned genotype dataset was performed using IMPUTE2 [31]. PRSs were calculated based on genome-wide association (GWAS) results for the following five traits: (1) Asthma, reported by Moffatt et al. in 2010 [32], (2) ASD, reported by the Cross-Disorder Group of the Psychiatric Genomics Consortium in 2013 [33], (3) AD, reported by Paternoster et al. in 2015 [13], (4) IBD, reported by Liu et al. in 2015 [34] and (5) RA reported by Stahl et al. in 2010 [35]. In order to evaluate the optimal P-value threshold (P_T) for discriminating cases and the reference set, we calculated additive PRSs on a broad range of P-value thresholds for SNP inclusion ($P_T < 0.001$, $P_T < 0.005$, $P_T < 0.01$, $P_T < 0.05$, $P_T < 0.1$, $P_T < 0.5$, and $P_T < 1$). Polygenic scores were calculated by multiplying the number of risk alleles (0, 1, 2) by the SNP effect size (beta or log-transformed odds ratio). This study aims to detect genetic predisposition for several other adverse immunological outcomes after suffering from cow's milk allergy per se, in relation to the general Dutch population. In this context, the age difference between cases and control was assumed not to be a limiting factor and we considered the reference set in this study to be appropriate. Prior to statistical analyses, PRS values were standardized to a mean of 0 and standard deviation of 1. In this study, we analysed common variants, implying that after this standardization the risk alleles in the reference set would be normally distributed centred around 0. In the former CMA patients, for any trait analysis, a deviation of the mean from 0 indicates enrichment of risk alleles. Detailed descriptions of genotyping, quality control, imputation, polygenic risk scoring and further statistical analyses can be found/have been included in Supplementary File S1. Genotypes of the used SNPs, including imputation score (RSQ), of all included cases and the reference set and the appropriate meta-data that were used in this study are available on reasonable request and, according to the Dutch privacy law, only after a data transfer agreement.

3. Results

Genotype sample: The mean call rates in CMA cases and the reference set were 97.1% and 97.7% respectively (Table 1). Although in both the cases and the reference set we observed a single sample with a call rate lower than 95%, subsequent data evaluation did not show any other deviation of standard (GenABEL) SNP array quality checks in these samples. Therefore, no samples were excluded in further analyses. Imputation to the 1000 Genomes reference panel (phase1, v3 (20101123, GRCh37 / hg19) yielded 27,507,174 genetic variants. Post-imputation quality control, based on minor allele frequency (MAF > 0.01) and INFO score (INFO > 0.9), yielded a total number of 6,932,124 variants present in all samples which were available for calculating PRS.

Table 1. Characteristics cow's milk allergic children (CMA) and the reference set.

Genotype Data	N	Female	Male	Age	CR (mean)	CR (min)	CR (max)
CMA	22	6	16	11.8 ± 4.9 *	97.1%	93.0%	98.0%
Reference set	307	130	177	53.3 ± 0.58 **	97.7%	94.0%	99.0%

* Mean age $\pm$ Standard deviation (months) at blood drawing. ** Mean age $\pm$ standard deviation (years) at blood drawing. CR: call-rate.

Polygenic risk scores (PRS): PRS analyses on imputed genotype data of CMA and the reference set were performed using SNP sets obtained for each trait. The number of SNPs that entered each PRS analysis is described in Table S1. We observed a significantly higher PRS in CMA cases compared to the reference set for all SNP set *P*-value thresholds for asthma, with the exception of the *P*-value threshold of $P_T < 0.005$ that showed a similar direction of effect, but did not yield a significant differential association (Figure 1, Table 2). The two largest mean asthma PRS differences (δ) between cases and the reference set were observed for the *P*-value thresholds $P_T < 0.005$ ($\delta = 0.41$, OR = 1.50, 95% CI = 0.98–2.32, $P = 0.065$) and $P_T < 0.01$ ($\delta = 0.58$, OR = 1.85 95% CI = 1.16–2.94, $P = 0.009$) (Figure 1, Table 2). For the ASD PRS, we observed a significantly higher PRS for the SNP set $P_T < 0.001$ in CMA cases compared to the reference set. However, for all other ASD SNP sets we found opposite trends in CMA cases of which the PRS SNP set $P_T < 0.1$ was significantly lower ($P = 0.001$) in CMA patients compared to the reference set (Figure 1). Although we found no significant differences between cases and the reference set for all P_T thresholds of the AD PRS analyses, our analyses showed a consistent trend to a negative PRS in CMA cases compared to the reference set. Similar findings were observed for Inflammatory bowel disease SNP sets, consistently showing lower PRS in CMA cases than in the reference set (Figure 1, Table 2), although not reaching significance. Finally, for RA we observed suggestively higher PRS in CMA patients compared to the reference set, with the exception of the largest SNP set of $P_T < 1$ (Figure 1, Table 2).

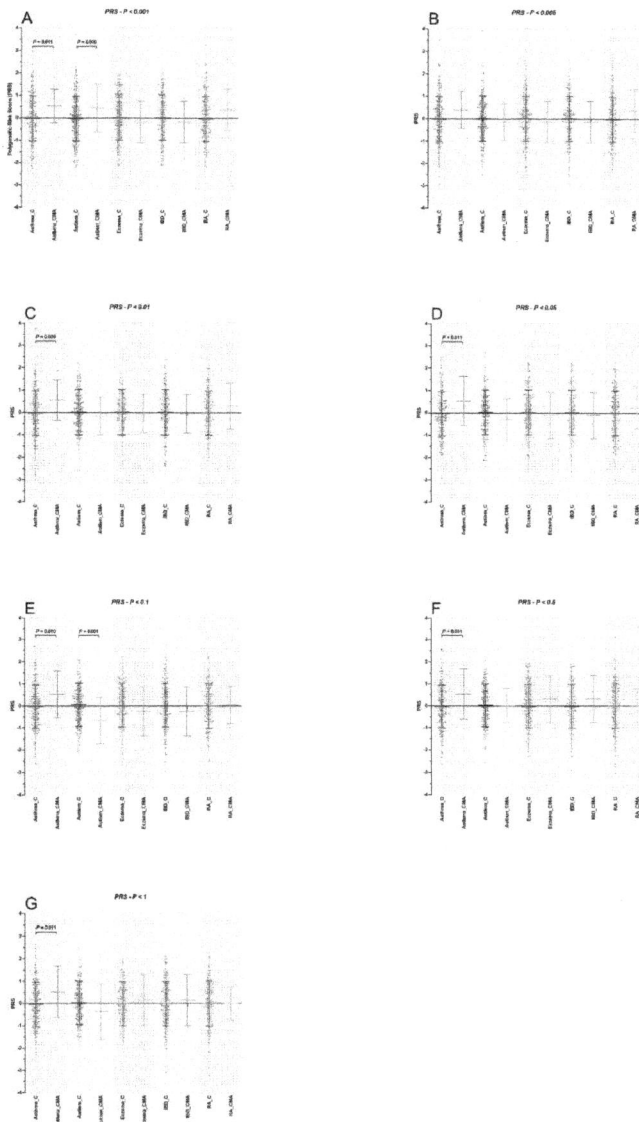

Figure 1. Polygenic risk scoring (weighted) for hypersensitive immune disorders asthma, autism spectrum disorder, atopic eczema, inflammatory bowel disease and allergic rhinitis (Prism 5.01, 2007). C: represents PRS in controls ($N = 307$), CMA represent PRS in (former) cow's milk allergic children ($N = 22$). (**A**) PRS scoring with (GWAS) cutoff of $P_T < 0.001$, (**B**) PRS scoring with (GWAS) cutoff of $P_T < 0.005$, (**C**) PRS scoring with (GWAS) cutoff of $P_T < 0.01$, (**D**) PRS scoring with (GWAS) cutoff of $P_T < 0.05$, (**E**) PRS scoring with (GWAS) cutoff of $P_T < 0.1$, (**F**) PRS scoring with (GWAS) cutoff of $P_T < 0.5$ and (**G**) PRS scoring with (GWAS) cutoff of $P_T < 1$. Parametric test (*t*-test) was performed to test for differences in means of the PRS between former CMA patients and healthy controls. We assumed a $P < 0.05$ statistically significant.

Table 2. Odds ratios polygenic risk score (PRS) analyses.

PRS Analysis	Asthma	ASD	AD	IBD	RA
$P_T < 0.001$	1.79 (1.14–2.84), **0.012**	1.59 (1.04–2.44), **0.032**	0.82 (0.53–1.28), 0.383	0.82 (0.53–1.28), 0.383	0.82 (0.53–1.28), 0.383
$P_T < 0.005$	1.50 (0.98–2.32), 0.065	0.85 (0.55–1.32), 0.481	0.86 (0.56–1.32), 0.483	0.86 (0.56–1.32), 0.483	0.86 (0.56–1.32), 0.483
$P_T < 0.01$	1.85 (1.16–2.94), **0.009**	0.83 (0.54–1.28), 0.410	0.94 (0.61–1.45), 0.780	0.94 (0.61–145), 0.780	0.94 (0.61–1.45), 0.780
$P_T < 0.05$	1.73 (1.13–2.27), **0.012**	0.73 (0.47–1.13), 0.161	0.89 (0.58–1.37), 0.596	0.89 (0.58–1.37), 0.596	0.89 (0.58–1.37), 0.596
$P_T < 0.1$	1.74 (1.13–2.66), **0.012**	0.45 (0.28–0.74), **0.002**	0.77 (0.50–1.18), 0.231	0.77 (0.50–1.18), 0.231	0.77 (0.50–1.18), 0.231
$P_T < 0.5$	1.74 (1.13–2.66), **0.011**	0.67 (0.43–1.05), 0.078	1.42 (0.91–2.23), 0.124	1.42 (0.91–2.23), 0.124	1.42 (0.91–2.23), 0.124
$P_T < 1$	1.73 (1.13–2.65), **0.012**	0.67 (0.43–1.04), 0.074	1.17 (0.76–1.81), 0.479	1.17 (0.76–1.81), 0.479	1.17 (0.76–1.81), 0.479

Odds ratio and the corresponding 95% confidence interval (95% CI) and corresponding *P*-value for all PRS analyses (logistic regression) based on standardized PRS values representing the effect for one standard deviation change in PRS. ASD: autism spectrum disorder; AD: atopic dermatitis; IBD: inflammatory bowel disease; RA: rheumatoid arthritis. P-values < 0.05 were assumed significant, annotated in bold.

Association of PRS with follow-up allergy related traits: Characteristics of the subset of CMA patients (N = 19, six females), obtained when they were around six years of age, are described in Table 3. In general, the highest incidence we observed concerned asthma-related and AD-related symptoms. In the latter context, we observed that diagnosed asthma, asthma medication use, AD and the use of topical steroids at the age of six years were present in one third of the former CMA patients (Table 3). Next, we performed association analysis on each symptom within the four prospective allergic disorder symptom groups, i.e., on asthma, AR, AD, and FA. Since our sample size was small, we limited the number of tests and performed these analyses only for a subset of the PRS outcomes, namely for asthma $P_T < 0.001$ and $P_T < 1$ and for ASD $P_T < 0.001$ and $P_T < 0.01$. We assumed that allergic disorders, as obtained in the prospective data, are not independent of each other, therefore, we assumed associations with a nominal *P*-value < 0.05 as significant. For the asthma PRS $P_T < 0.001$, we observed a positive PRS associated with nightly coughing (P = 0.02, Figure 2A and Table S2). However, it should be noted that both dyspnoea and clinical diagnosed asthma also showed a trend of a higher PRS. For all asthma PRS scores, with exception of $P_T < 0.005$, we have detected a significantly higher PRS in former CMA patients. In order to address the other end of the spectrum of SNP sets, we additionally analysed the asthma extreme SNP set of $P_T < 1$. Although not significant, this association showed similar trends with regard to the average PRS scores for the significantly associated symptoms in the asthma PRS $P_T < 0.001$. Furthermore, one of CMA patients that showed a positive IgE level within the first 2.5 years lifespan (Table 3) showed a value >1 for all the asthma PRSs, while the other IgE positive patient showed merely negative values. We observed that dyspnoea, nightly coughing and diagnosed asthma showed relatively high average PRS of 1.0, 1.0, and 0.8, respectively (Figure 2B and Table S2).

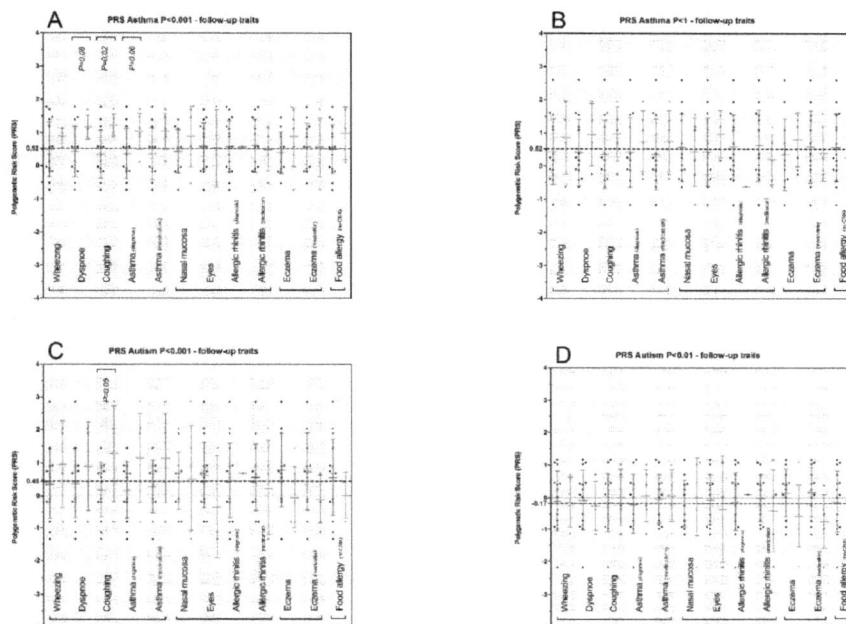

Figure 2. Polygenic risk scoring (weighted) for hypersensitive immune disorders asthma and autism spectrum disorder in relation to (prospectively obtained) hypersensitive immunological traits (Prism 5.01, 2007): (i) asthma related: wheezing; dyspnoe, (nightly) coughing; asthma clinically diagnosed; asthma medication (past 12 months), (ii) allergic rhinitis related: chronically irritated nasal mucosa, chronically irritated eyes, allergic rhinitis clinically diagnosed, Allergic rhinitis medication (past 12 months), (iii) atopic dermatitis: eczema; topical steroids use (past 12 months), (iiii) food allergy. (**A**) Asthma PRS scoring with (GWAS) cutoff of P_T <0.001 and (**B**) asthma PRS scoring with (GWAS) cutoff of P_T < 1. (**C**) ASD PRS scoring with (GWAS) cutoff of P_T < 0.001 and (**D**) ASD PRS scoring with (GWAS) cutoff of P_T < 0.01. ANOVA was performed to test for differences in means of the PRS between CMA patients scored for particular symptom or not. Red indicates no, green indicates yes. We assumed a P < 0.05 statistically significant.

Table 3. Six year follow-up characteristics former CMA patients.

General	
Total (*N*)	19
Female (*N*)	6
Male (*N*)	13
Age (mean ± SD)	7.0 ± 1.0 *
2 year IgE positive **	2
Asthma related	*N* (%)
Wheezing	4 (21.1%)
Dyspnoea	4 (21.1%)
Coughing at night	5 (26.3%)
Asthma diagnosed	6 (33.3%)
Asthma medication	6 (33.3%)
Allergic rhinitis related	*N* (%)
Irritated nasal mucosa)	6 (33.3%)
Eyes	3 (15.8%)
Allergic rhinitis diagnosed	1 (5.3%)
Allergic rhinitis medication	5 (26.3%)

Table 3. *Cont.*

General	
Atopic dermatitis related	*N* (%)
Eczema	7 (36.8%)
Topical steroids	6 (31.6%)
Allergy related	*N* (%)
Food allergy	3 (15.8%)

* Mean age ± Standard deviation (years) at the time of the follow study. ** Number of CMA patients that became IgE positive (IgE < 0.35 kU/L) during the first 2.5 years of life, data on IgE levels was available for 16 of 19 former CMA patients.

For ASD we selected *P*-value thresholds that were significant ($P_T < 0.001$ and $P_T < 0.01$) to test for association with prospective data. For the PRS based on $P_T < 0.001$, we observed significant association with nightly coughing ($P = 0.05$) and a trend ($P = 0.07$) with clinical diagnosed asthma (Figure 2C and Table S3). The PRS in the former CMA cases using the threshold of $P_T < 0.001$ showed positive associations, while in all other *P*-value thresholds we observed protective associations of the scores (negative δ). In the prospective data analyses, we found no evidence of any association of the PRS obtained from the SNP set of $P_T < 0.01$ with any symptom of the four disorders (Figure 2D and Table S3).

4. Discussion

To date, literature on the genetic architecture of common variants underlying CMA is limited to candidate gene/variant approaches, and the present study is the first to describe cross-trait association between a large number of common variants and CMA. Despite the fact that genome-wide surveys are absent and, therefore, the specific genetic architecture of CMA remains unclear, the phenotypic relation with other types of allergy have been studied widely [11,15,36]. Recently, our group and others have hypothesized that food allergy susceptibility, including CMA, involves an epigenetic component as well [37,38]. The latter hypothesis was strengthened by the fact that a genetic component cannot explain the vast increase of food allergy prevalence world-wide [7]. The way food is currently processed in comparison to the pre-industrialization period has been hypothesized to involve exposure to new antigens that might underlie the recent observed increase in allergic sensitization(s) [39].

On the other hand, it should be noted that clinical diagnosis of food allergies has also been improved dramatically over the last decades, which may have resulted in substantial lower number of misdiagnosis and, thus, an improved estimate of its prevalence. Moreover, given the fact that nowadays food allergy has been acknowledged as a public health issue, its diagnosis has become common practice compared to 50+ years ago, which probably also contributed to the increase of food allergies prevalence [40,41]. Furthermore, without excluding an epigenetic component contributing to the expression of food allergy, the genetic component especially for CMA might very well resemble or overlap with the genetic component of other more common types of later onset allergies, and might, therefore, represent a strong predictor for later onset hypersensitive immune disorders.

Phenotypically, this phenomenon has been very well covered by the "allergic march" hypothesis, which was recently studied extensively by Alduraywish et al. [42]. Genetic mechanisms underlying their observations were addressed in the present study, using PRS that included common variants detected in/for other types of hypersensitive immune disorders: asthma, AR and AD. Moreover, the use of prospective data on allergy and allergic symptoms in our CMA cohort can be considered as validation tool of our findings with regard to the risk of these later onset allergies. In this context, we were able to detect consistent significantly higher PRSs for asthma for virtually all SNP *P*-value thresholds compared to the reference set, but not for AR or AD. Moreover, our asthma based findings reflected relatively high odds ratios for these PRS. These findings strongly suggest that the genetic predisposition for later onset of asthma in children with a history of CMA involves a strong overlapping genetic component that is not reflected by positive IgE plasma levels within the first 2.5 years. In order to validate these findings on the asthma PRS analyses, the availability of prospective data is essential. Our former CMA patients were followed up after approximately six years. The timing of this follow-up with regard to the manifestation of asthma or asthmatic symptoms is in line with earlier reports on disease onsets covered by the allergic march hypothesis [42].

Although we cannot conclude that asthma and asthma-related symptoms are more common in former CMA patients (i.e., we did not include CMA screen negatives in this context), our data does indicate that the genetic load of asthma associated common variants is enriched in the former CMA patients that suffer from asthma and asthma-related phenotypes. In the context of a potential prediction of later onset asthma in CMA patients, our symptom association results, obtained with the *P*-value threshold of $P_T < 0.001$, are substantial. All former CMA cases that were diagnosed at a later age with asthma and/or reported nightly coughing and/or dyspnoea showed, without exception, relatively high PRS scores. The delta mean PRS difference between these patients was for dyspoea ($\delta = 0.74$), for nightly coughing ($\delta = 0.88$) and for asthma ($\delta = 0.68$). A considerable number of former CMA patients showed a high PRS, but showed so far not (yet) any asthmatic symptoms. Whether these patients will develop asthmatic symptoms at a later age, should be monitored in additional follow-up studies. In case such follow-ups report that these high PRS former CMA patients do develop symptoms, this would result in a strong specificity measure as well, implying that genetic profiling of CMA patients may be a powerful tool in predicting asthmatic symptoms in later life, possibly leading to preventive interventions in the future.

Accumulating evidence shows that a hypersensitive immune system, reflected by asthma or allergies, is associated with ASD [20]. Moreover, gastrointestinal dysfunction has also been associated with autoimmunity and ASD [19]. At the same time, gastrointestinal dysfunction, reflected for example by IBD, in relation to asthma has also been previously described [43]. In order to further explore this possible shared genetic component between allergies, asthma, ASD and IBD in our study, we performed additional PRS analyses based on/using SNP sets obtained from recently published ASD and IBD GWAS [34]. Our results yielded suggestive evidence of a shared genetic component of the smallest SNP set ($P_T < 0.001$) of ASD in former CMA patients. Surprisingly, the other ASD *P*-value threshold, especially $P_T < 0.01$, showed an opposite effect, reflecting an ASD favorable genetic component in former CMA patients (Figure 2C,D, Table S3). Although we cannot rule out the possibility that only a (small) subset of risk alleles may indeed be shared in CMA and ASD, these conflicting observations should be interpreted with caution. Further discussion on limitations and strengths of our study are described in Supplementary File S1.

Future studies should aim to include larger samples of CMA cases, which allow sufficient statistical power to uncover the genetic architecture of CMA. Identification of variants and genes related to CMA may aid in identifying overlapping risk loci between CMA, allergies and immune-related traits. In conclusion, to the best of our knowledge, this is the first study to investigate polygenic risk scores and cross-trait genetic liability in CMA patients, using genome-wide SNP sets from GWAS on immune-related traits. Our PRS analyses in former CMA patients show a significantly higher PRS of immune-related traits in patients compared to the reference set and, thus, a shared genetic susceptibility of CMA and asthma and asthmatic symptoms. Moreover, on the basis of the prospective data in our former CMA patients, we found a strong indication that our PRS of asthma might contribute to an accurate prediction of later onset asthma in these patients. Improving insight into the link between genetic predisposition for immune-related traits/disorders might lead to early clinical intervention in order to prevent the manifestation of asthma or limit its severity later in life.

Supplementary Materials: The following are available online at http://www.mdpi.com/2072-6643/10/11/1582/s1, Supplementary File S1: supplementary methods and supplementary discussion, Table S1. Number of included SNPs (MAF > 0.01) in polygenic risk score (PRS) analyses, Table S2: Association analyses of asthma *P* < 0.001 and *P* < 1 polygenic risk score (PRS) per follow-up symptom outcome, Table S3: Association analyses of autism *P* < 0.001 and *P* < 0.1 polygenic risk score (PRS) per follow-up symptom outcome.

Nutrients **2018**, *10*, 1582

Author Contributions: Authorship was assigned on the basis of the contributions of each author to the work described in the manuscript: Conceptualization: P.H., P.R.J., N.C.M.P. and D.P.; methodology: P.H., D.P. and P.R.J.; software: P.H., A.V. and P.R.J.; validation: N.C.M.P. and P.H.; formal analysis: P.H., A.V. and P.R.J.; investigation: P.H., P.R.J. and N.C.M.P.; resources: N.C.M.P. and A.B.S.; data curation: P.H., P.R.J., A.V. and N.C.M.P.; original draft preparation: P.H. and P.R.J.; review and editing: P.R.J., P.H., A.V., D.P., N.C.M.P., A.B.S. and M.M.A.M.M.; visualization: P.H.; supervision: P.H.; project administration: P.H.; funding acquisition: M.M.A.M.M., D.P. and A.B.S.

Funding: This research received no external funding.

Acknowledgments: We would like to thank the children, parents and all staff involved in clinical data collection for their participation. P.R. Jansen is supported by the Stichting Vrienden van het Sophia ('Friends of Sophia' foundation, project number: S14–27). We would like to thank Ir Michiel Adriaens, MACSBIO, Sciences, Faculty of Humanities and Sciences, Maastricht University, The Netherlands, for his assistance and guidance with the genotype imputations.

Conflicts of Interest: The authors declare no conflict of interest.

References

1. Gerrard, J.W.; MacKenzie, J.W.; Goluboff, N.; Garson, J.Z.; Maningas, C.S. Cow's milk allergy: Prevalence and manifestations in an unselected series of newborns. *Acta Paediatr. Scand. Suppl.* **1973**, *234*, 3–21. [CrossRef] [PubMed]
2. Petrus, N.C.; Schoemaker, A.F.; van Hoek, M.W.; Jansen, L.; Jansen-van der Weide, M.C.; van Aalderen, W.M.; Sprikkelman, A.B. Remaining symptoms in half the children treated for milk allergy. *Eur. J. Pediatr.* **2015**, *174*, 759–765. [CrossRef] [PubMed]
3. Chafen, J.J.; Newberry, S.J.; Riedl, M.A.; Bravata, D.M.; Maglione, M.; Suttorp, M.J.; Sundaram, V.; Paige, N.M.; Towfigh, A.; Hulley, B.J.; et al. Diagnosing and managing common food allergies: A systematic review. *JAMA* **2010**, *303*, 1848–1856. [CrossRef] [PubMed]
4. Schoemaker, A.A.; Sprikkelman, A.B.; Grimshaw, K.E.; Roberts, G.; Grabenhenrich, L.; Rosenfeld, L.; Siegert, S.; Dubakiene, R.; Rudzeviciene, O.; Reche, M.; et al. Incidence and natural history of challenge-proven cow's milk allergy in european children—Europrevall birth cohort. *Allergy* **2015**, *70*, 963–972. [CrossRef] [PubMed]
5. Tsai, H.J.; Kumar, R.; Pongracic, J.; Liu, X.; Story, R.; Yu, Y.; Caruso, D.; Costello, J.; Schroeder, A.; Fang, Y.; et al. Familial aggregation of food allergy and sensitization to food allergens: A family-based study. *Clin. Exp. Allergy* **2009**, *39*, 101–109. [CrossRef] [PubMed]
6. Benhamou, A.H.; Schappi Tempia, M.G.; Belli, D.C.; Eigenmann, P.A. An overview of cow's milk allergy in children. *Swiss Med. Wkly.* **2009**, *139*, 300–307. [PubMed]
7. Host, A.; Halken, S. A prospective study of cow milk allergy in danish infants during the first 3 years of life. Clinical course in relation to clinical and immunological type of hypersensitivity reaction. *Allergy* **1990**, *45*, 587–596. [CrossRef] [PubMed]
8. Host, A.; Halken, S.; Jacobsen, H.P.; Christensen, A.E.; Herskind, A.M.; Plesner, K. Clinical course of cow's milk protein allergy/intolerance and atopic diseases in childhood. *Pediatr. Allergy Immunol.* **2002**, *13* (Suppl. 15), 23–28. [CrossRef] [PubMed]
9. Morita, H.; Nomura, I.; Matsuda, A.; Saito, H.; Matsumoto, K. Gastrointestinal food allergy in infants. *Allergol. Int.* **2013**, *62*, 297–307. [CrossRef] [PubMed]
10. Saarinen, K.M.; Pelkonen, A.S.; Makela, M.J.; Savilahti, E. Clinical course and prognosis of cow's milk allergy are dependent on milk-specific ige status. *J. Allergy Clin. Immunol.* **2005**, *116*, 869–875. [CrossRef] [PubMed]
11. Bonnelykke, K.; Sparks, R.; Waage, J.; Milner, J.D. Genetics of allergy and allergic sensitization: Common variants, rare mutations. *Curr. Opin. Immunol.* **2015**, *36*, 115–126. [CrossRef] [PubMed]
12. Marenholz, I.; Esparza-Gordillo, J.; Ruschendorf, F.; Bauerfeind, A.; Strachan, D.P.; Spycher, B.D.; Baurecht, H.; Margaritte-Jeannin, P.; Saaf, A.; Kerkhof, M.; et al. Meta-analysis identifies seven susceptibility loci involved in the atopic march. *Nat. Commun.* **2015**, *6*, 8804. [CrossRef] [PubMed]
13. Paternoster, L.; Standl, M.; Waage, J.; Baurecht, H.; Hotze, M.; Strachan, D.P.; Curtin, J.A.; Bonnelykke, K.; Tian, C.; Takahashi, A.; et al. Multi-ancestry genome-wide association study of 21,000 cases and 95,000 controls identifies new risk loci for atopic dermatitis. *Nat. Genet.* **2015**, *47*, 1449–1456. [CrossRef] [PubMed]

14. Bunyavanich, S.; Schadt, E.E.; Himes, B.E.; Lasky-Su, J.; Qiu, W.; Lazarus, R.; Ziniti, J.P.; Cohain, A.; Linderman, M.; Torgerson, D.G.; et al. Integrated genome-wide association, coexpression network, and expression single nucleotide polymorphism analysis identifies novel pathway in allergic rhinitis. *BMC Med. Genom.* **2014**, *7*, 48. [CrossRef] [PubMed]

15. Henneman, P.; Petrus, N.C.; Venema, A.; van Sinderen, F.; van der Lip, K.; Hennekam, R.C.; Mannens, M.; Sprikkelman, A.B. Genetic susceptibility for cow's milk allergy in dutch children: The start of the allergic march? *Clin. Transl. Allergy* **2015**, *6*, 7. [CrossRef] [PubMed]

16. Ramasamy, A.; Curjuric, I.; Coin, L.J.; Kumar, A.; McArdle, W.L.; Imboden, M.; Leynaert, B.; Kogevinas, M.; Schmid-Grendelmeier, P.; Pekkanen, J.; et al. A genome-wide meta-analysis of genetic variants associated with allergic rhinitis and grass sensitization and their interaction with birth order. *J. Allergy Clin. Immunol.* **2011**, *128*, 996–1005. [CrossRef] [PubMed]

17. Ierodiakonou, D.; Garcia-Larsen, V.; Logan, A.; Groome, A.; Cunha, S.; Chivinge, J.; Robinson, Z.; Geoghegan, N.; Jarrold, K.; Reeves, T.; et al. Timing of allergenic food introduction to the infant diet and risk of allergic or autoimmune disease: A systematic review and meta-analysis. *JAMA* **2016**, *316*, 1181–1192. [CrossRef] [PubMed]

18. Alduraywish, S.A.; Standl, M.; Lodge, C.J.; Abramson, M.J.; Allen, K.J.; Erbas, B.; von Berg, A.; Heinrich, J.; Lowe, A.J.; Dharmage, S.C. Is there a march from early food sensitization to later childhood allergic airway disease? Results from two prospective birth cohort studies. *Pediatr. Allergy Immunol.* **2017**, *28*, 30–37. [CrossRef] [PubMed]

19. Brown, A.C.; Mehl-Madrona, L. Autoimmune and gastrointestinal dysfunctions: Does a subset of children with autism reveal a broader connection? *Expert Rev. Gastroenterol. Hepatol.* **2011**, *5*, 465–477. [CrossRef] [PubMed]

20. Lyall, K.; Van de Water, J.; Ashwood, P.; Hertz-Picciotto, I. Asthma and allergies in children with autism spectrum disorders: Results from the charge study. *Autism Res.* **2015**, *8*, 567–574. [CrossRef] [PubMed]

21. Theoharides, T.C. Is a subtype of autism an allergy of the brain? *Clin. Ther.* **2013**, *35*, 584–591. [CrossRef] [PubMed]

22. Visscher, P.M.; Brown, M.A.; McCarthy, M.I.; Yang, J. Five years of gwas discovery. *Am. J. Hum. Genet.* **2012**, *90*, 7–24. [CrossRef] [PubMed]

23. Eichler, E.E.; Flint, J.; Gibson, G.; Kong, A.; Leal, S.M.; Moore, J.H.; Nadeau, J.H. Missing heritability and strategies for finding the underlying causes of complex disease. *Nat. Rev. Genet.* **2010**, *11*, 446–450. [CrossRef] [PubMed]

24. Schizophrenia Working Group of the Psychiatric Genomics Consortium. Biological insights from 108 schizophrenia-associated genetic loci. *Nature* **2014**, *511*, 421–427. [CrossRef] [PubMed]

25. Wray, N.R.; Lee, S.H.; Mehta, D.; Vinkhuyzen, A.A.; Dudbridge, F.; Middeldorp, C.M. Research review: Polygenic methods and their application to psychiatric traits. *J. Child Psychol. Psychiatry* **2014**, *55*, 1068–1087. [CrossRef] [PubMed]

26. Dudbridge, F. Power and predictive accuracy of polygenic risk scores. *PLoS Genet.* **2013**, *9*, e1003348. [CrossRef]

27. Keil, T.; McBride, D.; Grimshaw, K.; Niggemann, B.; Xepapadaki, P.; Zannikos, K.; Sigurdardottir, S.T.; Clausen, M.; Reche, M.; Pascual, C.; et al. The multinational birth cohort of europrevall: Background, aims and methods. *Allergy* **2010**, *65*, 482–490. [CrossRef] [PubMed]

28. McBride, D.; Keil, T.; Grabenhenrich, L.; Dubakiene, R.; Drasutiene, G.; Fiocchi, A.; Dahdah, L.; Sprikkelman, A.B.; Schoemaker, A.A.; Roberts, G.; et al. The europrevall birth cohort study on food allergy: Baseline characteristics of 12,000 newborns and their families from nine european countries. *Pediatr. Allergy Immunol.* **2012**, *23*, 230–239. [CrossRef] [PubMed]

29. Lopuhaa, C.E.; Roseboom, T.J.; Osmond, C.; Barker, D.J.; Ravelli, A.C.; Bleker, O.P.; van der Zee, J.S.; van der Meulen, J.H. Atopy, lung function, and obstructive airways disease after prenatal exposure to famine. *Thorax* **2000**, *55*, 555–561. [CrossRef] [PubMed]

30. Roseboom, T.J.; van der Meulen, J.H.; Ravelli, A.C.; van Montfrans, G.A.; Osmond, C.; Barker, D.J.; Bleker, O.P. Blood pressure in adults after prenatal exposure to famine. *J. Hypertens.* **1999**, *17*, 325–330. [CrossRef] [PubMed]

31. Howie, B.; Marchini, J.; Stephens, M. Genotype imputation with thousands of genomes. *G3 (Bethesda)* **2011**, *1*, 457–470. [CrossRef] [PubMed]

32. Moffatt, M.F.; Gut, I.G.; Demenais, F.; Strachan, D.P.; Bouzigon, E.; Heath, S.; von Mutius, E.; Farrall, M.; Lathrop, M.; Cookson, W.O.; et al. A large-scale, consortium-based genomewide association study of asthma. *N. Engl. J. Med.* **2010**, *363*, 1211–1221. [CrossRef] [PubMed]

33. Cross-Disorder Group of the Psychiatric Genomics, Consortium. Identification of risk loci with shared effects on five major psychiatric disorders: A genome-wide analysis. *Lancet* **2013**, *381*, 1371–1379.

34. Liu, J.Z.; van Sommeren, S.; Huang, H.; Ng, S.C.; Alberts, R.; Takahashi, A.; Ripke, S.; Lee, J.C.; Jostins, L.; Shah, T.; et al. Association analyses identify 38 susceptibility loci for inflammatory bowel disease and highlight shared genetic risk across populations. *Nat. Genet.* **2015**, *47*, 979–986. [CrossRef] [PubMed]

35. Stahl, E.A.; Raychaudhuri, S.; Remmers, E.F.; Xie, G.; Eyre, S.; Thomson, B.P.; Li, Y.; Kurreeman, F.A.; Zhernakova, A.; Hinks, A.; et al. Genome-wide association study meta-analysis identifies seven new rheumatoid arthritis risk loci. *Nat. Genet.* **2010**, *42*, 508–514. [CrossRef] [PubMed]

36. Matsumoto, K.; Saito, H. Epicutaneous immunity and onset of allergic diseases—per-"eczema"tous sensitization drives the allergy march. *Allergol. Int.* **2013**, *62*, 291–296. [CrossRef] [PubMed]

37. Tsakok, T.; Du Toit, G.; Lack, G. Prevention of food allergy. *Chem. Immunol. Allergy* **2015**, *101*, 253–262. [PubMed]

38. Petrus, N.C.; Henneman, P.; Venema, A.; Mul, A.; van Sinderen, F.; Haagmans, M.; Mook, O.; Hennekam, R.C.; Sprikkelman, A.B.; Mannens, M. Cow's milk allergy in dutch children: An epigenetic pilot survey. *Clin. Transl. Allergy* **2016**, *6*, 16. [CrossRef] [PubMed]

39. Borad, S.G.; Kumar, A.; Singh, A.K. Effect of processing on nutritive values of milk protein. *Crit. Rev. Food Sci. Nutr.* **2017**, *57*, 3690–3702. [CrossRef] [PubMed]

40. Sicherer, S.H.; Sampson, H.A. Food allergy: A review and update on epidemiology, pathogenesis, diagnosis, prevention, and management. *J. Allergy Clin. Immunol.* **2018**, *141*, 41–58. [CrossRef] [PubMed]

41. Loh, W.; Tang, M.L.K. The epidemiology of food allergy in the global context. *Int. J. Environ. Res. Public Health* **2018**, *15*, 2043. [CrossRef] [PubMed]

42. Alduraywish, S.A.; Lodge, C.J.; Campbell, B.; Allen, K.J.; Erbas, B.; Lowe, A.J.; Dharmage, S.C. The march from early life food sensitization to allergic disease: A systematic review and meta-analyses of birth cohort studies. *Allergy* **2016**, *71*, 77–89. [CrossRef] [PubMed]

43. Park, J.H.; Jeong, D.Y.; Peyrin-Biroulet, L.; Eisenhut, M.; Shin, J.I. Insight into the role of tslp in inflammatory bowel diseases. *Autoimmun. Rev.* **2017**, *16*, 55–63. [CrossRef] [PubMed]

nutrients

MDPI

Article

Raw Cow's Milk Reduces Allergic Symptoms in a Murine Model for Food Allergy—A Potential Role for Epigenetic Modifications

Suzanne Abbring [1], Johanna Wolf [2], Veronica Ayechu-Muruzabal [1], Mara A.P. Diks [1], Bilal Alashkar Alhamwe [2,3], Fahd Alhamdan [2], Hani Harb [2], Harald Renz [2], Holger Garn [2], Johan Garssen [1,4], Daniel P. Potaczek [2,5] and Betty C.A.M. van Esch [1,4,*]

1   Division of Pharmacology, Utrecht Institute for Pharmaceutical Sciences, Faculty of Science, Utrecht University, 3584 CG Utrecht, The Netherlands
2   Institute of Laboratory Medicine, Member of the German Center for Lung Research (DZL) and the Universities of Giessen and Marburg Lung Center (UGMLC), 35043 Marburg, Germany
3   College of Pharmacy, International University for Science and Technology (IUST), Daraa 15, Syria
4   Danone Nutricia Research, 3584 CT Utrecht, The Netherlands
5   John Paul II Hospital, 31-202 Krakow, Poland
*   Correspondence: e.c.a.m.vanesch@uu.nl; Tel.: +31-625732735

Received: 14 June 2019; Accepted: 23 July 2019; Published: 25 July 2019

Abstract: Epidemiological studies identified raw cow's milk consumption as an important environmental exposure that prevents allergic diseases. In the present study, we investigated whether raw cow's milk has the capacity to induce tolerance to an unrelated, non-milk, food allergen. Histone acetylation of T cell genes was investigated to assess potential epigenetic regulation. Female C3H/HeOuJ mice were sensitized and challenged to ovalbumin. Prior to sensitization, the mice were treated with raw milk, processed milk, or phosphate-buffered saline for eight days. Allergic symptoms were assessed after challenge and histone modifications in T cell-related genes of splenocyte-derived CD4$^+$ T cells and the mesenteric lymph nodes were analyzed after milk exposure and after challenge. Unlike processed milk, raw milk decreased allergic symptoms. After raw milk exposure, histone acetylation of Th1-, Th2-, and regulatory T cell-related genes of splenocyte-derived CD4$^+$ T cells was higher than after processed milk exposure. After allergy induction, this general immune stimulation was resolved and histone acetylation of Th2 genes was lower when compared to processed milk. Raw milk reduces allergic symptoms to an unrelated, non-milk, food allergen in a murine model for food allergy. The activation of T cell-related genes could be responsible for the observed tolerance induction, which suggested that epigenetic modifications contribute to the allergy-protective effect of raw milk.

Keywords: epigenetics; farming effect; food allergy; histone acetylation; milk processing; raw milk

1. Introduction

Allergic diseases are a growing public health concern. In the previous decades, their prevalence has increased to such an extent that, nowadays, 20 to 30% of the world's population is suffering from some form of allergic disease [1]. With a severe impact on quality of life and extensive healthcare costs, the vast prevalence of allergic diseases has major socio-economic consequences [2]. Unfortunately, to date, there is neither a cure nor an effective and safe treatment. Allergy management focuses on allergen avoidance and symptomatic treatment with the self-administration of epinephrine in the case of systemic anaphylaxis upon accidental exposure.

Even though there are no effective preventive approaches for allergic diseases, there seems to be a natural solution. Several epidemiological studies have shown that children growing up on a farm have a reduced risk of developing asthma and allergies compared to children living in the same rural area but not growing up on a farm [3–7]. This protective 'farm effect' was demonstrated in many populations and it persisted into adult life [8]. Farm exposures that were associated with this allergy-protective effect appeared to be contact with livestock and animal feed, exposure to stables and barns, and consumption of raw, unprocessed, cow's milk [9–11]. Especially, the consumption of raw cow's milk is of importance, since its protective effect was found to be independent of farm status, giving it the potential to confer protection for a general, non-farming, population [9,10,12,13]. Recently, these epidemiological findings were confirmed by showing a causal relationship between raw cow's milk consumption and the prevention of allergic asthma in a murine model [14].

How raw cow's milk can be allergy protective is currently still unclear. Neither the protective raw milk constituents nor the underlying mechanisms are known. Heat-sensitive milk components, like immunoglobulins, lactoferrin, alkaline phosphatase, TGF-β, microRNAs, etc., are likely candidates, since epidemiological as well as preclinical studies have shown that milk processing, and particularly heating, abolishes the allergy-protective effect of raw cow's milk consumption [13–16]. However, the actual bioactive component(s) involved remain to be elucidated. Regarding the underlying mechanisms, several of the bioactive components that are present in raw milk are theoretically able to create a tolerogenic environment by, for example, promoting regulatory T cell development, enhancing epithelial barrier function and modulating the gut microbiome, however, none of these effects were actually investigated after drinking raw milk [17,18].

An emerging field is the contribution of epigenetic modifications in regulating the development of allergic diseases. Allergic diseases are the result of a complex interplay between the genes and environmental factors. These environmental factors can influence gene expression via epigenetic mechanisms, such as DNA methylation and histone modifications [19,20]. Epigenetic modifications are reversible and they affect the accessibility of the DNA to transcription enzymes, thereby regulating gene expression [19]. Environmental factors and components recently gaining interest in this regard are microbes, obesity, stress, and tobacco smoke, but it has also been suggested that nutrients might exert their effects through epigenetic mechanisms [19,21]. This indicates that epigenetic regulation might also be involved in the allergy-protective effect of raw cow's milk consumption.

Before certified raw cow's milk (raw cow's milk obtained from a farm that is legally allowed to sell raw milk [22]) can become part of a preventive approach for allergic diseases, compelling evidence that thoroughly investigates components and mechanisms that are involved is needed. As a first step, the many epidemiological studies showing an allergy-protective effect of raw cow's milk consumption need to be strengthened by causal evidence. In a previous study, we were able to show causality in a murine house dust mite-induced asthma model [14]. With the current research, we aimed to assess whether raw cow's milk has the capacity to induce tolerance to an unrelated, non-milk, food allergen. Besides, we studied the contribution of epigenetic regulation by assessing histone acetylation of T cell-related genes, as a potential mechanism underlying the protective effects.

2. Materials and Methods

2.1. Animals

Specific pathogen-free, three- to five-week-old, female C3H/HeOuJ mice were purchased (Charles River Laboratories, Sulzfeld, Germany) and were randomly allocated to the control or experimental groups. The mice were housed in filter-topped makrolon cages (one cage/group, *n* = 6–8/cage) with standard chip bedding, Kleenex tissues, and a plastic shelter on a 12 h light/dark cycle with unlimited access to food ('Rat and Mouse Breeder and Grower Expanded'; Special Diet Services, Witham, UK) and water at the animal facility of Utrecht University (Utrecht, The Netherlands). All animal procedures were approved by the Ethical Committee for Animal Research of the Utrecht University and were complied with the European Directive on the protection of animals used for scientific purposes (DEC 2014.II.12.107 & AVD108002015346).

2.2. Experimental Design—Tolerance Induction, Sensitization and Challenges

After an acclimatization period of one week, the mice were orally treated (i.e., intragastrically (i.g.) by using a blunt needle) with 0.5 mL certified raw, unprocessed, cow's milk (Hof Dannwisch, Horst, Germany), processed shop milk (full fat milk, 3.5%; EDEKA, Germany), or phosphate-buffered saline (PBS; as a control) for eight consecutive days (days −9 to −2). Following this oral tolerance induction period, mice were sensitized i.g. once a week for five weeks to the hen's egg protein ovalbumin (OVA; 20 mg/0.5 mL PBS; grade V; Sigma-Aldrich, Zwijndrecht, The Netherlands) while using 10 µg cholera toxin (CT; List Biological Laboratories, Campbell, CA, USA) as an adjuvant (days 0, 7, 14, 21, 28; *n* = 8/group). Sham-sensitized control mice (*n* = 6) received CT alone (10 µg/0.5 mL PBS). Five days after the last sensitization (day 33), all of the mice were intradermally (i.d.) challenged in both ear pinnae with 10 µg OVA in 20 µL PBS to determine acute allergic symptoms. Mice were subsequently i.g. challenged (7 h after the i.d. challenge) with 50 mg OVA in 0.5 mL PBS. Sixteen hours later (day 34), blood samples were taken via cheek puncture and mice were killed by cervical dislocation. The spleens were then collected for ex vivo analysis. Additional groups of mice (*n* = 6/group) were used in a follow-up experiment to assess the involvement of epigenetic regulation. These mice were killed by cervical dislocation either one day after the oral tolerance induction period (day −1) or one day after both challenges (day 34). Figure 1 shows a schematic representation of the experimental timeline.

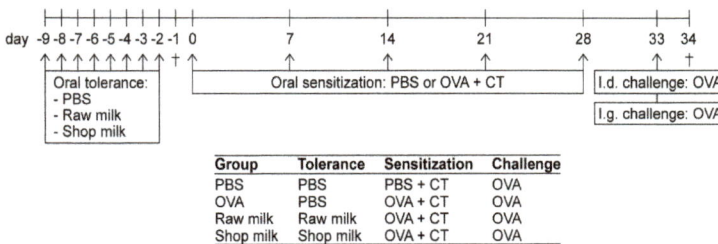

Figure 1. Schematic representation of the experimental timeline. For epigenetic measurements, additional groups of mice were killed after the tolerance induction period (day −1) and after both challenges (day 34; as indicated by †). PBS, phosphate-buffered saline; OVA, ovalbumin; CT, cholera toxin; i.d., intradermal; i.g., intragastric.

2.3. Assessment of the Acute Allergic Response

The acute allergic skin response, anaphylactic shock symptoms, and body temperature were evaluated by a researcher blinded to treatment upon i.d. challenge with OVA (10 µg OVA/20 µL PBS) in the ear pinnae of both ears to determine the severity of the acute allergic symptoms. The acute allergic skin response was measured as Δ ear swelling (µm) by subtracting the mean ear thickness before i.d. challenge from the mean ear thickness 1 h after i.d. challenge. Ear thickness at both of the timepoints was measured in duplicate for each ear using a digital micrometer (Mitutoyo, Veenendaal, The Netherlands). The mice were anesthetized using inhalation of isoflurane to perform the i.d. challenge as well as the ear measurements (Abbott, Breda, The Netherlands). The severity of anaphylactic shock symptoms was determined 30 min after i.d. challenge by using a previously described, validated, scoring table [23]. Body temperature was also measured 30 min after i.d. challenge (using a rectal thermometer) to monitor the anaphylactic shock-induced drop in body temperature.

2.4. Detection of OVA-Specific IgE and mMCP-1 in Serum

Blood was collected via cheek puncture 16 h after i.g. challenge, centrifuged at 10,000 rpm for 10 min, and the serum was stored at −20 °C until the analysis of OVA-specific IgE and mouse mast cell protease-1 (mMCP-1) levels by means of ELISA. OVA-specific IgE titers were detected, as described previously [24]. Levels are expressed in arbitrary units (AU), which were calculated based on a titration curve of pooled sera serving as an internal standard. The concentrations of mMCP-1 were determined while using a mMCP-1 Ready-SET-Go!® ELISA (eBioscience, Breda, The Netherlands), according to the manufacturer's protocol.

2.5. Ex Vivo OVA-Specific Stimulation of Splenocytes for Cytokine Measurements

Spleens were collected and homogenized while using a syringe and a 70 μm nylon cell strainer. The obtained single cell splenocyte suspensions were incubated with lysis buffer (8.3 g NH_4Cl, 1 g KHC_3O and 37.2 mg EDTA dissolved in 1 L demi water, filter sterilized) to remove the red blood cells and then resuspended in RPMI 1640 medium (Lonza, Verviers, Belgium), supplemented with 10% heat-inactivated fetal bovine serum (FBS; Bodinco, Alkmaar, The Netherlands), penicillin (100 U/mL)/streptomycin (100 μg/mL; Sigma-Aldrich), and β-mercaptoethanol (20 μM; Thermo Fisher Scientific, Paisley, Scotland). The splenocytes (8×10^5 cells/well) were cultured in U-bottom culture plates (Greiner, Frickenhausen, Germany), either with medium or with 50 μg/mL OVA for four days at 37 °C, 5% CO_2. The supernatants were collected and stored at −20 °C until cytokine analysis. The concentrations of IL-5 and IL-13 were measured by means of ELISA, as described elsewhere [25]. The concentrations of IFNγ, IL-10, and IL-17 were measured using a Cytometric Bead Array (CBA) Mouse Th1/Th2/Th17 Cytokine Kit (BD Biosciences, Alphen aan de Rijn, The Netherlands), according to the manufacturer's instructions. The results were obtained using FACS Canto II and analyzed with FCAP Array Software, version 3.0 (BD Biosciences, Alphen aan de Rijn, The Netherlands). Cytokine concentrations measured after medium stimulation were subtracted from cytokine concentrations measured after OVA stimulation to determine the OVA-specific cytokine response. A zero was entered when this resulted in a negative value.

2.6. Chromatin Immunoprecipitation to Determine Histone Acetylation Status in Splenocyte-Derived CD4+ T Cells and Mesenteric Lymph Nodes (MLN)

At day −1 (after the tolerance induction period) and at day 34 (after both challenges), CD4+ T cells were isolated from splenocytes of raw milk- and shop milk-treated mice using MACS, according to the manufacturer's instructions (Miltenyi Biotec, Leiden, The Netherlands).

Isolated CD4+ T cells were frozen with 15% dimethyl sulphoxide (DMSO; Sigma-Aldrich) in heat-inactivated FBS (Bodinco) and then stored in liquid nitrogen until further analysis. For the MLN, the entire tissue, containing a full population of the MLN cells, was frozen in 15% DMSO-FBS and stored in liquid nitrogen until further analysis. Detailed methodology of chromatin immunoprecipitation, followed by real-time polymerase chain reaction (ChIP-qPCR), along with its thoughtful validations, were previously described in detail [26]. In brief, the MLN tissues were first smashed through a mesh, washed with 1 mL of PBS (Sigma-Aldrich), and centrifuged at 8000 rpm for 5 min. at 4 °C. The pellet was then resuspended in 1 mL of warm PBS. The cross-linking of the cells was performed by incubating the cells with paraformaldehyde (PFA; Carl Roth GmbH, Karlsruhe, Germany) to a final concentration of 1% for 8 min at room temperature. The reaction was quenched by adding glycine to a final concentration of 125 mM (Carl Roth GmbH). After centrifugation at 8000 rpm for 5 min at 4 °C and washing with cold PBS, the samples were subjected to 20 min of incubation with lysis buffer I (Table S1) at 4 °C. Lysis buffer II (Table S1) was added with 1% sodium dodecyl sulfate (SDS; Carl Roth GmbH) for 5 min at 4 °C. Shearing of the DNA-protein complexes with the Bioruptor (Diagenode, Liège, Belgium) was conducted afterwards while using 30 cycles (30 s on, 30 s off) for CD4+ T cells and 40 cycles (40 s on, 40 s off) for MLN cells. Finally, the interfering debris was removed by centrifugation at 15,000 rpm for 15 min at 4 °C. Sepharose beads (GE Healthcare Bio-Sciences, Uppsala, Sweden) were first washed with lysis buffer II with 0.1% SDS. Following centrifugation at 3000 rpm for 2 min at room temperature, the beads were blocked with 1 mg/mL bovine serum albumin (BSA; Sigma-Aldrich) and 40 μg/mL salmon sperm DNA (Sigma-Aldrich) overnight at 4 °C. After washing the prepared beads with lysis buffer II with 0.1% SDS and centrifugation at 3000 rpm for 5 min at 4 °C, 30 μL of beads slurry per immunoprecipitation (IP) per number of samples were stored at 4 °C for the next day. To perform chromatin preclearing, 20 µL of beads slurry per antibody were added to the previously cross-linked chromatin samples, incubated with rotation for 2 h at 4 °C, and then centrifuged at 8000 rpm for 5 min at 4 °C. To the rest of the beads, 500 µL of lysis buffer II with 0.1% SDS and 1 µg of unspecific IgG (Abcam, Cambridge, UK) per sample were added and then incubated with rotation for 1 h at 4 °C. After washing three times with lysis buffer II with 0.1% SDS, 20 µL of the IgG-coupled beads were added to the precleared chromatin, incubated with rotation for 2 h at 4 °C, and then centrifuged at 8000 rpm for 5 min at 4 °C. Ten percent of the resulting supernatant containing chromatin were stored as the input control. The rest was divided into equal parts, to which 4 µg of either H3 or H4 (Millipore, Darmstadt, Germany) or 0.5 µg of IgG (Abcam) were added. The samples were then incubated at 4 °C overnight. Thirty microliter of the blocked beads slurry kept aside before, were added to each IP, and incubated for 2 h at 4 °C. After centrifugation at 8000 rpm for 5 min at 4 °C, the beads were washed twice with wash buffer I, twice with lysis buffer II, three times with wash buffer III (Table S1), and then twice with TE buffer with pH 8.0 (Table S1). The elution of the chromatin was performed by adding 500 µL of elution buffer (Table S1) to the sepharose beads, vortexing and incubating with rotation for 30 min After centrifugation at 8000 rpm for 2 min at 4 °C, the supernatants containing each IP, as well as the input controls, were mixed with 20 µL of 5 M NaCl, 10 µL of 0.5 M EDTA (Sigma-Aldrich), 20 µL of 1 M Tris-HCl (pH 7.2), 1 µL of Protease K (20 mg/mL; Sigma-Aldrich), and 1 µL of RNAse A (10 mg/mL; Sigma-Aldrich) per sample. All of the samples were incubated at 55 °C for 3 h and then at 65 °C overnight. Afterwards, DNA was purified while using the QIAquick PCR purification kit (Qiagen, Hilden, Germany). The purified DNA was subjected to qPCR that was performed with specific mouse gene promoter primers (Table S2) and Rotor-Gene SYBR Green PCR Kit (Qiagen), performed on Rotor-Gene Q (Qiagen). We were unfortunately unable to successfully amplify RORγ from H4-immunoprecipitated MLN DNA despite of two rounds of repetition, most probably due to the presence of a specific inhibition of this PCR in this batch of the samples. Percent enrichment to the input was calculated using the following formula: % enrichment $= 100 \times 2^{[(\text{CT input}-3.3)-\text{CT sample}]}$. Subsequently, the % enrichment of the isotype (IgG) control was subtracted from % enrichments that were obtained for specific antibodies. For final normalization, to further eliminate the variation caused by sample handling, such value obtained for each specific gene was divided by that of the positive control gene ribosomal protein L32 (RPL32) [26,27].

2.7. Statistical Analysis

Experimental results are expressed as mean ± standard error of the mean or as individual data points or box-and-whisker Tukey plots when the data were not normally distributed and analyzed using GraphPad Prism software (version 7.03, GraphPad Software, San Diego, CA, USA). Differences between pre-selected groups were statistically determined using one-way ANOVA, followed by Bonferroni's multiple comparisons test. Square root transformation was applied to mMCP-1 concentrations prior to ANOVA analysis. Anaphylactic shock scores and OVA-specific IgE levels were analyzed using the Kruskal–Wallis test for non-parametric data, followed by Dunn's multiple comparisons test for pre-selected groups. For histone acetylation and cytokine concentrations, differences between groups were statistically determined with an unpaired two-tailed Student's t-test. Welch's correction was used when the group variances were not equal. When data did not obtain normality, a Mann–Whitney test was performed. The results were considered to be statistically significant when $p < 0.05$.

3. Results

3.1. Raw Milk Reduces OVA-Induced Allergic Symptoms

Mice were orally treated for eight consecutive days with raw, unprocessed, cow's milk before being sensitized and challenged with OVA to assess whether raw cow's milk has the capacity to induce tolerance to an unrelated, non-milk, food allergen. Upon i.d. challenge with OVA, acute allergic symptoms were, as expected, increased in OVA-sensitized allergic mice when compared to

PBS-sensitized control mice. An increased acute allergic skin response, increased anaphylactic shock symptoms, and an anaphylactic shock-induced drop in body temperature illustrated this (Figure 2A–C). Treating mice with raw milk prior to OVA-sensitization reduced acute allergic symptoms when compared to PBS-treated allergic mice. The allergic skin response and anaphylactic shock symptoms were decreased and the body temperature of these mice remained high (Figure 2A–C). Mice were also treated with a processed, shop, milk to determine whether this allergy-suppressive effect is abolished upon milk processing. Treatment with this shop milk did not confer protection against allergic symptoms (Figure 2A–C).

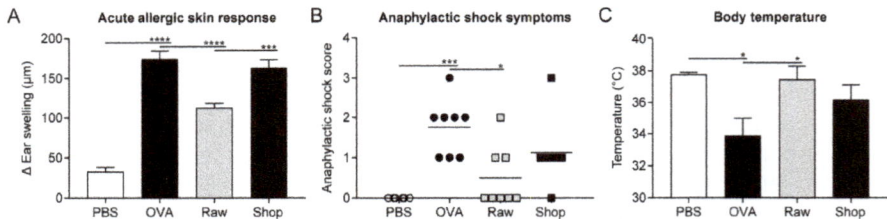

Figure 2. Reduced acute allergic symptoms upon ovalbumin (OVA) challenge in mice treated with raw milk. (**A**) The acute allergic skin response measured as Δ ear swelling 1 h after intradermal (i.d.) challenge. (**B**) Anaphylactic shock scores and (**C**) body temperature determined 30 min after i.d. challenge. Data are presented as mean ± standard error of the mean for the acute allergic skin response and body temperature and as individual data points for anaphylactic shock scores, $n = 6$ in PBS group and $n = 8$ in all other groups. * $p < 0.05$, *** $p < 0.001$, **** $p < 0.0001$, as analyzed with one-way ANOVA followed by Bonferroni's multiple comparisons test for pre-selected groups (**A**,**C**) or Kruskal–Wallis test for non-parametric data followed by Dunn's multiple comparisons test for pre-selected groups (**B**). PBS, phosphate-buffered saline; OVA, ovalbumin; raw, raw cow's milk; shop, shop milk.

3.2. OVA-Specific IgE Levels and Mucosal Mast Cell Degranulation Are Not Affected by Raw Milk Exposure

The effect of raw and shop milk on serum OVA-IgE levels was investigated since food allergens mainly induce type I hypersensitivity reactions, which are characterized by the production of allergen-specific IgE antibodies. Serum OVA-IgE levels were elevated in OVA-sensitized mice when compared to PBS-sensitized mice (Figure 3A). Even though OVA-IgE levels were not significantly affected by exposure to both milk types, they did follow a similar pattern as the acute allergic symptoms, with low OVA-IgE levels in the raw milk group and higher levels in the shop milk group (Figure 3A). In addition, serum mMCP-1 concentration, as a marker for mucosal mast cell degranulation, was measured. mMCP-1 concentrations were increased in the OVA-sensitized mice when compared to PBS-sensitized mice, but were unaffected by treatment with raw or shop milk (Figure 3B).

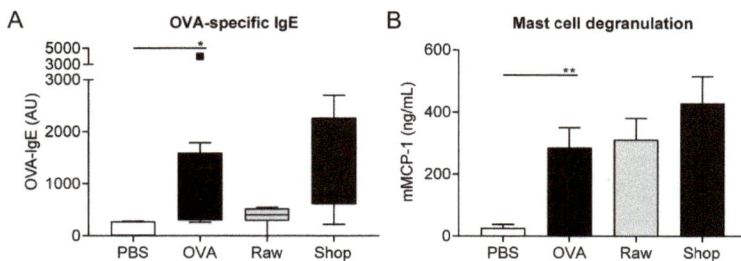

Figure 3. Raw milk treatment did not affect ovalbumin (OVA)-specific IgE levels and mouse mast cell protease-1 (mMCP-1) concentrations. (**A**) OVA-specific IgE levels and (**B**) mMCP-1 concentrations measured in serum 16 h after intragastric challenge. Data are expressed as box-and-whisker Tukey plot (in which outliers are shown as separately plotted points) for OVA-specific IgE levels and as mean ± standard error of the mean for mMCP-1 concentrations, $n = 6$ in PBS group and $n = 8$ in all other groups. * $p < 0.05$, ** $p < 0.01$ as analyzed with Kruskal–Wallis test for non-parametric data followed by Dunn's multiple comparisons test for pre-selected groups (**A**) or one-way ANOVA followed by Bonferroni's multiple comparisons test for pre-selected groups (**B**). PBS, phosphate-buffered saline; OVA, ovalbumin; AU, arbitrary units; raw, raw cow's milk; shop, shop milk; mMCP-1; mucosal mast cell protease-1.

3.3. Raw Milk Treatment Initially Increases Histone Acetylation of Several T Cell Subset Genes, While after Both Challenges It Specifically Reduces Th2-Related Gene Acetylation

Environmental factors might interact with genes that are involved in allergy development via epigenetic regulation. Histone acetylation (associated with higher gene expression) at selected Th1-, Th2-, Th17-, and regulatory T cell (Treg)-specific genes of splenocyte-derived CD4$^+$ T cells was assessed to determine whether epigenetic modifications contribute to the allergy-protective effect of raw cow's milk consumption. Surprisingly, histone H4 acetylation of Th2-related genes (GATA3, IL-4, IL-5, and IL-13) was higher after eight days of raw milk exposure when compared to shop milk exposure (day −1; Figure 4A). Raw milk exposure also increased the histone acetylation of T-bet and tended to increase the histone acetylation of FoxP3 (day −1), which indicated a type of general immune stimulation (Figure 4D). After both challenges (day 34), this general immune stimulation that was induced by raw milk was resolved and the histone acetylation of Th2 genes was lower as compared to shop milk (Figure 4B,E). Furthermore, the histone acetylation pattern of Th2-related genes is visualized by the raw milk/shop milk ratio, which shifted from in favor of raw milk after tolerance to in favor of shop milk after challenge (Figure 4C). A similar pattern was observed for IL-17, whereas the raw milk/shop milk ratio for Th1- and Treg-specific genes remained in favor of raw milk throughout the experiment (Figure 4F). For histone H3, the acetylation patterns were comparable (Figure S1).

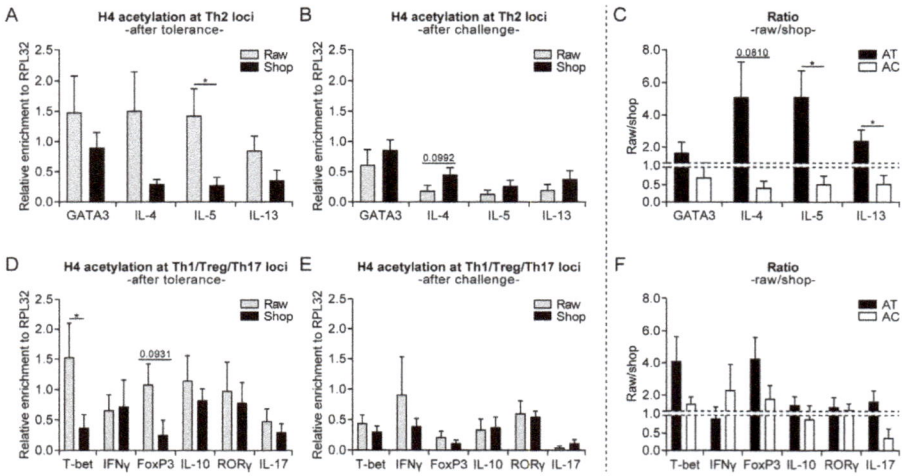

Figure 4. Increased histone acetylation of several T cell subset genes directly after raw milk exposure, while only Th2-related gene acetylation was reduced in raw milk-treated mice after both challenges. (**A**) Histone H4 acetylation at Th2 loci after the tolerance induction period (day −1), (**B**) after both challenges (day 34) and (**C**) the raw milk/shop milk ratio. (**D**) Histone H4 acetylation at Th1/Treg/Th17 loci after the tolerance induction period (day −1), (**E**) after both challenges (day 34) and (**F**) the raw milk/shop milk ratio. Histone H4 acetylation status was determined by means of chromatin immunoprecipitation in CD4$^+$ T cells derived from splenocytes of raw milk and shop milk-treated mice. Results are expressed as relative enrichment after normalization to ribosomal protein L32 (RPL32) as mean ± standard error of the mean, $n = 6$/group. * $p < 0.05$ as analyzed with an unpaired two-tailed Student's *t*-test. A Mann–Whitney test was used for T-bet, IFNγ, FoxP3, RORγ (after tolerance), T-bet, IL-17 (after model), and T-bet, IFNγ, RORγ (ratio raw/shop) since data did not obtain normality. Raw, raw cow's milk; shop, shop milk; AT, after tolerance; AC, after challenge.

3.4. Systemically Observed Acetylation Profile of Th2-Related Genes Induced by Raw Milk also Visible Locally

MLN were analyzed to determine whether the systemically observed alterations in histone H4 acetylation of T cell genes induced by raw milk are also visible locally. Despite being less strong, the shift in acetylation of Th2-related genes was also evident in the MLN (Figure 5A–C). Raw milk exposure for eight days led to higher acetylation of Th2-related cytokine genes (IL-4, IL-5, and IL-13) when compared to shop milk (day −1), while a lower acetylation of these genes was observed after both challenges (day 34; Figure 5A,B). For GATA3, histone acetylation was lower in the raw milk group after tolerance, as well as after the challenges (Figure 5A,B). The general immune stimulation, as observed after tolerance in CD4$^+$ T cells derived from the spleen of raw milk-treated mice, was not observed in the MLN. No significant differences were found between raw milk and shop milk in histone acetylation levels at Th1, Th17, and Treg loci (Figure 5D). After the challenges, histone acetylation of T-bet was increased in shop milk-treated mice when compared to raw milk-treated mice (Figure 5E), which resulted in a shift in the raw milk/shop milk ratio towards more favorable in shop milk after challenge (Figure 5F). A similar shift was observed for IL-10 (Figure 5F). Histone H3 acetylation was also assessed for MLN, but no significant differences between the groups were observed (Figure S2).

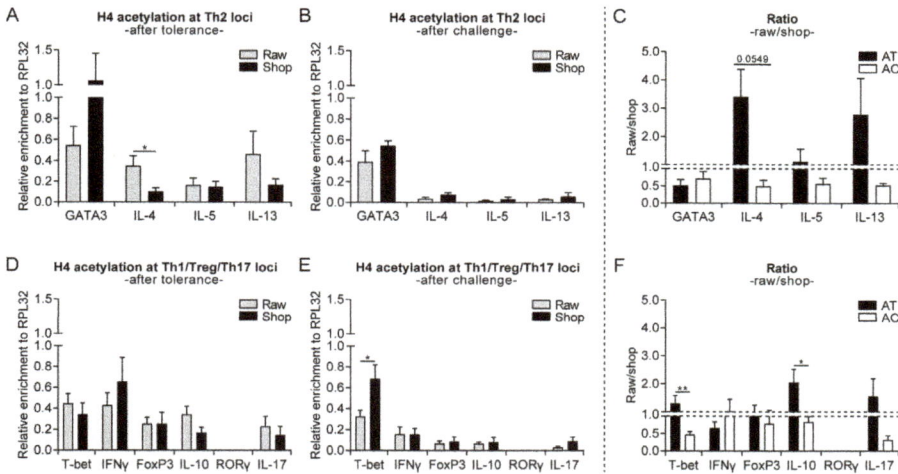

Figure 5. Raw milk-induced acetylation pattern of Th2-related genes observed in splenocyte-derived CD4$^+$ T cells also visible locally in mesenteric lymph nodes (MLN). (**A**) Histone H4 acetylation at Th2 loci after the tolerance induction period (day −1), (**B**) after both challenges (day 34) and (**C**) the raw milk/shop milk ratio. (**D**) Histone H4 acetylation at Th1/Treg/Th17 loci after the tolerance induction period (day −1), (**E**) after both challenges (day 34) and (**F**) the raw milk/shop milk ratio. Histone H4 acetylation status was determined by means of chromatin immunoprecipitation in MLN of raw milk- and shop milk-treated mice. The results are expressed as relative enrichment after normalization to ribosomal protein L32 (RPL32) as mean ± standard error of the mean, *n* = 4–6/group. * *p* < 0.05, ** *p* < 0.01 as analyzed with an unpaired two-tailed Student's *t*-test. A Mann–Whitney test was used for GATA3, IL-10 (after tolerance), IL-10 (after model) and GATA3 (ratio raw/shop) since data did not obtain normality. Raw, raw cow's milk; shop, shop milk; AT, after tolerance; AC, after challenge; MLN; mesenteric lymph nodes.

3.5. Cytokine Production by OVA-Stimulated Splenocytes Corresponds to Histone Acetylation

Cytokine production upon ex vivo stimulation of splenocytes with OVA was measured since differences in histone acetylation levels of cytokine genes do not necessarily result in differences in actual cytokine production. To be able to look at the OVA-specific cytokine response, the concentrations were only measured after both challenges (day 34). Concentrations were low for the Th2-related cytokines IL-5 and IL-13 (Figure 6A,B). However, the tendency towards a reduced IL-5 production in raw milk-treated mice is of interest when compared to shop milk-treated mice (Figure 6A), which corresponds to the lower IL-5 acetylation in splenocyte-derived CD4$^+$ T cells that were observed in histones H4 and H3 (Figure 4B and Figure S1B). IFNγ and IL-17 concentrations also correspond with the observed acetylation patterns, although no significant difference between the milk groups was observed (Figures 4E and 6C,E). In the case of IL-10, the cytokine concentration did not resemble gene acetylation, since the reduced IL-10 production in raw milk-treated mice was not observed in IL-10 gene acetylation (Figures 4E and 6D). Ex vivo stimulation of MLN with OVA did not result in measurable cytokine production.

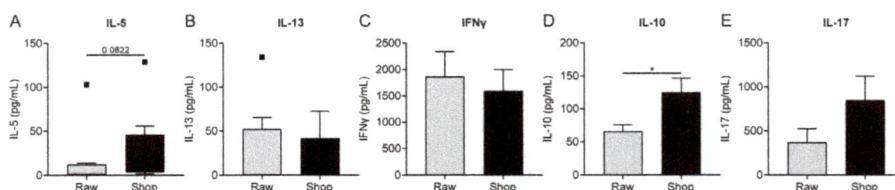

Figure 6. Cytokine concentrations produced by ovalbumin (OVA)-stimulated splenocytes corresponded with observed histone acetylation. (**A**) IL-5, (**B**) IL-13, (**C**), IFNγ, (**D**) IL-10 and (**E**) IL-17 concentrations measured in supernatant after ex vivo stimulation of splenocytes with OVA for four days (37 °C, 5% CO_2). Data are presented as box-and-whisker Tukey plot (in which outliers are shown as separately plotted points) for IL-5 and IL-13 concentrations and as mean ± standard error of the mean for IFNγ, IL-10 and IL-17 concentrations after subtracting baseline cytokine levels, n = 8/group. * $p < 0.05$ as analyzed with a Mann-Whitney test (**A**,**B**) or an unpaired two-tailed Student's *t*-test (**C**–**E**). OVA, ovalbumin; raw, raw cow's milk; shop, shop milk.

4. Discussion

After showing causality in a murine house dust mite-induced asthma model [14], the present study demonstrates that raw, unprocessed, cow's milk is also protective in a murine model for food allergy. Raw milk induced oral tolerance to a non-milk, food allergen, by reducing acute allergic symptoms after intradermal challenge with OVA. This protective effect was not observed when a processed, shop milk was used to treat the mice. Looking at epigenetic modifications, raw milk exposure for eight days prior to sensitization led to higher histone acetylation of Th1-, Th2-, and Treg-related genes of splenocyte-derived CD4$^+$ T cells when compared to shop milk exposure. At the end of the study, after the induction of allergic symptoms, this general immune stimulation was resolved and histone acetylation of Th2-related genes was lower when compared to shop milk. A similar, but less strong, pattern was locally visible, in the MLN. These results suggest that epigenetic regulation plays a role in the allergy-protective effect of raw milk.

Food allergies are thought to occur due to the failure to develop or the loss of oral tolerance [28]. Oral tolerance is the phenomenon of local and systemic immune hyporesponsiveness to ingested food proteins [29]. Actively inducing or restoring oral tolerance is an interesting approach for preventing or treating food allergies. For this, research has mainly focused on specific immunomodulation while using the allergen. Both inducing oral tolerance by allergen exposure in early life and restoring oral tolerance via various types of allergen-specific immunotherapy are frequent topics of immunological research [30,31]. However, using the intact allergen for oral tolerance induction might also trigger sensitization or allergic symptoms in high-risk patients [32,33].

Instead of specific immunomodulation, generic immunomodulation does not use the allergen to induce oral tolerance, preventing the risk of severe side effects. Generic immunomodulation is based on using beneficial immunomodulatory components that can create an environment that favors oral tolerance induction [34]. Mainly dietary components, such as, probiotics, prebiotics, synbiotics, and n-3 polyunsaturated fatty acids (PUFAs) have proven to be beneficial in this respect [35].

Several epidemiological studies already suggested that raw, unprocessed, cow's milk may have the capacity to prevent allergic diseases by inducing tolerance via generic immunomodulation. Raw cow's milk consumption was, for example, shown to be inversely associated with asthma, which indicated protection in the absence of the allergen [13]. In a murine house dust mite-induced asthma model, we confirmed these findings by showing a causal relationship between raw cow's milk consumption and the prevention of allergic asthma [14]. In the current study, raw cow's milk induced tolerance to OVA, an unrelated, non-milk, food allergen, which further substantiates this hypothesis.

Strikingly, processed, shop milk was not able to induce tolerance to OVA. This confirms earlier findings, which showed that milk processing abolishes the allergy-protective effect of raw milk [13–16].

The milk processing chain consists of various steps to preserve milk along the supply chain. Each of these steps (e.g., machine milking, skimming, homogenization, heat treatment, storage, and packaging) induces changes in the composition of the milk, which makes it hard to pinpoint one particular raw milk constituent that is responsible for the protective effects [36]. Even though comparing a raw milk with a shop milk (consumed by most people) was a logical first step in our opinion, future research should focus on testing milk from the same milk source that only differs in one processing step (skimmed milk, pasteurized milk, ultra-high temperature processing milk, etc.). Besides elucidating the raw milk component(s) involved, this will give the opportunity to look into the cellular mechanisms inducing tolerance in more depth.

Epigenetic regulation might be one of the mechanisms by which raw cow's milk exerts its allergy-protective effect. Since environmental factors are known to be able to modulate gene expression through epigenetic mechanisms, we wondered whether this also applied to raw milk. Epigenetic mechanisms can modify the accessibility of genes for transcription without altering the DNA nucleotide sequence which means that they can modulate the phenotype without affecting the genotype [19]. In this way, epigenetic mechanisms are key in the plasticity of gene expression. They are essential for developmental processes, like cellular differentiation, contributing, for example, to the flexibility among CD4$^+$ T cell subsets [37]. The classical epigenetic mechanisms comprise DNA methylation and histone modifications, including histone acetylation, methylation, phosphorylation, and ubiquitination [20].

We assessed histone acetylation at the promoter regions of Th1-, Th2-, Th17-, and Treg-related genes of splenocyte-derived CD4$^+$ T cells and MLN to determine the role of epigenetic mechanisms in the allergy-protective effect of raw milk. During histone acetylation, an acetyl group is added to a lysine residue at the N-terminal tail of a histone (mainly histones H3 and H4). This removes the positive charge on the histones that are involved, resulting in a decreased interaction with the negatively charged DNA. Consequently, the DNA is less tightly wrapped around the histones, which makes it more accessible to the transcriptional machinery. Therefore, higher histone acetylation usually results in higher gene transcription, while the opposite is true for reduced histone acetylation [19].

In line with the protective effects that were observed on acute allergic symptoms and IgE, histone acetylation of Th2-related genes (GATA3, IL-4, IL-5, and IL-13) of splenocyte-derived CD4$^+$ T cells after allergy induction was lower in raw milk-treated mice than in shop milk-treated mice. The strongest effects were observed on histone H4 acetylation at Th2 cytokine genes. Since histone acetylation substantially contributes to and is an important marker for an open chromatin structure [19,20], we assessed whether the acetylation levels positively correlated with cytokine production. Unfortunately, Th2 cytokine concentrations were low, but the tendency towards a reduced IL-5 production in raw milk-treated mice as compared to shop milk-treated mice suggests that there is indeed a positive correlation. Several other studies already confirmed that differences in H4 acetylation levels at Th2 cytokine genes indeed correlate with cytokine production [26,38]. Affecting epigenetic marks on Th2 cytokine genes might be an interesting preventive approach since type 2 cytokines play a predominant role in allergic diseases by directing the effector phase of an allergic response [39].

After allergy induction, the histone acetylation of Th1-, Th17-, and Treg-related genes did not differ between raw milk- and shop milk-treated mice. Although, here, histone acetylation patterns were reflected in cytokine production. The only cytokine for which the production did not correspond to gene acetylation was IL-10, which suggested that histone H3/H4 acetylation is not a main driver of IL-10 synthesis. Furthermore, we observed that IL-10 production was reduced in raw milk-treated mice as compared to shop milk-treated mice. This seems to be in contrast with the observed allergy protection, since IL-10 is known as a regulatory cytokine. However, in a murine model for OVA-induced food allergy, it was shown that IL-10 could also have proinflammatory effects. IL-10 was demonstrated to be essential for the development of food allergy by inducing mucosal mast cell expansion and activation [40]. This indicates that lowering IL-10 concentrations in a murine OVA-induced food allergy model might be beneficial. Besides systemically looking at splenocyte-derived CD4$^+$ T cells, we also locally assessed histone acetylation in the MLN. Here, similar effects were observed, although less strong. This might have to do with the fact that the whole tissue was used for ChIP analysis, rather than the isolated T cells. This may have resulted in weaker effects, as other cell types might also express the genes measured.

In addition to looking at histone acetylation patterns at the end of the study (after allergy induction), we also directly assessed histone acetylation after the eight days of milk exposure. Surprisingly, histone acetylation of the Th2-related genes of splenocyte-derived CD4$^+$ T cells was higher in the raw milk group as compared to the shop milk group. However, histone acetylation of T-bet and FoxP3 was also increased, which suggested a kind of general immune stimulation. Whether this general immune stimulation induced by raw milk is responsible for the observed allergy protection at the end of the study we do not know yet. Previously, however it has been demonstrated that acquiring tolerance in food allergic children involves epigenetic regulation of the FoxP3 gene [41]. Furthermore, epidemiological studies have shown that raw cow's milk consumption was associated with increased DNA demethylation of FoxP3 and increased numbers of Tregs [42]. Unfortunately, we did not look at Treg numbers in our study, but since active suppression by Tregs is considered to be one of the main effector mechanisms for oral tolerance [43], the observed increase in histone acetylation of the FoxP3 gene might contribute to the allergy-protective effect. Inhibiting de novo histone acetylation with histone acetyltransferase inhibitors might be an interesting approach for further investigating the role of histone acetylation in the allergy-protective effect of raw milk.

How raw milk affects epigenetic marks on T cell-related genes is currently unclear, but there are some indications. Microbes that were derived from farm dust, known to prevent allergic asthma, were, for example, shown to operate via epigenetic mechanisms [44], which suggested that microbes that are present in raw milk might have similar effects. Furthermore, raw milk contains higher levels of n-3 PUFAs than industrially processed milk [15]. These n-3 PUFAs reduce the risk of developing allergic diseases and they have been shown to lower the acetylation of IL-13 genes [45,46]. In addition, raw milk contains components, like lactoferrin, which can promote the growth of *Bifidobacteria* and *Lactobacilli* in the gut [17,18]. These bacteria are potent producers of short-chain fatty acids and these short-chain fatty acids are known for their capacity to inhibit histone deacetylases, thereby increasing gene transcription. Whether the above-mentioned components in the concentrations present in raw milk can influence epigenetic mechanisms and subsequently contribute to the allergy-protective effect of raw milk should be clarified in future studies. The possible involvement of the epigenetic mechanisms should also be investigated in the case of the anti-allergic effects of human breast milk consumption [47].

5. Conclusions

In conclusion, we show the potency of raw cow's milk to induce tolerance to a non-milk, food allergen. This allergy-protective effect was abolished by industrial milk processing, emphasizing the importance of minimally processed milk. The allergy-protective constituents of raw milk remain elusive and it should be investigated in follow-up studies. In addition, we showed that raw milk is able to modulate gene expression through epigenetic mechanisms. Raw milk might have induced oral tolerance by targeting histone marks on T cell-related genes. Whether this is a cause–effect relationship and whether effects are more pronounced with longer raw milk exposure should be assessed in future research. Nevertheless, our data suggest that the consumption of certified raw cow's milk can contribute to allergy prevention and epigenetic regulations, especially histone modifications, might be one of the underlying mechanisms.

Supplementary Materials: The following are available online at http://www.mdpi.com/2072-6643/11/8/1721/s1, Figure S1: Acetylation patterns of histone H3 were comparable to histone H4 in splenocyte-derived CD4$^+$ T cells, Figure S2: No differences between groups observed for histone H3 acetylation in MLN, Table S1: Buffers used for ChIP, Table S2: Primers used for qPCR.

Author Contributions: Conceptualization, S.A., J.W., V.A.-M., H.H., D.P.P. and B.C.A.M.v.E.; Funding acquisition, H.R., J.G., D.P.P. and B.C.A.M.v.E.; Investigation, S.A., J.W., V.A.-M., M.A.P.D., B.A.A., F.A. and H.H.; Methodology, S.A., J.W., V.A.-M., B.A.A., F.A., H.H., H.G., D.P.P. and B.C.A.M.v.E.; Project administration, S.A., D.P.P. and B.C.A.M.v.E.; Supervision, S.A., H.G., J.G., D.P.P. and B.C.A.M.v.E.; Visualization, S.A., J.W., D.P.P. and B.C.A.M.v.E.;

Writing–original draft, S.A.; Writing–review & editing, J.W., V.A.-M., M.A.P.D., B.A.A., F.A., H.H., H.R., H.G., J.G., D.P.P. and B.C.A.M.v.E.

Funding: This research was financially supported by Danone Nutricia Research.

Acknowledgments: The authors would like to thank T. Baars (Research Institute of Organic Agriculture, Frick, Switzerland) for providing the raw cow's milk used in this study.

Conflicts of Interest: J.G. and B.C.A.M.v.E. are (partly) employed at Danone Nutricia Research. All other authors report no potential conflict of interest.

References

1. World Allergy Organisation (WAO) White Book on Allergy: Update 2013. Available online: https://www.worldallergy.org/UserFiles/file/WhiteBook2-2013-v8.pdf (accessed on 4 February 2019).
2. Zuberbier, T.; Lotvall, J.; Simoens, S.; Subramanian, S.V.; Church, M.K. Economic burden of inadequate management of allergic diseases in the European Union: A GA(2) LEN review. *Allergy* **2014**, *69*, 1275–1279. [CrossRef] [PubMed]
3. Braun-Fahrlander, C.; Gassner, M.; Grize, L.; Neu, U.; Sennhauser, F.H.; Varonier, H.S.; Vuille, J.C.; Wuthrich, B. Prevalence of hay fever and allergic sensitization in farmer's children and their peers living in the same rural community. SCARPOL team. Swiss Study on Childhood Allergy and Respiratory Symptoms with Respect to Air Pollution. *Clin. Exp. Allergy* **1999**, *29*, 28–34. [CrossRef]
4. Kilpelainen, M.; Terho, E.O.; Helenius, H.; Koskenvuo, M. Farm environment in childhood prevents the development of allergies. *Clin. Exp. Allergy* **2000**, *30*, 201–208. [CrossRef] [PubMed]
5. Von Ehrenstein, O.S.; Von Mutius, E.; Illi, S.; Baumann, L.; Bohm, O.; von Kries, R. Reduced risk of hay fever and asthma among children of farmers. *Clin. Exp. Allergy* **2000**, *30*, 187–193. [CrossRef]
6. Riedler, J.; Eder, W.; Oberfeld, G.; Schreuer, M. Austrian children living on a farm have less hay fever, asthma and allergic sensitization. *Clin. Exp. Allergy* **2000**, *30*, 194–200. [CrossRef]
7. Alfven, T.; Braun-Fahrlander, C.; Brunekreef, B.; von Mutius, E.; Riedler, J.; Scheynius, A.; van Hage, M.; Wickman, M.; Benz, M.R.; Budde, J.; et al. Allergic diseases and atopic sensitization in children related to farming and anthroposophic lifestyle–the PARSIFAL study. *Allergy* **2006**, *61*, 414–421. [CrossRef]
8. Von Mutius, E.; Vercelli, D. Farm living: Effects on childhood asthma and allergy. *Nat. Rev. Immunol.* **2010**, *10*, 861–868. [CrossRef] [PubMed]
9. Riedler, J.; Braun-Fahrlander, C.; Eder, W.; Schreuer, M.; Waser, M.; Maisch, S.; Carr, D.; Schierl, R.; Nowak, D.; von Mutius, E.; et al. Exposure to farming in early life and development of asthma and allergy: A cross-sectional survey. *Lancet* **2001**, *358*, 1129–1133. [CrossRef]
10. Waser, M.; Michels, K.B.; Bieli, C.; Floistrup, H.; Pershagen, G.; von Mutius, E.; Ege, M.; Riedler, J.; Schram-Bijkerk, D.; Brunekreef, B.; et al. Inverse association of farm milk consumption with asthma and allergy in rural and suburban populations across Europe. *Clin. Exp. Allergy* **2007**, *37*, 661–670. [CrossRef]
11. Ege, M.J.; Frei, R.; Bieli, C.; Schram-Bijkerk, D.; Waser, M.; Benz, M.R.; Weiss, G.; Nyberg, F.; van Hage, M.; Pershagen, G.; et al. Not all farming environments protect against the development of asthma and wheeze in children. *J. Allergy Clin. Immunol.* **2007**, *119*, 1140–1147. [CrossRef]
12. Perkin, M.R.; Strachan, D.P. Which aspects of the farming lifestyle explain the inverse association with childhood allergy? *J. Allergy Clin. Immunol.* **2006**, *117*, 1374–1381. [CrossRef]
13. Loss, G.; Apprich, S.; Waser, M.; Kneifel, W.; Genuneit, J.; Buchele, G.; Weber, J.; Sozanska, B.; Danielewicz, H.; Horak, E.; et al. The protective effect of farm milk consumption on childhood asthma and atopy: The GABRIELA study. *J. Allergy Clin. Immunol.* **2011**, *128*, 766–773 e4. [CrossRef]
14. Abbring, S.; Verheijden, K.A.T.; Diks, M.A.P.; Leusink-Muis, A.; Hols, G.; Baars, T.; Garssen, J.; van Esch, B. Raw Cow's Milk Prevents the Development of Airway Inflammation in a Murine House Dust Mite-Induced Asthma Model. *Front. Immunol.* **2017**, *8*, 1045. [CrossRef]
15. Brick, T.; Schober, Y.; Bocking, C.; Pekkanen, J.; Genuneit, J.; Loss, G.; Dalphin, J.C.; Riedler, J.; Lauener, R.; Nockher, W.A.; et al. Omega-3 fatty acids contribute to the asthma-protective effect of unprocessed cow's milk. *J. Allergy Clin. Immunol.* **2016**, *137*, 1699–1706 e13. [CrossRef]
16. Brick, T.; Ege, M.; Boeren, S.; Bock, A.; von Mutius, E.; Vervoort, J.; Hettinga, K. Effect of Processing Intensity on Immunologically Active Bovine Milk Serum Proteins. *Nutrients* **2017**, *9*, 963. [CrossRef]

17. Abbring, S.; Hols, G.; Garssen, J.; van Esch, B. Raw cow's milk consumption and allergic diseases—The potential role of bioactive whey proteins. *Eur. J. Pharmacol.* **2019**, *843*, 55–65. [CrossRef]
18. Van Neerven, R.J.; Knol, E.F.; Heck, J.M.; Savelkoul, H.F. Which factors in raw cow's milk contribute to protection against allergies? *J. Allergy Clin. Immunol.* **2012**, *130*, 853–858. [CrossRef]
19. Potaczek, D.P.; Harb, H.; Michel, S.; Alhamwe, B.A.; Renz, H.; Tost, J. Epigenetics and allergy: From basic mechanisms to clinical applications. *Epigenomics* **2017**, *9*, 539–571. [CrossRef]
20. Alhamwe, B.A.; Khalaila, R.; Wolf, J.; von Bulow, V.; Harb, H.; Alhamdan, F.; Hii, C.S.; Prescott, S.L.; Ferrante, A.; Renz, H.; et al. Histone modifications and their role in epigenetics of atopy and allergic diseases. *Allergy Asthma Clin. Immunol.* **2018**, *14*, 39. [CrossRef]
21. Acevedo, N.; Frumento, P.; Harb, H.; Alashkar Alhamwe, B.; Johansson, C.; Eick, L.; Alm, J.; Renz, H.; Scheynius, A.; Potaczek, D.P. Histone Acetylation of Immune Regulatory Genes in Human Placenta in Association with Maternal Intake of Olive Oil and Fish Consumption. *Int. J. Mol. Sci.* **2019**, *20*, 1060. [CrossRef]
22. Verordnung über die Güteprüfung und Bezahlung der Anlieferungsmilch (Milch-Güteverordnung). Available online: http://www.gesetze-im-internet.de/milchg_v/index.html (accessed on 9 January 2019).
23. Li, X.M.; Schofield, B.H.; Huang, C.K.; Kleiner, G.I.; Sampson, H.A. A murine model of IgE-mediated cow's milk hypersensitivity. *J. Allergy Clin. Immunol.* **1999**, *103*, 206–214. [CrossRef]
24. Abbring, S.; Ryan, J.T.; Diks, M.A.P.; Hols, G.; Garssen, J.; van Esch, B. Suppression of Food Allergic Symptoms by Raw Cow's Milk in Mice is Retained after Skimming but Abolished after Heating the Milk-A Promising Contribution of Alkaline Phosphatase. *Nutrients* **2019**, *11*, 1499. [CrossRef]
25. Abbring, S.; Kusche, D.; Roos, T.C.; Diks, M.A.P.; Hols, G.; Garssen, J.; Baars, T.; van Esch, B. Milk processing increases the allergenicity of cow's milk-Preclinical evidence supported by a human proof-of-concept provocation pilot. *Clin. Exp. Allergy* **2019**, *49*, 1013–1025. [CrossRef]
26. Harb, H.; Amarasekera, M.; Ashley, S.; Tulic, M.K.; Pfefferle, P.I.; Potaczek, D.P.; Martino, D.; Kesper, D.A.; Prescott, S.L.; Renz, H. Epigenetic Regulation in Early Childhood: A Miniaturized and Validated Method to Assess Histone Acetylation. *Int. Arch. Allergy Immunol.* **2015**, *168*, 173–181. [CrossRef]
27. Haring, M.; Offermann, S.; Danker, T.; Horst, I.; Peterhansel, C.; Stam, M. Chromatin immunoprecipitation: Optimization, quantitative analysis and data normalization. *Plant Methods* **2007**, *3*, 11. [CrossRef]
28. Chinthrajah, R.S.; Hernandez, J.D.; Boyd, S.D.; Galli, S.J.; Nadeau, K.C. Molecular and cellular mechanisms of food allergy and food tolerance. *J. Allergy Clin. Immunol.* **2016**, *137*, 984–997. [CrossRef]
29. Pabst, O.; Mowat, A.M. Oral tolerance to food protein. *Mucosal. Immunol.* **2012**, *5*, 232–239. [CrossRef]
30. Palmer, D.J.; Prescott, S.L. Does early feeding promote development of oral tolerance? *Curr. Allergy Asthma Rep.* **2012**, *12*, 321–331. [CrossRef]
31. Kostadinova, A.I.; Willemsen, L.E.; Knippels, L.M.; Garssen, J. Immunotherapy - risk/benefit in food allergy. *Pediatr. Allergy Immunol.* **2013**, *24*, 633–644. [CrossRef]
32. Bellach, J.; Schwarz, V.; Ahrens, B.; Trendelenburg, V.; Aksunger, O.; Kalb, B.; Niggemann, B.; Keil, T.; Beyer, K. Randomized placebo-controlled trial of hen's egg consumption for primary prevention in infants. *J. Allergy Clin. Immunol.* **2017**, *139*, 1591–1599 e2. [CrossRef]
33. Palmer, D.J.; Metcalfe, J.; Makrides, M.; Gold, M.S.; Quinn, P.; West, C.E.; Loh, R.; Prescott, S.L. Early regular egg exposure in infants with eczema: A randomized controlled trial. *J. Allergy Clin. Immunol.* **2013**, *132*, 387–392 e1. [CrossRef]
34. Gourbeyre, P.; Denery, S.; Bodinier, M. Probiotics, prebiotics, and synbiotics: Impact on the gut immune system and allergic reactions. *J. Leukoc. Biol.* **2011**, *89*, 685–695. [CrossRef]
35. Vonk, M.; Kostadinova, A.; Kopp, M.; Van Esch, B.; Willemsen, L.; Knippels, L.; Garssen, J. Dietary Interventions in Infancy. *Allergy Immun. Toler. Early Child. First Steps Atopic March* **2015**, *261*, 261–284.
36. Michalski, M.C.; Januel, C. Does homogenization affect the human health properties of cow's milk? *Trends Food Sci. Technol.* **2006**, *17*, 423–437. [CrossRef]
37. Suarez-Alvarez, B.; Rodriguez, R.M.; Fraga, M.F.; Lopez-Larrea, C. DNA methylation: A promising landscape for immune system-related diseases. *Trends Genet.* **2012**, *28*, 506–514. [CrossRef]
38. Harb, H.; Raedler, D.; Ballenberger, N.; Bock, A.; Kesper, D.A.; Renz, H.; Schaub, B. Childhood allergic asthma is associated with increased IL-13 and FOXP3 histone acetylation. *J. Allergy Clin. Immunol.* **2015**, *136*, 200–202. [CrossRef]

39. Gandhi, N.A.; Bennett, B.L.; Graham, N.M.; Pirozzi, G.; Stahl, N.; Yancopoulos, G.D. Targeting key proximal drivers of type 2 inflammation in disease. *Nat. Rev. Drug Discov.* **2016**, *15*, 35–50. [CrossRef]
40. Polukort, S.H.; Rovatti, J.; Carlson, L.; Thompson, C.; Ser-Dolansky, J.; Kinney, S.R.; Schneider, S.S.; Mathias, C.B. IL-10 Enhances IgE-Mediated Mast Cell Responses and Is Essential for the Development of Experimental Food Allergy in IL-10-Deficient Mice. *J. Immunol.* **2016**, *196*, 4865–4876. [CrossRef]
41. Paparo, L.; Nocerino, R.; Cosenza, L.; Aitoro, R.; D'Argenio, V.; Del Monaco, V.; Di Scala, C.; Amoroso, A.; Di Costanzo, M.; Salvatore, F.; et al. Epigenetic features of FoxP3 in children with cow's milk allergy. *Clin. Epigenetics* **2016**, *8*, 86. [CrossRef]
42. Lluis, A.; Depner, M.; Gaugler, B.; Saas, P.; Casaca, V.I.; Raedler, D.; Michel, S.; Tost, J.; Liu, J.; Genuneit, J.; et al. Increased regulatory T-cell numbers are associated with farm milk exposure and lower atopic sensitization and asthma in childhood. *J. Allergy Clin. Immunol.* **2014**, *133*, 551–559. [CrossRef]
43. Weiner, H.L. Oral tolerance: Immune mechanisms and the generation of Th3-type TGF-beta-secreting regulatory cells. *Microbes Infect.* **2001**, *3*, 947–954. [CrossRef]
44. Brand, S.; Teich, R.; Dicke, T.; Harb, H.; Yildirim, A.O.; Tost, J.; Schneider-Stock, R.; Waterland, R.A.; Bauer, U.M.; von Mutius, E.; et al. Epigenetic regulation in murine offspring as a novel mechanism for transmaternal asthma protection induced by microbes. *J. Allergy Clin. Immunol.* **2011**, *128*, 618–625 e1-7. [CrossRef]
45. D'Vaz, N.; Meldrum, S.J.; Dunstan, J.A.; Lee-Pullen, T.F.; Metcalfe, J.; Holt, B.J.; Serralha, M.; Tulic, M.K.; Mori, T.A.; Prescott, S.L. Fish oil supplementation in early infancy modulates developing infant immune responses. *Clin. Exp. Allergy* **2012**, *42*, 1206–1216. [CrossRef]
46. Harb, H.; Irvine, J.; Amarasekera, M.; Hii, C.S.; Kesper, D.A.; Ma, Y.; D'Vaz, N.; Renz, H.; Potaczek, D.P.; Prescott, S.L.; et al. The role of PKCzeta in cord blood T-cell maturation towards Th1 cytokine profile and its epigenetic regulation by fish oil. *Biosci. Rep.* **2017**, *37*. [CrossRef]
47. Rajani, P.S.; Seppo, A.E.; Jarvinen, K.M. Immunologically Active Components in Human Milk and Development of Atopic Disease, With Emphasis on Food Allergy, in the Pediatric Population. *Front Pediatr.* **2018**, *6*, 218. [CrossRef]

MDPI

St. Alban-Anlage 66

4052 Basel

Switzerland

Tel. +41 61 683 77 34

Fax +41 61 302 89 18

www.mdpi.com

Nutrients Editorial Office

E-mail: nutrients@mdpi.com

www.mdpi.com/journal/nutrients

www.ingramcontent.com/pod-product-compliance
Lightning Source LLC
Chambersburg PA
CBHW051838210326
41597CB00033B/5699